SECURITY, INSECURITY AND MIGRATION IN EUROPE

Research in Migration and Ethnic Relations Series

Series Editor:
Maykel Verkuyten, ERCOMER
Utrecht University

The Research in Migration and Ethnic Relations series has been at the forefront of research in the field for ten years. The series has built an international reputation for cutting edge theoretical work, for comparative research especially on Europe and for nationally-based studies with broader relevance to international issues. Published in association with the European Research Centre on Migration and Ethnic Relations (ERCOMER), Utrecht University, it draws contributions from the best international scholars in the field, offering an interdisciplinary perspective on some of the key issues of the contemporary world.

Forthcoming

The Ethnic Penalty
Migration, Networks and Labour Markets in Canada and China
Reza Hasmath
ISBN 978 1 4094 0211 4

Other titles in this series at back of book

ERCOMER

**EUROPEAN RESEARCH CENTRE
ON MIGRATION & ETHNIC RELATIONS**

Security, Insecurity and Migration in Europe

Edited by

GABRIELLA LAZARIDIS
University of Leicester, UK

Routledge
Taylor & Francis Group
LONDON AND NEW YORK

First published 2011 by Ashgate Publishing

Published 2016 by Routledge
2 Park Square, Milton Park, Abingdon, Oxfordshire OX14 4RN
711 Third Avenue, New York, NY 10017, USA

First issued in paperback 2016

Routledge is an imprint of the Taylor & Francis Group, an informa business

British Library Cataloguing in Publication Data
Security, insecurity and migration in Europe. – (Research in migration and ethnic relations series)
1. European Union countries – Emigration and immigration – Social aspects. 2. European Union countries – Emigration and immigration – Economic aspects. 3. National security – European Union countries. 4. Asylum, Right of – European Union countries. 5. Immigrants – Civil rights – European Union countries. 6. Immigrants – Europe Union countries – Social conditions. 7. Women immigrants – European Union countries. 8. European Union countries – Ethnic relations.
I. Series II. Lazaridis, Gabriella.
325.4–dc22

Library of Congress Cataloging-in-Publication Data
Lazaridis, Gabriella.
Security, insecurity, and migration in Europe / by Gabriella Lazaridis.
p. cm. – (Research in migration and ethnic relations series)
Includes index.
ISBN 978-1-4094-0920-5 (hardback: alk. paper) – ISBN 978-1-4094-0921-2 (ebook) 1. Europe – Emigration and immigration. 2. Security systems – Europe. I. Title.
JV7590.L39 2010
355'.03304–dc22

2010038953

ISBN 13: 978-1-138-24669-0 (pbk)
ISBN 13: 978-1-4094-0920-5 (hbk)

In fond loving memory of my brother Stephanos Lazaridis and my niece Nicoletta Roussos, and to Lin Wymer (the hoixe who made me want to see Greece from her camera lenses)

Contents

List of Tables

Acknowledgements

This book stems from papers delivered at an international conference on Security, Insecurity and Migration in Europe which took place at the University of Leicester, UK, on 18/19 September 2009. Reflecting the interdisciplinary nature of the conference themes, participants represented a rich variety of approaches: from politics, sociology, geography, education and demography. The conference was organized by Gabriella Lazaridis (University of Leicester) and the Hellenic Institute of Migration (IMEPO), which also funded the conference. The editor would like to thank Vassiliki Papakosta from IMEPO, and Jane Russell, Sally Marsden, Carol Adjai and Thomas Gill from the University of Leicester, whose efforts were crucial to the planning and smooth running of the conference. At the Leicester end, where the editing of the book took place, the editor is most grateful to Nick James for his excellent editorial assistance and for indexing the book so efficiently. I am also grateful for Lianne Sherlock and Neil Jordan's valuable support. Finally the editor would like to thank the contributors for their willingness to revise their chapters against a tight schedule and to accept our editorial input with such good grace.

Gabriella Lazaridis

List of Contributors

Giovanna Campani is Professor of Intercultural Education and Gender Anthropology in the Faculty of Education at the University of Florence, Italy. She has worked extensively on migration, ethnic relations and gender issues. Her most recent book is *Dalle minoranze agli immigrati: il pluralismo culturale e religioso in Italia*, 2008, Unicoepli, Milan.

Georgios Karyotis is a Lecturer in International Relations at the University of Strathclyde. His research interests include international security theory, European migration policy, terrorism and South-Eastern Europe politics. His most recent article on 'Religion, Securitization and Anti-Immigration Attitudes' was published in the Journal of Peace Research in 2010.

Anna Maria Konsta is an Adjunct Professor of European Politics and Law at the American College of Thessaloniki. She has published in the field of European and Comparative Social Law.

Khalid Koser is Associate Dean, Head of the New Issues in Security Programme, and Director of the New Issues in Security Course (NISC) at the Geneva Centre for Security Policy. He has also been Non-Resident Fellow in Foreign Policy Studies at the Brookings Institution, Research Associate at the Graduate Institute of International and Development Studies in Geneva, Visiting Fellow at the Lowy Institute in Sydney, and Associate Fellow at Chatham House. He is chair of the UK's Independent Advisory Group on Country Information. He is co-editor of the Journal of Refugee Studies and on the editorial board for Global Governance, Ethnic and Racial Studies and Forced Migration Review.

Anna Krasteva is director of the Department of Political Sciences and of CERMES (Centre for Refugees, Migration and Ethnic Studies) at the New Bulgarian University in Sofia, and editor-in-chief of the journal, *South Eastern Europe*, published by Brill. She is currently a fellow at the Institute for Advanced Studies in Nantes, France. She is author of more than 70 publications in 13 countries.

Gabriella Lazaridis is a Senior Lecturer in the Department of Politics and International Relations, University of Leicester. She has published extensively in the fields of gender, ethnicity and migration in Europe, and the implementation of European policies in southern Europe. Her latest book, *Women's Work and Lives in Rural Greece*, was published by Ashgate in 2009. She is currently working on two monographs, one on Gender and Migration to be published by Springer and one on International Migration in Europe to be published by Ashgate.

Branka Likic-Brboric is a political economist and an economic historian. She is a Research Fellow at the Department of Government, Uppsala University and a Guest Lecturer at REMESO, Institute for Migration, Ethnicity and Society, Linköping University. Her research focuses on the configuration of the global governance regime and its normative base, in terms of discourses of global social justice, migrants' rights and the 'decent work' agenda. Her special concern is the interplay between EU enlargement, new migration landscapes, informal economy and the concomitant reconfiguration of European social models.

Mark Maguire lectures in the Department of Anthropology, National University of Ireland Maynooth. His ethnographic research is on programme refugee, asylum, elite migration, and new security technologies. He is author of *Differently Irish* (2004). In 2008 he held the 2nd Fulbright Advanced Research and University Lecturing Award in Stanford University, California, where he will again be a Visiting Professor in 2011.

Kees Nagtegaal is a political scientist working at Leiden University – Montesquieu Institute in The Hague, where he organizes the educational programs. He also works as a lecturer in the Institute of Public Administration, Faculty of Social Sciences, Leiden University, Leiden.

Mojca Pajnik works as a scientific counsellor at the Peace Institute, Institute for Contemporary Social and Political Studies in Ljubljana, Slovenia, and as assistant professor at the Faculty of Social Sciences, University of Ljubljana. Her recent publications include *Prostitution and Human Trafficking: Gender, Labor and Migration Aspects* (PI, 2008). She is co-author of *Where in the Puzzle: Trafficking from, to and through Slovenia* (IOM, 2003), and of *Immigrants, Who Are You? Research on Immigrants in Slovenia* (PI, 2001).

Bernd Parusel is a political scientist. He currently works as a Researcher for the European Migration Network (EMN) at the Federal Office for Migration and Refugees in Nuremberg, Germany.

Annie Phizacklea is Emeritus Professor in Sociology at Warwick University, UK. Her research and writing has focussed on gender, migration and employment.

Natalia Ribas-Mateos is Ramón y Cajal Fellow at ESOMI, Universidad de A Coruña, Spain. Her latest book is *The Mediterranean in the Age of Globalization. Migration, Welfare and Borders* (Transaction Publishers, 2005).

Liza Schuster is a Senior Lecturer in the Department of Sociology, City University. Her research interests include migration, in particular forced and undocumented migration, racism and the links between the two. Her publications include 'The Use and Abuse of Asylum in Britain and Germany', 'Dublin II and Eurodac: The (un)intended consequences(?)' in *Gender, Place and Culture* (forthcoming 2011) and 'Turning Refugees into "Illegal" Migrants', *Ethnic and Racial Studies* (forthcoming 2011).

Burcu Toğral is a PhD candidate at the University of Hamburg. Currently, she is conducting fieldwork at the Foundation Jose y Ortega in Madrid. She also engages in the activities of Hamburg Refugee Council of Hamburg.

Anastassia Tsoukala is Professor of Criminology at the University of Paris 11 and Senior Researcher at Paris 5-Sorbonne University. Her research centres on the design and implementation of security policies in Europe, and on the social construction of threat. Recent publications: *Football Hooliganism in Europe*, Palgrave Macmillan, 2009; *Terror, Insecurity and Liberty*, Routledge, 2008 (co-edited with D. Bigo).

Introduction

Gabriella Lazaridis

Europe, the latest enlargement of the European Union (EU), globalization, and increasing instability in the developing world, have generated new, complex forms of population movement, and have reconfigured refugee and asylum issues. The forces that determine these movements are powerful (for example, growing disparities in the level of prosperity and human security experienced by different societies in the developed migrant destination countries and the developing world), and in the current economic climate, the ability to modify them is limited. The national states and supra-national institutions of Europe have been presented with massive challenges as they have sought to grapple with the evolution of new forms of mobility and migration within and into the EU and its western satellites. These have unfolded against the continuing redefinition of relations with the non-European world. These complex and shifting forms of population movement lead to questions about established notions of identity and citizenship, and create challenges for different scales of governance in both sending and receiving states in the light of emerging divergences between the interests of migrants and the states trying to control their movement and entry, as well as between the interests of governments, NGOs and civil societies.

Since the events of 9/11, security concerns have gained unprecedented dominance on western governments' national and international political agendas, a fact evidenced by the adoption of intense 'securitization' policies and measures to control cross-border movements. Migration issues, having often been framed in terms of security concerns, have simultaneously given rise to issues of insecurity: on the one hand, security of borders, political, societal and economic security/insecurity in the host country; while on the other hand, concerns about human security, that is social, legal, economic and other conditions of (in)security for EU citizens and for those migrants who enter or have entered Europe. In terms of state security, migration is a core target of increasingly globally networked surveillance capabilities garnered in 'the war against terror'. In terms of human security, it exposed the gap between the protections that migrants formally enjoy under international law and the realities they experience as they travel and work across different countries, and hence a human rights deficit. Stringent passport controls, the militarization of borders, heightened visa requirements, compulsory, indefinite detention, deportation and other measures implemented in destination countries run counter to the maintenance of the human security and safety of citizens (including established ethnic minorities) and of migrants entering or residing in the EU and other western countries, and often lead to the economic and/or sexual exploitation of particularly vulnerable groups (see Balzacq 2009; Balzacq & Carrera 2008; Catz 2003; Carter & Merrill, 2007; Ceyhan & Tsoukala 2002; Lorentzen 2003). Muslims (as migrants or members of ethnic minority communities) are particularly subjected to severe surveillance and security measures. The intensification of surveillance raises questions of legitimate state power, civil liberties and democratic accountability. Evidence also suggests that migrant and ethnic minority populations suffer disproportionate levels of insecurity due to the global economic recession, both the lack of opportunities this creates and the hostility it fuels within western societies against 'outsiders' seen as competitors for scarce resources (see Awad 2009; OECD 2009).

Hence, irregular migration poses difficult challenges for states as well as exposing migrants to insecurity and vulnerability. This book addresses both the concerns of states in this respect and the need to protect the fundamental rights of migrants. Against this background, the book explores a number of related themes:

- The dilemmas involved in framing policy-making on immigration in Europe.
- The implications of the securitization of migration and migrants.
- The extent to which irregular migration constitutes a real or perceived threat to state security and human security.
- Migration, economic in/securities and the informalization of European labour markets.
- Migration, societal in/securities and the rise of racism and xenophobia in Europe.
- The living conditions and security needs of migrants in European countries.
- Processes of inclusion and exclusion.
- Gender and migration.
- The role of social capital in negotiating spaces of control.
- Human trafficking as a threat to security; who is insecure?
- Migrants as objects of globally networked surveillance.
- The implications of state structures and institutional capabilities for the migration-security balance sheet.

Typically, the literature on the securitization of migration approaches it as a 'top-down' process, in which various political, societal and security elites present migration as an existential threat to fundamental values of European societies and states. Chapter 1 by Georgios Karyotis provides a framework in which to study the consequences of this process. However, unlike most normative critiques of securitization, this chapter does not focus on the detrimental impact of securitization on the human rights of migrants, who are treated as being inherently dangerous or inferior. Instead, it examines securitization from the perspective of the actors that facilitate the dominant interpretation of migration as a security threat. In doing so, it seeks to provide a pragmatic rather than an ethical assessment of the political practice of linking migration to security in Europe.

Karyotis claims that although elites are goal-maximizing actors, they fail to recognize the unintended consequences and hidden costs of securitizing migration. Securitization, he argues, is socially constructed and a disproportionate response to the threats posed by migration. It is also a counter-productive policy frame for managing migration, as it exacerbates rather than reduces actual threats to identity and public order. Finally, securitization creates a vicious circle of supply and demand for security, which is unattainable and ultimately hampers elites' ability to support contradictory aims, such as those relating to labour needs for immigration. The overall analysis in this chapter demonstrates that the securitization of migration serves only-short term needs, while harming in the long-term other interests, including of those who constructed the security frame in the first place.

In Chapter 2 Mark Maguire outlines some of the major developments in biometric security during the past decade, from 'smart' borders to the securitization of access to spaces and privileges. This chapter explores biometrics by taking a number of different access routes. Firstly, it outlines the history of biometrics in the age of European imperialism, when technologies for recognition, such as fingerprinting, and technologies for recording and storage, such as those used by police forces, were first developed. Secondly, it explores in detail two contemporary examples of biometric security, those of Ireland and Qatar. Maguire tracks the developments in Ireland, and the voices calling for debate, in order to suggest that what is at stake is not simply a question of civil liberties or rights (narrowly conceived) but rather the progressive instantiation of new ways of governing through securitization. He also explores the possible trajectory of current developments by considering the example of the Gulf State of Qatar – a state with considerable investment in biometrics, and with graduated forms of residence and citizenship.

The central argument in this chapter is that a full understanding of the implications of biometric securitization in Europe may only be achieved by attending to historical developments and to developments outside Europe. Studying the history of biometrics reveals the specific problematizations of identity and governance that biometrics such as fingerprinting were deployed to deal with. The history of biometrics also shows that colonial and European governments were, even in the nineteenth and early twentieth centuries, wrestling with questions of security and liberty, and debating universality and the targeting of specific populations. The contemporary examples of Ireland and Qatar highlight the globalized nature of the biometric security industry, the range of applications, and areas of deployment. The image that emerges, however fractured, is of Europe facing a future of securitization in which individual identity verification will be used to grant or deny access and privileges.

In Chapter 3 Natalia Ribas-Mateos argues that the complex Mediterranean context, characterized by its sea border, and the US–Mexico region with its extended land border, may well represent the two regions of the world where the processes of integration-interaction between asymmetric countries are currently most in evidence, and therefore constitute key sites in which to study border reinforcement and the vulnerability of human rights.

Her hypothesis is that questions related to the evolution of new forms of mobility, issues of border security and conditions of insecurity (specifically vulnerability), can be examined in terms of human (in)security in highly specific places, namely borders, as key research sites. Here, the gap between the protection migrants formally enjoy under international law, national laws, and the actual experiences of individuals reveal a marked absence of human rights. It is also at such strategic sites that we encounter many issues specifically mentioned in this book: securitization and migration, risk security, irregular migration, racism, gender issues, human trafficking, globally networked surveillance etc. In this chapter, Ribas-Mateos looks at the Mediterranean region in terms of the securitization pattern that has been exported from other areas. Indeed, many authors have referred in particular to the Mexico–US region as a laboratory of the postmodern security-insecurity contradictions of the world. The question posed here is: can such a model be easily exported to Mediterranean Europe? And why are such border areas examples of extreme security yet at the same time also of extreme insecurity for migrants?

In the first part of Chapter 4 Khalid Kosner assesses the impacts of the recent global economic and financial crisis on international migration patterns, processes, and policies in Europe. These impacts include a reduction in labour migration, rising unemployment among foreign workers, deteriorating working and living conditions, increasing returns, and reduction in irregular migration flows, but an increase in irregular migration stocks, and a reduction in remittances. The evidence for impacts on other migration flows, ranging from student to humanitarian migration, is less clear. The overwhelming policy response by European states has consisted of increasing restrictions, achieved through cutting the number of work permits for foreigners, offering incentives for migrants to return, reducing shortage occupation lists and reinforcing labour market tests, introducing changes in visa levels and entry requirements, limiting possibilities to change status or renew work permits, and tightening controls on family inflows.

The second part of this chapter considers some of the security implications of these changes, bearing in mind reservations about the 'securitization' of migration. There are possible implications for national security in certain European countries where stocks of irregular migrants have risen substantially and quickly, where they enter into competition with legal migrants and even local populations for work, turn to petty crime to subsist, or place an extra burden on scarce resources. Equally, irregular migrants may become a catalyst for anti-immigration violence and racist attacks. The growth in irregular migrant stocks also has the potential to undermine public confidence in government policy on migration and integration. Another implication for national security is a possible knock-on effect on future migration, if the recent decline in labour migration is sustained, with the potential to restrain technological and economic progress and inhibit cultural innovation.

Equally the impacts of the crisis have implications for the human security of the migrants involved, arising from growing unemployment, deteriorating working conditions, the transition from legal to irregular status, the rising incidence of violent attacks, and for some the prospect of returning home to even greater poverty. There may also be an impact on their families as family reunion becomes more difficult, or remittances decrease.

Chapter 5 by Branka Likic-Brboric addresses the viability of EU strategies for the configuration of a transnational social space that places at its core a European Social Model and a rights-based migration regime, particularly considering the EU's current responses to the global crisis. It explores the cumulative impact of the EU's commitment to globalization, asymmetric policies of EU enlargement and the ongoing securitization of the EU migration regime on the configuration of the EU-wide political economy of inequality, new forms of labour casualization–*cum*-informalization and new migratory landscapes. It is claimed in this chapter that the unfolding process of EU enlargement has not successfully addressed the inherent 'liberal paradox' of the EU project, and that the EU's responses to the current financial crisis reveal regionally differentiated reaffirmation of economic austerity measures. Coupled with the securitization of migration, this reinforced neo-liberal policy trend discredits social dialogue and generates a decay of democratic politics and European social citizenship. The circular migration policies informed by the combined logic of neo-liberal economic policy and securitized asylum *acquis* not only undermine migrants' human rights and create a vicious circle of enduring 'statelessness' but also emasculate social citizenship and the labour rights of citizens in the enlarged EU. The configuration of the EU-level institutional framework for Freedom, Security and Justice, has reinforced a new EU-wide landscape of precarization and insecurity that has been shaped by market fundamentalism. Moreover, the EU's target to combat irregular migration through stricter border control, neighbourhood and development policies, can be seen as a regional strategy of neo-liberalization of citizenship and re-commodification of labour that creates a 'useful migrant' and an 'employable citizen'. These mutually reinforcing frames of security and economic austerity breed multiple paradoxes that can only be challenged by an institutional rebalancing of EU policies in terms of an alternative that promotes European solidarities, social protection, 'decent work' and a rights-based mobility/migration regime.

Chapter 6 by Liza Schuster looks at the insecurities experienced by asylum seekers. She argues that although the number of asylum seekers in Europe is decreasing, the asylum policy of most EU member states continues to be driven by a fear of a potential increase. As a result, these states refuse to address the growing number of people caught between Eurodac and Dublin II. These includes the many thousands who transit Greece and Spain, where in theory they should claim asylum, but where in reality they have little chance of being able to make a claim, and almost no chance of having it examined properly, much less being actually granted asylum. This chapter addresses two arguments: firstly, that the concern of EU member states is actually to reduce the number of those able to apply for asylum to an absolute minimum, rather than to ensure access to this status to anyone who might need it. In other words, the rhetoric of abuse is a device to evade responsibility for providing protection to those in need, and secondly, it is argued that the assumption underlying these different provisions, that there is no right to choose where one makes a claim, is evidence of a failure to recognize asylum seekers as autonomous human beings (or human beings *tout court*), and the reason why the policies don't work even from a state's perspective.

In Chapter 7 Kees Nagtegaal shows that whereas present immigration policies seem to have been increasingly dominated by security issues since 9/11, security issues are not new in the context of immigration. Since the beginning of the twentieth century, the security of nation states has been linked with immigration policies, and the security of refugees suffers as a consequence. To illustrate this, Nagtegaal takes an historical approach, and gives an overview of the refugee policies of various Dutch coalition governments in the twentieth century by referring to a number of refugee crises. He shows that the basic framing of refugee policies in The Netherlands has always been within a framework of threat to the nation state and its citizens.

International treaties stimulated the formalization of the rights of refugees. Yet state sovereignty is all important, since it gives national authorities the power to decide who will be allowed to enter and which criteria will be used as a filter to judge applicants. The responsibility governments feel towards their citizens, and the duty they have, once they have signed international treaties and, based on these, created national legislation, requires them to balance interests. Many countries, including The Netherlands, have been party to these treaties. Yet authorities prefer to make their own policies, although they claim to follow the spirit of these treaties. The balance between the stronger nation states and their security on the one hand and the weaker refugees and their security, based on international treaties and national legislation, on the other hand is one between two very unequal parties.

Chapter 8 by Bernd Parusel deals with the entry into, and reception of, unaccompanied minor migrants in the EU. It provides some basic data on the number of unaccompanied children and youth under the age of 18 who have applied for international protection in selected EU Member States, as well as information on their main countries of origin and motivations for seeking protection. This Chapter also looks into some immigration patterns observed by individual EU Member States in relation to such minors.

The central argument is that unaccompanied minors in Europe are exposed to two conflicting policy frameworks. On the one hand, migration and asylum policies both in an EU context and within the individual Member States are based on the principle that migration flows need to be controlled and restricted. These policies have created situations in which unaccompanied migrant children are exposed to insecurity and vulnerability. This becomes apparent when unaccompanied minors attempt to enter EU territory, in contacts between foreign minors and government authorities such as border police or asylum officials, and during the difficult, and sometimes impossible, process of obtaining legal residence, education and care.

On the other hand, the EU and its Member States are bound to international commitments to protect minors and to respect their security needs. State actors have to apply the 'best interest of the child' principle as a criterion for actions taken by public authorities in relation to all children and young persons, irrespective of their legal status or nationality. Yet, national policies and practices vary significantly across the EU, and the well-being of children and their personal security needs are not always a primary concern by any means. Parusel concludes that, because of international human rights standards, advocacy work and the involvement of non-governmental actors in policy planning and in the development of new legal instruments, there are certain chances, although limited, for an improvement of existing policies towards greater respect for the needs and rights of unaccompanied children and youth.

In Chapter 9, Giovanna Campani argues that a 'security-linked' migration threat is constantly evoked in anti-immigrant political rhetoric in many European countries. This means that the concept of 'security', which is already unclear from the academic point of view because of its multiple definitions (such as 'national security', 'human security', etc.) is blurred by an ideological discourse, in which perceived cultural, economic and physical safety threats, border control and terrorism are put together for political reasons.

Campani attempts to show how the gender approach can contribute to the deconstruction of the 'security rhetoric' and to a better analysis of the quantitative data and the 'security' policies. She argues that the anti-immigrant rhetoric on security is largely gendered: irregular migration of men is considered a threat to national security (in terms of raising crime, terrorism, black work), while that of women is strongly associated with questions of morality and sexuality (prostitution). Women can also be seen as victims, especially trafficking victims. This narrative of victimhood reserved to women produces a rhetoric that considers irregular migrant women as forced to migrate (mainly to work in prostitution) and irregular migrant men as a threat to the society (and as potential criminals). The fact that irregular migration (as a process provoked by extremely restrictive migratory policies) exposes all migrants to insecurity and vulnerability is hidden. The gender approach illustrates the fact that the largely accepted statement that irregular migrants represent a special threat to security should be questioned by the study of gendered trajectories. Campani argues that the 'security' rhetoric in political discourse, and 'security' measures,, harsher border control and the pressure on irregular migrants, rarely play a positive role in assuring better human security for either immigrants or natives.

In questioning usual analyses of media discourses on migration-related threats, which only highlight the amplifying aspects of social problem construction, Chapter 10 by Anastassia Tsoukala seeks to capture the way media representation of migratory threat evolves according to an array of social, political and economic interests. To unravel the interactions that take place between the framing of the issue and certain prevailing vested interests at a given moment, it rests upon analysis of immigration- and asylum-related French and Greek press discourses in the late 2000s as compared to media representation of the issues during the 1980s and 1990s.

Analysis of relevant articles uncovers the fact that the initially broadly shared securitization of the issue is closely associated with a desecuritization process as regards immigration-related economic and demographic threats. In this context, desecuritization is seen as the outcome of a move that seeks to protect certain domestic interests that are consensually accepted as vital for the well-being of host societies. The importance of this process, which is arguably attributed to acknowledgement of the growing dependence of domestic economies and societies on a foreign workforce, highlights the complexity of the framing of longstanding 'social problems'. Far from always being structured in binary terms, threat-fuelling discourses appear as flexible communication tools likely to adjust themselves to the specificities of evolving social and political fields.

This ongoing interplay between desecuritizing and securitizing discourses, and the ensuing decreasing visibility of immigration-related economic and demographic threats, do not, however, imply any significant changes in the policy-making realm. Migrants continue to be associated with many different security and cultural threats which eventually legitimize increasingly restrictive migration and asylum policies. Consequently, desecuritization is mainly to be understood as a rhetorical move.

Chapter 11 by Anna Krasteva is predicated on the assumption that the new security concept, centred on the individual rather than on the state, is attacked by new figures of insecurity, at the micro rather than the macro level, concerning identities rather than the borders. The objective is to analyse the political logic of the production of gendered insecurity on the base of the mixture of Islam, radicalism and gender. A case study of the French public debate on burqa is analysed.

When combined, gender, Islam and radicalism produce an explosive mixture. The objective of this chapter is to analyse the political logic of the production of gendered insecurity by the French public discourse on the basis of this mixture, an amalgam of Islam and radicalism on the one hand, and national identity and laïcité on the other.

During the period of adoption of the national security concept the central actor of difference is the terrorist, the one who succeeds in crossing the borders and attacking the very core of the state. This actor has also religious and gender characteristics – he is a man and a Muslim.

The new stage in the focus on human security needs a new figure of otherness, and one such has been immediately constructed: the veiled woman. She also has clear gender and religious characteristics: female and Muslim.

This case study enables understanding of the reasons for, and the mechanisms involved in creating figures of otherness, which challenge the republican political culture and threaten the collective identity. This new threat, reified by the political discourse, is a favourable ground for a further securitization of policy. Eventually, the ultimate end, that of human security, is not strengthened, rather a very securitarian vision and policy is reproduced and enhanced.

Chapter 12 by Burcu Toğral addresses how the securitization of migration and 'new racism' have converged under the 'auspice' of the cultural difference-as-a-threat narrative in Europe. It is done through delineating the transformation of discourses surrounding racism and migration, and revealing how this transformation provides new means of justification for the racialized, securitarian policies and discourses. In doing this, this chapter does not intend to delve into the 'real' reasons behind these discourses and policies, but it seeks to demonstrate that articulation of migrants' 'cultural differences' in terms of threat has become a very seductive strategy to mask and depoliticize the real issues at stake. More precisely, 'cultural differences' of certain groups of migrants, especially those of Muslims and Arabs, have been constructed and reified as a striking destabilizing factor, a threat to the so-called cultural homogeneity and national identity of the member states as well as to European 'values'. In this context, securitization of migration in the thematic field of culture embodies the same appeals as the 'new racism'; a racism which was built upon cultural discourses rather than biological ones. On closer scrutiny, these processes and convergence have manifested themselves in the integration discourses and policies across the EU. Through mandatory and discriminatory integration policies, cultural difference-as-a-threat narrative has been materialized and reproduced. These policies are based on the premise that the differences of particular groups of migrants are a threat that must be decontaminated and tamed. They draw a framework through which these migrants and their 'inferior' culture can be brought in line with the 'superior' culture and values of member states. Indeed, this binary grouping and growing focus on the culture has decisive political utility. They help depoliticize political problems such as racial discrimination and the subordination of migrants in the economic, political and social fabric of European societies. In the context of these arguments, this chapter calls for critical reflection on integration, multiculturalism and similar culturalist strategies, and asserts that idealized representation of integration and multiculturalism hides the structural problems and sustains the domination and subordination of migrant groups through the 'innocent' appeal to the culture.

In Chapter 13 Mojca Pajnik argues that productivity rates in the aging societies of the EU member states are increasingly only possible to maintain due to the migrant labour force filling positions in undervalued and low-paid work, especially in construction, manufacturing, and the feminized sectors of care and domestic work. Migrants tend to increasingly occupy precarious jobs, often related to black market economies with no or low levels of security, and with poor recognition of their skills. Also those officially employed on the basis of various types of work permit find themselves in the grips of the flexibilization of the market reality, being forced to accept vulnerable, short-term, low-paid and low skilled jobs.

The actual demand for migrant labour remains poorly reflected in state policies, which, in turn, define and regulate the migrants' positions primarily in terms of limitations and closure. The chapter argues that current migration and labour market policies sustain migrants in 'rightless' positions in which their security is not respected . Ensuring migrants the 'right to have rights' (Arendt, H. [1948]1976) is thematized as a precondition for any serious attempt to claim migrants' integration as a politics of security and of equality, not of subordination. Drawing on the concept of social citizenship, the chapter questions the contemporary rationale of migration and labour policies, and problematizes what appear as 'strategies of circular conditioning' of migrants' lives. Migrants' narratives obtained through biographical narrative interviews in Slovenia are used to pinpoint the various condition-setting mechanisms (work permits, the quota system, residence permits, access to social services etc.) that restrain migrants' rights. Furthermore, the chapter debates social citizenship both as utopian imagery and as a point of reference that can facilitate the development of the 'needs discourse' pertaining to migrant populations. It explores how social citizenship could function as a concept which accommodates the irreversibility of social rights that have been in decline since the 'crisis' of the welfare state.

Chapter 14 by Gabriella Lazaridis and Anna Maria Konsta focuses on the experiences of Albanian migrant women, making a distinction between 'ethnic-Greek-Albanians' (originating from Northern Epirus) and 'other' Albanians. It shows that different members of the Albanian migrant population in Greece are positioned differently in relation to various dimensions of citizenship, according to their gender, ethnicity, legal residence status etc. Lazaridis and Konsta claim that there is a *plastic citizenship* emerging for these women, in which boundaries are blurred and processes of becoming or not becoming are fluid, changing over time and influenced by notions of who should belong and who should not, who is entitled to what rights and who is not; at the same time, they argue that this *plastic citizenship* leads to a transformation of their subjectivities into *plastic subjectivities*, which can enable resistance and contain all kinds of possibilities to enable them to 'wiggle' and escape from the rigidity of the societal structure.

New transnational forms of female subjectivities emerge in Europe today which are regulated by immigration, family and labour laws. Greek and European legal regulations prove to be insufficient to cope with the needs of the transcultural female legal subject. National and EU citizenship rights, if interpreted in a narrow sense, do not apply to these migrant women. The transcultural female legal subject finds herself excluded from the host society. In this respect migrant women are transformed by law and through law from legal *subjects* into 'legal *abjects*'. As shown in this chapter, the process of *abjectification* is not linear and differs according to one's differential inclusion and/or exclusion in the host country, and involves differing forms of human (in)securities.

In her concluding remarks, Annie Phizacklea pulls together the arguments put forward in the chapters of this book, and asks whether it is possible for the already limited access of third country nationals to protection be safeguarded. She argues that 'there has to be a real commitment in practice to a rights based mobility/ migration regime in Europe rather then the lip service that is paid to such rights in day to day realities. Given the record of practice over the last 40 years this amounts to a very uphill struggle'.

This book brings to the forefront the unresolved dilemma: whether to approach the issue from a state or human-centric perspective and whether, or to what extent, these divergent approaches can be reconciled. The questions flowing from this dilemma are important not only for academics but also for policy-makers and practitioners. It is essential that questions of the 'human rights deficit', the violation of civil liberties, economic precarity, racism and hostility, and the exploitation of vulnerable migrants, form the focus of academic critiques and alternative policy approaches.

The themes covered in the chapters which follow do not exhaust the list of possible topics for analysis. The issues of security, insecurity and migration remain a potentially fertile area for further investigation.

References

Arend, H. ([1948]1976) *The Origins of Totalitarianism*, New York: Harcourt

Awad, I. (2009) *The Global Economic Crisis and Migrant Workers: Impact and Response*. Geneva, ILO, accessed: 14 January 2010 at: www.ilo.org/global/About_the_ILO/Media_and_public_information/Feature_stories/lang--en/WCMS_112537/index.htm

Balzacq, T. (ed.) (2009) *The External Dimension of EU Justice and Home Affairs: Governance, Neighbours, Security*, Basingstoke: Palgrave Macmillan

Balzacq, T. and Carerra, S. (eds) (2008) *Security Versus Freedom?: A Challenge for Europe's Future*. Aldershot: Ashgate

Carter, D. and Merrill, H. (2007) 'Bordering Humanism: Life and Death on the Margins of Europe', in *Geopolitics*, 12/2: 248–64

Catz, P. (2003) 'United Kingdom: Withdrawing from International Human Rights Standards' in Brouwer, E., Catz, P. and Guild, E. (eds) *Immigration, Asylum, Terrorism: A Changing Dynamic of European Law,* Nijmegen: Instituut voor Rechtssociologie/Centrum voor Migratierecht

Ceyhan, A. and Tsoukala, A. (2002) 'The Securitization of Immigration in Western Societies: Ambivalent Discourses and Policies', in *Alternatives*, No. 27 (suppl.): pp. 21–39

Lorentzen, J. (2003), 'Keep Out! Protectionism, Migration Control and Globalization' in Tiersky, R. (ed.) *Europe Today: National Politics, European Legislation and European Security*, Lanham, MD: Rowman & Littlefield

Organization for Economic Cooperation and Development (2009) International Migration Outlook, Special Focus: Managing the Labor Migration Beyond the Crisis, Paris: OECD

Chapter 1
The Fallacy of Securitizing Migration: Elite Rationality and Unintended Consequences

Georgios Karyotis

Introduction

It has become commonplace to argue that migration in Europe and beyond is intimately linked to questions of security. As mentioned in the introduction to this book, migration is associated with a range of threats covering the whole socioeconomic and political spectrum. Immigrants and asylum seekers are often seen as a threat to public order and stability. They are also believed to be 'plotting' to exploit national welfare provisions and available economic opportunities at the expense of citizens. Above all, they are seen as a threat to the identity of societies and thus as a challenge to the very existence of a traditional pattern of living. Attempting to capitalize on such fears, Nick Griffin (2003), chairman of the British National Party (BNP) claimed in an official statement, that while not racist, his party and society as a whole 'must not become multi-racist either', a message repeated by right-wing parties across Europe.

The horrific attacks of 11 September 2001, as well as subsequent terrorist incidents in Madrid and London, exacerbated public anxiety towards migrants in Europe (Faist, 2002; Bigo, 2006). In all cases, the perpetrators matched a specific ethnic profile, while some of those involved in the U.S. attacks, had previously lived in Hamburg for years. Nevertheless, these events did not cause the insecurities, uncertainties, ambiguities and complexities that characterize migration policies at both domestic and European Union (EU) level. Rather, they strengthened and legitimized the security logic that has dominated asylum and immigration policies in Europe since the late 1970s (Huysmans, 2000, 2006; Geddes, 2003; Karyotis, 2007).

Lavenex (2001) refers to this logic as the 'realist policy frame', which contains a dominant interpretation of migration as a security problem. Framing generally involves making some aspects of a perceived reality more salient in discourse 'in such a way as to promote a particular problem definition, causal interpretation, moral evaluation, and/or treatment recommendation' (Entman, 1993: 52). In the case of migration, the realist frame is rooted in a state-centric philosophy, emphasizing the need to secure borders, restrict migration and homogenize all categories of migrants into a single policing-repression scheme. In contrast, the liberal frame focuses on the individual. It is primarily concerned with the protection of migrant human rights and the reduction of barriers to labour migration, which is considered beneficial to the economy.

Once a frame has gained prominence, it becomes established as the 'correct' or the 'standard' way to define an issue. The construction of the security frame in particular is known as 'securitization', where an issue enters the realm of security, not because of the objective threats it poses but because it is presented as such (Buzan *et al.*, 1998). Securitization is therefore a process through which elites, with 'the most effective means of public persuasion and the best resources for suppressing or marginalizing alternative opinions' succeed in defining an issue as an existential threat to fundamental values of society and the state (Van Dijk, 1993: 45).

The implication is that the elites' decision to securitize migration is a deliberate and calculated one. While both civil society and scholars have widely criticized the moral bankruptcy of the realist frame for its impact on migrant rights in particular (McSweeney, 1996; Lohrmann, 2000; Thouez, 2002; Guild, 2003), the consequences of securitization on the state and its elites has received scarce attention. The aim of this chapter is to address this imbalance and explore securitization from the perspective of securitizing elites. In doing so, it seeks to provide a pragmatic, rather than an ethical assessment of the political practice of linking migration to security in Europe. Findings highlight the fallacies of securitizing migration, arguing that as a policy option, it is more costly than often assumed.

The argument is developed progressively over three interconnected parts. The first section addresses the philosophical and normative aspects of securitization, which are underpinned by a belief that security justifies any sacrifice, particularly at the expense of migrants. The second part questions whether the decision to securitize is in fact ever based on rational calculations. The final part explores the consequences of the securitization of migration, which it argues serves only short-term needs, while harming in the long-term other interests, including those that supported the security discourse in the first place.

Security values and trade-offs

Many of the normative criticisms of securitization can be located in a long-standing philosophical debate concerning the importance of security and the cost and potential trade-offs associated with its pursuit. In its broadest formulation, this involves a question of whether security, as a value and policy goal, justifies the compromise of other key values, such as freedom or justice. This section will briefly address this and in doing so, draw attention to the ethical dilemmas associated with the securitization of migration.

Traditionally, the realm of security has been dominated by realist ideas, which consider the state as the only referent object of security, i.e. the actor that is to be secured. Accordingly, during the Cold War when realism was at its prime, security became synonymous with 'national security' (Yergin, 1997). The survival of the state at any cost was the agenda of security studies, while the state was simultaneously the focus and provider of security. This reasoning echoes Hobbes's state of nature, described in Leviathan as 'a war of all men against all men'. The sovereign state comes into existence to provide order and security, 'while in the absence of authority, unrestricted competition driven by quite rational individual calculation brings about unwished for and disastrous outcome for all concerned' (Hampsher-Monk, 1992: 25).

In this light, security and liberty are assumed to be set in a zero-sum game, in which more of one is taken to mean less of the other (Zedner, 2009: 135). This dichotomy is not inescapable, since the two could be seen as 'interrelated, mutually reinforcing goods' (Dinh, 2002: 400; Huysmans, 2006). By controlling dangers, the state enables individuals to realize their freedom, while, in turn, the lack of oppression contributes to the maintenance of order within the state. Nonetheless, the presence of imminent threat, real or perceived, stimulates the antagonistic framing of security and other key values, such as freedom.

If we were to accept that such a trade-off exists, then, according to realism, we must prioritize security and allow the state enough power to ensure it. As Hobbes (1985: 225) puts it, 'there can be no peace without subjection' and therefore the state can legitimately constrain civil liberties and individual values in order to establish order and unity. Even classical liberal thinkers such as John Stuart Mill (1991: 190–91), who opposed unlimited state control, recognized security as 'the most vital of all interests', because in its absence 'we would have no ability to look forward with any degree of confidence to the future'.

However, placing security above all other values is ethically questionable. Benjamin Franklin famously noted that 'they that can give up essential liberty to obtain a little temporary safety deserve neither liberty nor safety' (Dinh, 2002: 399). Drawing on this, and in contrast to the realist approach, critical security scholars have sought to re-orientate security away from the state and toward 'human security' (Ayoob, 1997; McSweeney, 1996). Introduced in the United Nations Development Programme (UNDP) report of 1994, human security is a concept calling for the pursuit of the 'security of individuals, not just the security of their nations', in other words, it is about the 'security of people, not just security of territory'. This is based on a core, moral belief that the only appropriate referent object of security is the individual, whose welfare, wellbeing and freedom cannot be compromised in order to safeguard the state as a political unit. After all, Dworkin (2005: 86) maintains, 'among the most fundamental of all moral principles is the principle of shared humanity: that every human life has a distinct and equal inherent value.'

An overemphasis upon statist security at the expense of individual freedoms is, according to Ken Booth (1991: 320), not only immoral but also illogical. To illustrate this, he draws an analogy between a house and its inhabitants:

> A house requires upkeep, but it is illogical to spend excessive amounts of money and effort to protect the house against flood, dry rot and burglars if this is at the cost of the well-being of the inhabitants. There is obviously a relationship between the well-being of the sheltered and the state of the shelter, but can there be any question as to whose security is primary?

In this critique, Booth concludes that states should be the means of security rather than its end, whereas individuals, citizens or not, should be its primary focus. To extend Booth's analogy (1991: 319), what would the value of security really be, if its pursuit required the imposition of such 'physical and human constraints', which would stop the inhabitants of the house 'from carrying out what they would otherwise freely choose to do'?

Relating these to migration, the two contrasting approaches correspond to what Lavenex (2001) referred to as the realist and liberal policy frames. The former sees migration as a vulnerability to state security, while the latter examines it primarily through a humanitarian perspective. At the European Union level, this translates to an observed tension between the policy objective of managing the perceived threats from migration on the one hand and the protection of human rights on the other (Balzacq and Carrera, 2006). The conclusion reached by most analysts is that the realist imperative is the one that has been driving policy developments, at the expense of humanitarian or other considerations (Geddes, 2003; Guiraudon, 2003; Huysmans, 2006). In other words, as Didier Bigo (2006: 35) notes, the EU's headlining goals of promoting freedom, security and justice is 'infiltrated and contaminated' by an overemphasis on strengthening security. While this security bias is not new (Huysmans, 2000), the war on terror has further strengthened it, with the overzealous application of various security practices and border controls in Europe and beyond (see chapters by Maguire and Ribas-Mateos in this volume).

The dominance of the realist frame on migration points towards an updated reformulation of the philosophical debate between security and freedom: the suggested trade-off is in fact between the 'liberties of the few against the security of the majority' (Waldron, 2003: 194). In other words, the common claim is that *our* security justifies limitations to *their* rights. The securitization of migration is a crucial manifestation of this dichotomy, since it legitimizes repressive measures against migrants, particularly those that match a given ethnic, religious or political profile. This attempt to mobilize a 'we' against a supposedly threatening 'them' is not only a central tenant of social identity theory and self-categorization (Tajfel and Turner, 1986) but also, according to Carl Schmitt, the essence of politics. As he put it, an act or antithesis 'transforms into a political one if it is sufficiently strong to group human beings effectively according to friend and enemy' (Schmitt, 1996: 37; Huysmans, 1998).

The conceptual link of securitization with the Schmittian legacy of 'realpolitik' unsurprisingly adds fuel to the critics' concerns about its ethical integrity and responsibility (McSweeney, 1996). Proponents of the liberal policy frame and human security in particular reject the focus on the state and the pursuit of security at the expense of other values. Further to the philosophical arguments discussed above, this rejection reflects unease with the exploitation and violation of the human rights of migrants, who, in a securitized frame, are perceived as inherently inferior and/or dangerous (Guild, 2003; Thouez, 2002; Lohrmann, 2000). For instance, Togral in her contribution to this volume explains how the securitization of migration results in the masking of a new form of racism in Europe.

Notably, proponents of securitization theory themselves recognize the dangers of securitizing societal issues like migration, and argue that 'security' should not be idealized but seen as a failure to deal with issues as 'normal politics' (Buzan *et al.*, 1998: 29). Therefore, according to Waever (2000: 6–7), 'the ideal of the securitization approach is – ceteris paribus – desecuritization, that issues are not lifted above normal politics with an urgency and 'necessity' that has often antidemocratic effects.' The implication is that securitizing actors, typically political and security elites, should have a heightened sense of responsibility when they talk security. However, the limitation of any normative calls to overturn the security frame on migration, advocates the need to take a closer look at the rationality of the elites that supported the securitization of migration in the first place, an issue examined in the next section.

Elite rationality and securitization

One key assumption about security decision making is that it differs form other policy areas in the degree of centralization and pluralism. Security 'is a structured field' and only those with the societal currency, knowhow and status are able to write legitimate security discourses. Threat perceptions are constructed and appropriate responses are determined by those in positions of power that allow them to be generally accepted as voices of security (Buzan *et al.*, 1998: 31). Therefore, although in principal nobody is excluded from becoming a securitizing actor, the field of security is biased in favour of political elites and 'security professionals' (Bigo, 1994).

Securitization occurs when securitizing actors, speaking or acting in the name of a referent object, succeed in convincing a relevant section of society that exceptional measures are needed in response to an existential threat. Any issue can through this mechanism become a security one, not necessarily because of the nature or the objective importance of the threats it poses but because it is presented as such (on the role of 'speech acts' see Waever, 1995). Accordingly, securitization, Huysmans explains (1998: 571), becomes a governmental technique 'which retrieves the ordering force of the fear of the violent death by a mythical replay of the Hobbesian state of nature. It manufactures a rupture in the routinized, everyday life by fabricating an existential threat which provokes experiences of the real possibility of violent death'.

The motives of securitizing actors in framing an issue in security terms remain surprisingly undertheorized in securitization research (Balzacq, 2005; Karyotis,

2007b: 275). Do elites support the security frame on migration out of a genuine concern for the existential threats they believe it poses? Are their discourse and policies based on a cost-benefit assessment of a range of possible responses? What other factors impact on their decision to make a securitizing move? Although it is obviously not possible to get inside elites' heads to fully understand their positioning on an issue, understanding the migration-security nexus, let alone attempting to untangle it, requires a closer investigation of security decision making.

As discussed, securitization is considered a purposeful, orchestrated, elite-driven process. Implicitly, elites are assumed to be acting rationally, in terms of both personal interests and their professional responsibility. The tangible consequences of successful securitization, the theory goes, are an increased urgency to deal with the issue, with additional resources and exceptional means outside the formal and established procedures of politics (Buzan *et al.*, 1998). If the decision to securitize was to be seen as rational, it should theoretically lead to a better handling of an issue, while promoting the vested interests of political elites, security professionals and the mass media, all of which are deemed to benefit from the securitization of migration in particular (Boswell, 2008; Bigo, 2002).

However, if by 'rational' we refer to the classic expected-utility model, then the assumption of rationality in securitization instances is false for at least three reasons. First, information overload, uncertainty and complexity 'make it almost impossible' for policy makers 'to live up to the ideal of rational method' (Hill, 2003: 102). In the case of irregular migration, this is very relevant, since statistics about its size and impact are notoriously slippery and unreliable (Brochmann and Hammar, 1999). Second, psychological factors, influenced by a particular cultural and social context, also limit pure rationality. These include pre-existing beliefs, emotions and ideas about their own role and the values that need to be protected (e.g. identity), which shape their 'operational environment' (Sprout and Sprout, 1969). Third, path-dependencies and historical conditions associated with a perceived threat can either facilitate or restrict policy-makers' ability to define it in security or alternative terms (Hay, 2006).

These do not suggest that decision makers are irrational (Gigerenzer and Selten, 2002). Indeed, 'they want to make rational decisions, but they cannot always do so' (Jones, 1999: 298), since the constraints discussed above make it impossible to reach the optimized solution to a policy problem. This 'bounded rationality' (Simon, 1995) leads them to accept the first outcome which approximates their preferences, rather than strive for the best option, which may be costly and unattainable (Hill, 2003). In turn, once a particular way of viewing or dealing with an issue becomes established and institutionalized, it is likely to resist change, even if the social power relations that facilitated its emergence have changed (Coleman, 1998; Karyotis and Patrikios, 2010).

The crucial question then is how to reverse securitization and, in our case, the security frame which has come to dominate migration policy in Europe. The problem for the analyst is that that '[e]ven when one writes about security with the aim of achieving de-securitization or to sensitise everybody to the problems of securitization, one securitizes by way of putting these issues in security terms' (Waever, 2000: 15; Huysmans, 1998). As for policy makers, their bounded rationality that contributed to the securitization of migration in the first place is likely to make more liberal frames and policies appear risky and costly. This is despite any ethical or moral considerations, which, when security is at stake, are sidelined and perceived as expendable. Therefore, instead of focussing on moral or normative calls for desecuritization, a possibly more fruitful avenue would be to attempt to convince elites that the securitization of migration is counterproductive and damaging.

Securitization aims and unintended consequences

Elites that support the security-migration nexus do so because they believe they are serving their country's, and in some cases, their own interests. Nobody wants to feel that their ideas and policies are unjust or immoral, which is why even right-wing parties attempt to legitimize and justify their hostile position towards migrants. For instance, Ian Cobain (2006) from *The Guardian*, who had worked undercover in the extreme-right BNP for seven months, reported that members were explicitly instructed to avoid using racist or anti-semitic language in public, in an attempt to clean the party's image. To pursue electoral gain, one BNP member was quoted as saying, 'people must stop seeing us as ogres.'

The main reasons that are commonly used to support the need for securitization are critically examined in the first part of this section. Drawing on these, the second part looks at the unintended consequences of securitization, while the final part assesses the extent to which securitization actually promotes the self-interests of the elites that support it.

Deconstructing the security-migration nexus

Migration is perceived as a threat for reasons that cut through a range of societal, criminological and economic arguments. The conviction that migration poses existential threats to European states and to the EU appears to be the driving force behind their restrictive policies. This is not surprising according to Ceyhan and Tsoukala (2002: 22) since, the natural tendency is to fear the 'different, the alien, the undocumented migrant, the refugee, the Muslim, the non European'. However, when assessing the objective significance of the threats that migration in Europe is deemed to be posing, the legitimation for securitization becomes more difficult to defend.

The overarching, underlying concern is that migration is, above all, a threat to societal security. Waever (1993: 23) notes that societal security 'concerns the ability of a society to persist in its essential character under changing conditions and possible or actual threats'. The sacred values that are deemed to be threatened and need to be protected relate to 'traditional patterns of language, culture, association, and religious and national identity and custom' (Waever, 1993: 23). Cultural norms, which define belonging in a community, can be seen to be violated by 'unwanted migrants' (Weiner, 1992). Therefore, elites in Europe, who often see themselves as defenders of national purity and societal security, may feel that their role demands they deal with immigrants and asylum seekers as a threat to communal harmony and cultural homogeneity (e.g. Ibryamova, 2002).

The reality however is that identity can never be seen as frozen or monolithic, as such securitizing discourses suggest (McSweeney, 1996). European societies are not static entities but have been evolving and developing in a rather harmonious way, despite the influx of migrants. After all, all European societies are the result of 'multiple migration and crossbreeding processes' and their culture is 'deeply influenced, even sometimes determined by migrant cultures' (Ceyhan and Tsoukala 2002: 29). Notably, perspective securitizing actors, the likes of Jörg Heider in Austria or Jean-Marie Le Pen in France to mention but a few high-profile examples, should be reminded of the traumatic warnings from recent European history about misusing the terms 'ethnic purity' and 'racial unity'. As Lohrmann (2000: 9) succinctly puts it 'the fact that receiving countries are confronted with immigrants with different cultural backgrounds does not represent a threat in itself. Rather, it is the political exploitation of these cultural differences that confers a security dimension to immigration.'

A discourse related to the identitarian axis sees migration as a threat to public order and internal security. The 'criminal migrant' thesis is based on the demonization of the 'other' and the creation of an artificial continuum between migration, crime, drug-trafficking and terrorism. This 'security continuum' is not just a rhetorical one but also reflected in EU institutional structures (Bigo, 1994). To be precise, starting in 1975, the EU's migration policy evolved gradually within intergovernmental fora, such as the TREVI Groups, the Ad hoc Group on Asylum and Immigration, the Schengen Treaties and Europol, all of which were preoccupied with internal security, thus strengthening the defensive and repressive logic of migration management (Karyotis, 2007a).

However, studies assessing the links between migration and crime have challenged the prejudicial stereotyping of migrants as inherently prone to deviant behaviour. Despite some legitimate security concerns, for instance about the links of migration to organized crime and human trafficking, the overall impact of migration on the crime rate and on the internal security of host countries in Europe is grossly exaggerated (Lohrmann, 2000). For instance, the Association of Chief Police Officers in the UK confirmed that there is no evidence of a higher rate of criminality among refugees and asylum seekers, who are in fact more likely to be the victims of crime in the UK (ACPO, 2001). Furthermore, the participation of migrants in serious criminality does not appear to be dramatic compared to citizens (Karydis, 1998), while where discrepancies are observed, these can typically be attributed to socioeconomic characteristics (e.g. age, education and income of migrants), rather than ethnicity (Hatton and Williamson, 2007). Finally, crimes committed by migrants tend to receive greater -often biased and inaccurate- coverage, while stories about occurrences of racism are much rarer (Triandafyllidou, 2002; Buonfino, 2004).

Economic concerns too, add to the prevailing insecurity towards immigrants and asylum seekers, who are seen as 'free riders', 'scroungers' or 'bogus', plotting to exploit the socio-economic fabric of host European societies. The economic burden they pose includes increasing unemployment, 'straining housing, education, and transportation facilities' (Weiner, 1992: 114) and overburdening 'already dilapidated welfare systems' (Held *et al.*, 1999: 313). The local resentment generated leads to what Huysmans (2000: 767) describes as welfare chauvinism, whereby 'immigrants and asylum-seekers are not simply rivals but *illegitimate* recipients of socio-economic rights'.

Yet, few dispute anymore that migration has a largely positive impact on European economies. Exaggerated concerns about migration lowering wages, causing unemployment or damaging the welfare system are unfounded. For instance, Home Secretary David Blunkett noted in 2001 that immigrants contributed £2.5 billion more in taxes than they consumed in tax-supported services in the UK (cited in Karyotis, 2007a:11). Immigrant communities bore the heaviest brunt of the restructuring of European economies in the 1980s (Held *et al.*, 1999: 325) and will have an even greater role to play in the future. This is because, as a report by the European Commission (2000: 21) explains, there are 'growing shortages of labour at both skilled and unskilled levels', amplified by the 'declining and ageing populations in Europe', which make migrants' contribution 'to the labour market, to economic growth and to the sustainability of social protection systems' of crucial importance. Indeed, if the current demographic trend continues, the United Nations (2000) predicted, European economies will need 700 million immigrants for the fifty years to come in order to sustain growth and support their social security systems.

From the above it can be concluded that the dominant belief that migration poses existential threats to society and the state is a fallacy that can be convincingly refuted, if a cost-benefit assessment was to be conducted. Yet, to explain why political elites, in many cases continue to reproduce fabricated truisms that heighten public anti-immigration attitudes, we need to recognize that their rationality is bounded. Preconceived ideas about their responsibility to protect the symbolic boundaries of the nation, in all its manifestations, provide a speculative answer for their persistence on securitization. An additional reason can be sought in the historical and operational context in Europe which saw receiving countries adopt highly restrictive policies since the 1970s. Since the security frame was subsequently fully institutionalized and adopted by the EU and even new immigration countries, policy makers may be hesitant to challenge it, despite the weight of the evidence against it.

Unintended consequences of securitization

Deriving from the above, a second related fallacy is that securitization is the best option to manage perceived or real threats from migration. Occasionally, securitization of certain issues can indeed bear fruitful results for society and the state (e.g. see counterterrorism policies in Greece in Karyotis, 2007b). With regards to migration however, securitization is a counter-productive management strategy, even if we were to accept that the threats discussed merit urgent attention. The unintended consequences of securitization are explored in this section.

First, securitization as a response to perceived threats to the identity of the host nation has the opposite of the desirable effect. Typically, migration in Europe has been short-term, with the majority of economic migrants opting to eventually return to their country of origin. For instance, despite increased migration movements from Central and Eastern Europe during the 1990s, permanent migration declined substantially in the same period (Grabbe, 2001). Paradoxically, it is the very restrictive policies advocated by the security frame that are more likely to lead migrants into settlement. This is because not only do restrictionism and inflexible barriers to entry encourage irregular movements but also they discourage migrants from investing and keeping strong ties with their own countries, in order to secure their access to work in Europe. This was the case for example in Germany after the oil crises in the 1970s, when the restrictive policies introduced encouraged family reunification and ultimately increased Turkish settlement (Entzinger, 1985). As Harris (2002: 31) explains 'preventing people working so that they would not become citizens forced them to become citizens in order to work'.

Second, the securitization of migration also leads to an increased rather than a reduced possibility of physical threats to public order. The scapegoating of migrants and reproduction of the criminal-migrant discourse, amplified by misinformation given by the media and politicians poses a major obstacle to their inclusion in European societies. In turn, as sociologist Robert Agnew (1992) explains, an increased intensity and frequency of strain experiences for migrants, e.g. through discrimination and presentation of negatively valued stimuli, is likely to trigger migrant anger, aggression and criminal behaviour, not reduce it. Examples of such racial tensions include the 2001 Bradford riots in Britain, the 2005 civil unrest in France, and the 2010 riots in Rosarno, Southern Italy. In the last case, hundreds of mostly African immigrants clashed with police during a demonstration, blaming racism and their atrocious living conditions for the violence and carrying placards which read 'We are not animals'.

Furthermore, justifying restrictive policies with reference to the threat of terrorism, particularly since the Seville European Council in 2002, has the unfortunate effect of blurring all types of migrants and incorporating illegal immigrants, labour immigrants and asylum-seekers into a single policing-repression scheme (Statham, 2003; Karyotis, 2007a). For instance, newspapers, such as Britain's biggest selling tabloid *The Sun* play on public fears, by claiming that 'asylum in Britain is now a Trojan Horse for terrorism' (20 January 2003) and that 'terrorists are using Britain's asylum shambles to sneak into the UK and go into hiding' (19 July 2005). Stories like the former, with headlines instructing readers to 'Read this and get angry' inevitably fuel social tensions and insecurities, while being legitimized with reference to official police statistics. Nevertheless, the 2005 London attacks would not have been prevented by stricter immigration and asylum policies or a closing down of borders, which would instead only make it more difficult to scrutinize and screen those that would subsequently attempt to enter European states through illegal routes.

In sum, the assumption that by securitizing migration elites are promoting their country's greater good is also a fallacy. Securitization does not create a safer society but one that lives in permanent fear from real or perceived threats. Paradoxically, it exacerbates negative effects on societal homogeneity and harmony through its distractive unintended consequences. Thus, even if we if we were to accept that migration poses existential threats to identity and public order, securitization does not appear to be conductive to a better way of managing it.

The political legitimacy trap

The discussion so far suggests that securitization is an excessive and ineffective response to the need for migration management. Does it at least serve the interests of the main securitizing actors that support it? In some cases, security professionals and law enforcement agencies involved in the provision of internal security may benefit from the securitization of migration, which may allow them to attract more resources (Bigo, 1994; 2002). However, since migration is 'a highly institutionalized field with a relatively weak level of civil society engagement', it is political elites that are best placed to shape public attitudes and determine 'in a relatively autonomous way' policy outcomes (Statham and Geddes, 2006: 248). This section explores the unintended consequences of securitization to political elites that support it in the first place.

Other than a threat to societal security, migration represents a direct threat to the legitimacy of political elites and the systems of government of the state (Buzan, 1991: 19). Determining who belongs in a community –commonly in an adversarial way- and controlling access to its territory is a defining function of the state, one which, in the final instance, is always determined by its elites. Since migration calls into question these symbolic boundaries of belonging, political elites use securitization in order to maintain a certain myth of control and thus safeguard their legitimacy (Bigo, 1998). A soft stance on migration, elites worry, may prove costly in electoral terms or 'lead to xenophobic popular sentiments and to the rise of anti-migrant political parties that could threaten the regime' (Weiner, 1992: 114). Under these circumstances, elites may sustain the security frame on migration to cement their power positions and prevent public reactions.

Securitization is, in the short term, a convenient and easy way of shifting blame and responsibility for all society's ills and their own failings (Guiraudon, 2003). The climate of crisis it creates, however, disguises its longer-term consequences.

Paradoxically, Bhagwati (2003: 99) notes, 'the ability to control migration has shrunk as the desire to do so has increased. The reality is that borders are beyond control and little can be done to really cut down on immigration.' A vicious circle is the outcome: The 'decisively restrictionist stance' of elites (Statham and Geddes, 2006: 248) creates unattainable public expectations for effective 'defence' against the perceived existential threats that migration poses. Inability to deliver on their promises leaves governing elites and security professionals responsible for controlling migration susceptible to scrutiny and public criticism, which is exploited by anti-migrant parties and other political opponents (Boswell, 1998). This in turn, makes political elites sustain the security frame, even verging in some cases on outright xenophobia, since they feel that 'their policy proposals must compete for this political territory' (Statham, 2003: 167).

These suggest that while securitization may protect the political legitimacy of elites in the short-term, it ends up undermining it in the medium and long term. Securitization hampers elites' ability to support contradictory aims, such as those relating to labour needs for immigration. This is because, as discussed above, under conditions of securitization, the distinctions between 'desirable' economic migrants, asylum seekers and irregular migrants become muddled in the public mind. As a result, construction of the security frame from the top-down creates demand for more securitization from the bottom-up, thus constraining political action and choices.

Conclusion

The analysis in this chapter suggests that securitizing migration is not just a question of ethics and humanity, relating for instance to the rights of migrants and to Europe's pursuit of justice and freedom. It is also not just a philosophical debate between proponents of security vs. liberty, which effectively translates to limitations in the freedom of migrants, who are seen as inferior and dangerous, for the benefit of the security of citizens. It is above all a question of consequences. Since securitization is a top-down process, orchestrated by goal-maximizing elites, it is their inherently bounded rationality that should be questioned in order to explore ways of untangling the security–migration nexus.

Through assessing the consequences of securitization, a number of contradictions can be traced and exemplified. First, the assumption that migration poses such grave dangers that legitimize its securitization and the adoption of exceptional measures by the state is a fallacy. As discussed, many studies have demonstrated that these threat perceptions are socially constructed and grossly exaggerated. Similarly, the conjecture that securitization reduces the level of threat and thus it results in better management of migration is equally false. In the end, securitization has unintended consequences that reduce not only the security of migrants and asylum seekers but also that of those it seeks to protect, by exacerbating threats to identity and public order. Finally, securitization creates a vicious circle of supply and demand for security, which is unattainable and ultimately harms the political legitimacy of securitizing actors and constraints them from pursuing contradictory goals in the area of migration. Elites should therefore seek ways to overturn the hegemony of the security frame, which, it is argued, is damaging to the state as well as to their own interests in the long term.

References

ACPO (2001) 'Guide to Meeting the Policing Needs of Asylum Seekers and Refugees', accessed at: http//www.acpo.police.uk/policies/index.html

Agnew, R. (1992) 'Foundation for a General Strain Theory', in *Criminology*, Vol. 30, No. 1, pp. 47–87

Ayoob, M. (1997) 'Defining Security: A Subaltern Realist Perspective' in Krause, K. and Williams, M. C. (eds) *Critical Security Studies*, Minneapolis: University of Minnesota Press

Balzacq, T. (2005) 'The Three Faces of Securitization: Political Agency, Audience, Context', in *European Journal of International Relations*, Vol. 11, No. 2, pp. 171–201

Balzacq, T. and S. Carrera (2006) *Security Versus Freedom? A Challenge for Europe's Future*, Aldershot: Ashgate

Bhagwati, J. (2003) 'Borders Beyond Control', in *Foreign Affairs*, Jan/Feb, pp. 98–104

Bigo, D. (1994) 'The European Internal Security Field: Stakes and Rivalries in a Newly Developing Area of Police Intervention' in Anderson, M. and den Boer, M. (eds) *Policing across National Boundaries*, London: Pinter, pp. 161–73

Bigo, D. (1998) 'Frontiers and Security in the European Union: The Illusion of Migration Control' in Anderson, M. and Bort, E. (eds) *The Frontiers of Europe*, London: Pinter, pp. 148–64

Bigo, D. (2002) 'Security and Immigration: Toward a Critique of the Governmentality of Unease', in *Alternatives*, Vol. 27, pp. 63–92

Bigo, D. (2006) 'Liberty, Whose Liberty? The Hague Programme and the Conception of Freedom' in Balzacq, T. and Carrera, S. (eds) *Security versus Freedom? A Challenge for Europe's Future*, Aldershot: Ashgate

Booth, K. (1991) 'Security and Emancipation', in *Review of International Studies*, Vol. 17, No. 4, pp. 313–27

Boswell, C. (2008) 'Migration Policy Debates in Europe after 9/11: Securitization, Embedded Liberalism, or the Quest for Legitimation?' in Givens, T. E. Freeman, G. P. and Leal, D. L. (eds) *Immigration Policy and Security: U.S., European, and Commonwealth Perspectives,* New York: Routledge

Brochmann, G. and Hammar, T. (eds) (1999) *Mechanisms of Immigration Control: A Comparative Analysis of European Regulation Policies*, Oxford: Berg Publishers

Buonfino, A. (2004) ‘Between Unity and Plurality: the Politicization and Securitization of the Discourse of Immigration in Europe’, in *New Political Science*, Vol. 26, No. 1, pp. 23–49

Buzan, B. (1991) *People, States and Fear: An Agenda for International Security Studies in the Post-Cold War Era*, 2nd edn, Boulder: Lynne Rienner

Buzan, B., Waever, O and de Wilde, J. (1998) *Security: A New Framework for Analysis*, Boulder: Lynne Reinner

Ceyhan, A. and Tsoukala, A. (2002) ‘The Securitization of Migration in Western Societies: Ambivalent Discourses and Policies’, in *Alternatives*, No. 27 (suppl.), pp. 21–39

Cobain, I. (2006) ‘Exclusive: Inside the Secret and Sinister World of the BNP’, in *The Guardian*, 21 December 2006

Coleman, W. D.(1998) ‘From Protected Development to Market Liberalism: Paradigm Change in Agriculture’, in *Journal of European Public Policy*, Vol. 5, No. 4, pp. 632–51

Dinh, V. D. (2002) ‘Freedom and Security after September 11’, in *Harvard Journal of Law and Public Policy*, Vol. 25, No. 2, pp. 399–406

Dworkin, R. (2005) ‘Terror and the Attack on Civil Liberties’ in Rockmore, T., Margolis, J. and Marsoobian, A. T. (eds) *The Philosophical Challenge of September 11*, Oxford: Oxford University Press

Entman, R. M. (1993) ‘Framing: Toward Clarification of a Fractured Paradigm’, in *Journal of Communication*, Vol. 43, No. 4, pp. 51–8

Entzinger, H. (1985) ‘Return Migration in Western Europe’, in *International Migration*, Vol. 23, No. 2, pp. 263–90

European Commission (2000) ‘Communication from the Commission to the Council and the European Parliament on a Community Immigration Policy’, COM (2000) 757 final, Brussels, 22 November 2000

Faist, T. (2002). ‘Extension Du Domaine de la Lutte’: International Migration and Security before and after 11 September 2001’, in *International Migration Review*, Vol. 36, No. 1, pp. 7–14

Geddes, A. (2003) *The Politics of Migration and Immigration in Europe*, London: Sage

Gigerenzer, G. and Selten, R. (2001) *Bounded Rationality: The Adaptive Toolbox*, Cambridge, MA: MIT Press

Grabbe, H. (2001) *The Sharp Edges of Europe: Security Implications of Extending EU Border Policies Eastwards*, Western European Union: Occasional Paper 13

Griffin, N. (2003) ‘The BNP: Anti-asylum Protest, Racist Sect or Power-Winning Movement?’, 2004, BNP website, accessed at: http://web.archive.org/web/20071010043702/http://www.bnp.org.uk/articles/race_reality.htm

Guild, E. (2003) ‘International Terrorism and EU Immigration, Asylum and Borders Policy: The Unexpected Victims of 11 September 2001’, in *European Foreign Affairs Review*, Vol. 8, No. 2, pp. 331–46

Guiraudon, V. (2000) 'European Integration and Migration Policy: Vertical Policy-Making as Venue Shopping', in *Journal of Common Market Studies*, Vol. 38, No. 2, pp. 251–71

Hampsher-Monk, I. (1992) *A History of Modern Political Thought: Major Political Thinkers from Hobbes to Marx*, Oxford: Blackwell Publishers

Harris, N. (2002) *Thinking the Unthinkable: The Immigration Myth Exposed*, London: I. B. Tauris

Hatton, T. and Williamson, J. (2007) *Global Migration and the World Economy: Two Centuries of Policy and Performance*, Cambridge, MA: MIT Press

Hay, C. (2006) 'Constructivist Institutionalism' in Rhodes, R. A. W., Binder, S. A., Rockman, Bert A. (eds) *The Oxford Handbook of Political Institutions*, Oxford: Oxford University Press, pp. 56–74

Held, D. *et al.* (1999) *Global Transformations: Politics, Economics and Culture*, Cambridge: Polity Press

Hill, C. (2003) *The Changing Politics of Foreign Policy,* Basingstoke: Palgrave Macmillan

Hobbes, T. (1985) *Leviathan*, London: Penguin [first published 1651]

Huysmans, J. (1998) 'The Question of the Limit: Desecuritisation and the Aesthetics of Horror in Political Realism', in *Millennium: Journal of International Studies*, Vol. 27, No. 3, pp. 569–89

Huysmans, J. (2000) 'The European Union and the Securitization of Migration', in *Journal of Common Market Studies,* Vol. 38, No. 5, pp. 751–77

Huysmans, J. (2006) *The Politics of Insecurity: Fear, Migration and Asylum in the EU*, London: Routledge

Ibryamova, N. (2002) 'Migration from Central and Eastern Europe and Societal Security in the European Union', The Jean Monnet Chair, University of Miami: Florida

Jones, B. D. (1999) 'Bounded Rationality', in *Annual Review of Political Science,* Vol. 2, pp. 297–32

Karydis, V. (1998) 'Criminality or Criminalization of Migrants in Greece? An Attempt at Synthesis' in Ruggiero, V., South, N., Taylor, I. (eds) *The New European Criminology*, London and New York: Routledge, pp.357–9

Karyotis, G. (2007a) 'European Migration Policy in the Aftermath of September 11: The Security–Migration Nexus', in *Innovation: The European Journal of Social Science Research*, Vol. 20, No. 1, pp. 1–17

Karyotis, G. (2007b) 'The Securitization of Greek Terrorism and the Arrest of the 'Revolutionary Organization November 17', in *Cooperation and Conflict*, Vol. 42, No. 3, pp. 271–93

Karyotis, G. and Patrikios, S. (2010) 'Religion, Securitization and Anti-Immigration Attitudes: The Case of Greece', in *Journal of Peace Research*, Vol. 47, No. 1, pp. 43–57

Lavenex, S. (2001) 'Migration and the EU's New Eastern Border: Between Realism and Liberalism', in *Journal of European Public Policy*, Vol. 8, No. 1, pp. 24–42

Lohrmann, R. (2000) 'Migrants, Refugees and Insecurity: Current Threats to Peace', in *International Migration*, Vol. 38, No. 4, pp. 3–22
McSweeney, B. (1996) 'Identity and Security: Buzan and the Copenhagen school', in *Review of International Studies*, Vol. 22, No. 1, pp. 81–93
Mill, J. S. (1991) 'Utilitarianism' in Gray, J. (ed.) *On Liberty and Other Essays*, Oxford: Oxford University Press [first published 1861]
Schmitt, C. (1996) *The Concept of the Political*, Chicago: University of Chicago Press [first published 1936]
Simon, H. A. (1995) 'Rationality in Political Behavior', in *Political Psychology*, Vol. 16, pp. 45–61
Sprout, H. and Sprout, M. (1969) 'Environmental Factors in the Study of International Politics' in Rosenau, J. N. (ed.) *International Politics and Foreign Policy*, New York: The Free Press
Statham, P. (2003) 'Understanding the Anti-Asylum Rhetoric: Restrictive Politics or Racist Publics?', in *Political Quarterly*, Vol. 74, No. 1, pp. 163–77
Statham, P. and Geddes, A. (2006) 'Elites and Organized Publics: Who Drives British Immigration Politics and in Which Direction?', in *West European Politics*, Vol. 29, No. 2, pp. 248–69
Tajfel, H. and Turner, J. (1986) 'The Social Identity Theory of Inter-group Behavior' in Worchel, S. and Austin, W. (eds) *Psychology of Intergroup Relations*, Chicago: Nelson-Hall
Thouez, C. (2002) 'Migration and Human Security' Paper prepared by the International Migration Policy Programme, for the Commission on Human Security, accessed at: www.humansecurity-chs.org
Triandafyllidou, A. (2002) 'Greece' in ter Wal, J. (ed.) *Racism and Cultural Diversity in the Mass Media: An Overview of Research and Examples of Good Practice in the EU Member States, 1995–2000*, Vienna: EUMC, pp. 149–72
UNDP (1994) *Human Development Report*, New York: Oxford University Press
United Nations (2000) *Replacement Migration: Is It a Solution to Declining and Ageing Populations?*, United Nations, Population Division, Economic and Social Affairs, New York
van Dijk, T. (1993) *Elite Discourse and Racism*, Newbury Park, Ca: Sage Publishers
Waever, O. (1993) 'Societal Security: the Concept' in Waever, O. *et al. Identity, Migration and the New Security Agenda in Europe*, London: Pinter
Waever, O. (2000) 'Security Agendas Old and New, and How to Survive Them', Paper prepared for the Workshop on 'The Traditional and New Security Agenda: Inferences for the Third World', Universidad Torcuato Di Tella, Working Paper No. 6, Buenos Aires
Waldron J. (2003) 'Security and Liberty: The Image of Balance', in *Journal of Political Philosophy*, Vol. 11, No. 2, pp. 191–210
Weaver, O. (1995) 'Securitization and Desecuritization' in Lipshutz, R. D. (ed.) *On Security*, New York: Columbia University Press, pp. 46–86

Weiner, M. (1992) 'Security, Stability and International Migration', in *International Security*, Vol. 17, No. 3, pp. 91–126
Yergin, D. (1997) *Shattered Peace: The Origins of the Cold War and the National Security State*, Boston: Houghton Mifflin Company
Zedner, L. (2009) *Security*, Key Ideas in Criminology Series, London: Routledge

Chapter 2
Vanishing Borders and Biometric Citizens

Mark Maguire

Introduction

In an influential 2004 essay Giorgio Agamben explains his refusal to travel to the USA and thereby submit to biometric screening. Biometrics, he argues, are at the vanguard of recent attempts to

> accustom citizens to supposedly normal and humane procedures and practices that had always been considered to be exceptional and inhumane. Today's electronically enhanced possibilities of the state to exercise control over its citizens [...] were unimaginable in the past. But there is one threshold in the control and manipulation of bodies, the transgression of which would [...] equal a next step towards what Foucault has referred to as the progressive animalization of man through extremely refined techniques. The electronic registration of finger prints, the subcutaneous tatoo [*sic*] and other such practices must be located on that threshold (Agamben, 2004: 1).

Agamben's stance and much of his evocative terminology are acting as the fashionable anchor tenant for a growing agglomeration of projects. Several commentators (Amoore, 2006; Dillon, 2007) follow Agamben by politicizing biometric technologies in the context of the post-9/11 securitization of migration, a process that seems to target 'potentially risky travellers [...] ranked according to their nationality or race' (Epstein, 2007: 159). However, much of this work over-theorizes biometric security prior to analysing the conditions of its possibility or the diverse cultures in which design, user experience and modes of reception are situated. This chapter represents a call for a more open, historically-informed and comparative discussion of biometric security.

Rather than seeing biometric security as part of current state-led attempts to normalize and render humane exceptional and inhumane practices, the history of biometrics shows its origins in colonialism and the government of civil life. Furthermore, in defiance of approaches that set out zero sum game in which state-led surveillance is on one side and civil liberties are on the other, much may be learned by looking away from the dominant legal structures and debates in Europe and North America and more towards parts of the world characterized by a different working through of security. These 'latitudes' (Ong, 2006) often function as workshops where new forms of convergence between governmentality and technology are hammered out.

This chapter briefly outlines the dimensions of the contemporary revolution in biometric security. There follows a discussion of the first biometric revolution in the late nineteenth century. This historical discussion builds upon earlier work (Maguire, 2009), and here I am retracing the history of biometrics in order to unpick the cultural conditions that shaped modes of reception, paying particular attention to security and liberty. Finally, I connect my analysis of nineteenth century developments to the contemporary reception of biometrics in Ireland and the Middle East. To borrow from Paul Rabinow (2007), I am interested here in the history of the present and the anthropology of the contemporary.

Biometric security today

Biometric security refers to the recognizing of humans and the verification of identities on the basis of physical characteristics which are taken to be unique to individuals. Some biometrics, such as fingerprints, have long been used, but with recent advances in imaging technology, algorithms and database systems, automated recognition on the basis of the iris, face, hand geometry, vascular patterns, the retina, DNA, and even one's body odour is now possible. Biometric security also denotes bio-behavioural biometrics such as handwriting, voice and keystroke patterns, together with experimental fields such as heart rhythm analysis and gait recognition (Jain, Flynn and Ross, 2008).

Biometric security follows a standardized set of criteria – the so-called seven pillars – with which to judge the suitability and efficacy of particular characteristics and systems (EC-DG JRC-IPTS, 2005; Jain, Bolle and Pankanti, 1999). These pillars include: (1) *universality*, i.e. that all humans share this characteristic, (2) *distinctiveness*, that for each person these characteristics are unique to a significant degree, and (3) that the characteristics are *permanent* and will therefore remain largely unchanged throughout life. The pillars also include standardized ratings according to (4) *collectability*, i.e. that the biometric may be efficiently collected, (5) *performance,* or the accuracy of matching, (6) the *acceptability* of the system to users, and, finally, (7) the degree to which the system is open to *circumvention* or 'spoofing'. Some biometrics, such as fingerprints, rank very highly; other biometrics may score well in terms of some of the pillars but not others. Retinal scanning, for example, scores well in terms of universality and distinctiveness but it involves stopping users, disconcerting recognition systems, and it raises ethical issues on account of its medical diagnostic potential.

Biometric security involves characteristics being imaged, 'enrolled' in a database and compared with stored data, while performance systems test for accuracy and filter out variables such as movement or lighting. Therefore, while developments in imaging technology and database systems are crucial to the current revolution in biometric security, the core driver has always been statistics.

There is never an exact match *per se*; rather, biometric security involves identifying tolerable margins of error. To take face recognition as an example, the variability of the human face, together with issues such as rotation, makeup and lighting, presents significant challenges to performance, and false acceptance (FAR) and false rejection rates (FRR) remain high for these systems.

Most people are familiar with biometric security as a result of passing through international airports or other borders. Several commentators have set out to document the uncomfortably proximate relationship between the biometric reading of the human body and racialization. This growing body of literature has shown the ways in which biometric systems rest easily alongside broader processes of racialization (Cole, 2001). However, recent research has also argued that biometric systems encode racism within their design (Browne, 2009; Magnet, 2009). 'Race' plays a complex role in biometric systems, but increasingly it plays a very simple role in what are called 'soft biometrics': systems that screen persons on the basis of crude descriptors such as height or 'ethnicity' before hard biometrics (such as fingerprints) are imaged, in order to increase efficacy (Jain, Flynn and Ross, 2008; Maguire, 2010). Moreover, as the discussion of the origins of biometrics herein will show, 'race' has a long history of association with biometric security.

Beyond areas such as border control, people are experiencing biometric security in previously virgin territory, from luxury cars to biometrically-enabled police handguns, and from employee credentialing to systems for identifying infants in large maternity hospitals. Benefit and welfare disbursements are also a fertile ground for market growth, as are the broader areas of e-government. Much could also be written about secure logical access, especially for online transactions – indeed, with the UK government claiming approximately £1.2 billion in losses per annum due to identity theft, this is becoming one of the major markets.

Naturally, some countries are moving faster than others. The United States, as one might expect, has a biometrics industry considerably pumped since the post-9/11 security concerns gave it a shot in the arm. The Federal Government is the big spender, but curiosities include the fact that over the past decade Walt Disney World has been in second place. This pattern is evident in many parts of the world: while the largest markets for biometric security solutions appear to be 'states', closer inspection reveals that a broader field of 'government' is both the market for and driver of biometrics.

The ways in which new forms of security are shaping and being shaped by an interstitial governmental field is clearly evident in many Middle Eastern countries. In a recent interview with *Arabian Business* a lead consultant for LogicaCMG discussed the United Arab Emirates' new joined-up ID system:

> A national ID card is the first step to enabling a smart society and its citizens to use their national ID cards on a daily basis to carry out daily transactions. Those daily transactions could be, for example, opening a bank account, applying for a credit card, paying for books and services over the internet, and paying for your

> parking – rather than putting a coin into the machine, you top up the chip on your ID card and put your ID card into the machine (Walid, 2007: 1).

According to this consultant, and to many of the experts I have interviewed, the conceptual challenge is to think in terms of society as being composed of differentiated access to spaces and privileges – a series of graduated e-gates – wherein 'the software is a good means to make sure that you don't have any people that you do not want in a particular area' (ibid.). From 2008 onwards, I attended numerous biometric security events, showcases and conferences in Europe and North America and have often been struck by what one might term the cultural imaginary of industry experts. Whether they are working on using biometrics in hospital settings or for border control, industry experts often see their 'solutions' as creating new forms of sociality, altering the way people will live in and move through the world.

Security 'solutions' are developed, marketed and deployed within a vast global industry, and companies located in countries such as Ireland often have clients throughout the world. One must recognize, therefore, that biometric security deployments in the Middle East, a region often described as the 'test bed', have direct connections to developments in Europe and especially to developments in Europe's border zones, which are, like the US-Mexico border, becoming laboratories for the testing of new forms of security (see chapter by Ribas-Mateos in this volume). New security thinking, potential corporate, civil or military markets, and even reflexive approaches to issues of data privacy and liberties, are held together in an industry without borders. Consequently, any analysis of contemporary biometrics must develop from a multi-sited research imaginary. Moreover, while 'bleeding edge' research in database management, encryption and imaging are often strikingly innovative, core aspects of contemporary biometric security emerged in the nineteenth century. The history of biometrics is certainly one of diffuse developments in scattered locations, but it is nonetheless important to understand the sites of problematizations, the political mentalities, and the actual interventions.

The revolution in biometric identification

Current scholarship holds that biometrics identification emerged in the nineteenth century within a broader identity revolution. The body entered systems of securitization when efforts by detectives and administrators intersected with more 'scientific' fields such as anthropometry. Biometrics, denoting the measurement of life, involved examining the human body for signs of hereditary forces, deviance or 'race'. However, key figures such as Francis Galton, Parisian detective Alphonse Bertillon and colonial administrator William Herschel realized that biometrics could identify persons. In nineteenth century London and Paris, and in the colonies, identity was a site of problematization.

The conditions for the possibility of biometric identification are easy to detect in nineteenth-century London. The image of the city as a home to 'outcasts', an unidentifiable and mobile population, is certainly available in Friedrich Engels's *The Condition of the Working-Class* (1892) and Henry Mayhew's *London Labour and the London Poor* (1861). But that image is sharper still in mainstream commentaries. Even as late as 1901, Walter Besant could describe the East End of London as a dark continent within the metropolis:

> The population is greater than that of Berlin and Vienna. [...] In the streets there are never seen any private carriages; there is no fashionable quarter. [...] People, shops, houses, conveyances – all together are stamped with the unmistakable seal of the working class [...] perhaps the strangest thing of all is this: in a city of two million people there are no hotels! That means, of course, that there are no visitors (quoted in Stedman Jones, 1971: 14).

Nineteenth-century Paris was also imagined as a city in which rich and poor were dangerously proximate, and fears of unidentifiable populations loomed large (Gillis, 1989; Fuchs and Moch, 1990; Pinkney, 1953; Thompson, 1997). The demographic historian Louis Chevalier has argued that rumours and paranoia were accompanied by the spectacular inscription of justice onto the bodies of offenders. In 1828, *Journal des Débats* covered a case of pillory outside the Palais de Justice during which the crime was branded onto an offender's body with a red-hot iron (Chevalier, 1973: 4). Punishments such as branding, the cropping of ears and tattooing – permanent, visible and irreversible marking of the body – were also methods of identification (cf. Agamben, 2004) and new and more efficient methods were an urgent requirement. The problem of mobile identities took shape in the figures of the recidivists, '*Habitués* and prison scum, mangy black sheep that threaten the whole heard with contagion', according to Prosecutor Charles Bertheau (Nye, 1984: 73; de Marsangy, 1844).

For much of the nineteenth century, however, European criminology did not examine the criminal body for individual identity but rather for anthropological type. The Italian 'positive school' of Cesare Lombroso imagined crime to be hereditary and atavistic, and accordingly the underlying moral state broke out as degenerative stigmata. As David Horn (2003) has shown, the positivists' gaze included physiognomic typologies, behavioural metrics such as gait or handedness, and measurements from a bewildering array of instruments – tachianthropometers, auricular goniometers, even olfactometers. Accordingly, to gaze upon the individual body-as-specimen meant to recognize that which was at one and the same time beneath its surface and running through and beyond it. French criminologists, recognizing the fallacy (*cum hoc ergo propter hoc*), derided their Italian counterparts by comparing their output to coffee – stimulating, but no nutritional value. Moreover, from distant colonies to the metropolitan heartlands of London and Paris the efficient administration of populations required knowledge of individuals and on this criminal anthropology was largely silent.

Prior to the 1870s, efforts were made to identify recidivists in Paris using unsystematic photographs – sometimes even silhouettes – accompanied by unverifiable names. This changed when Alphonse Bertillon joined the police in 1879 and turned anthropometric science into a tool for individual identification. Bertillon's anthropometric measurements and systematic descriptions of the body were augmented by standardized photographs (and eventually fingerprints). The resulting data was encoded on cards and meticulously filed. The system, essentially, identified through the process of elimination by working away from the normal distribution (Bertillon, 1893: 23). But *Bertillonage* failed to reach the bar of conclusive identification. It also failed as an export, proving insensitive to the administrative and physiological conditions of the colonies. Bertillon's legacy is certainly important – the mug shot, forensic photography and database-like filing, for example – but there is also much to be learned from *Bertillonage* as a cultural artefact. His primary innovation was to turn the biometric science of anthropometry towards the individual units of population rather than types. Perhaps just as importantly, his vision was for a world of securitized identities in which the police gazed through an already encoded eye upon a population composed of individuals with second lives as files.

Contemporary biometric security, however, owes more to Charles Darwin's cousin Francis Galton. He laid the foundations of modern statistics, popularized the notional division between nature and nurture, and became the 'father' of eugenics. Heidegger (1982: 28) once wrote, 'That period we call modern [...] is defined by the fact that Man becomes the centre and measure of all beings'. In the new province of knowledge opened by Darwin, Man was placed amongst all other beings, and Galton measured obsessively. He studied the populations in most of the institutions that characterized Victorian society, from prisons and hospitals to schools. His quest took him from experiments in composite photography (Galton, 1877: 344–7) to the then respected science of anthropometry, and even to early attempts at face recognition (Galton, 1910: 127). He always searched for the racial types that were, in a sense, always already there.

Throughout the 1880s Galton was collecting and analyzing data from his Anthropometric Laboratory in Kensington. His work suggested applications in socio-medical surveillance, such as the screening of candidates for the military and civil services (Galton, 1890: 8). He was also aware that biometrics could potentially verify individual identity, and he kept a careful eye on Bertillon's efforts in Paris. In 1880 Galton's attention was called to the possibilities of fingerprinting, first by the missionary Henry Faulds and then by East India Company administrator and later civil servant William Herschel. While both men played a role in the 'discovery' of fingerprinting, Herschel's contribution was crucial in that he had amassed samples and observations over several decades in India. Moreover, Herschel had actually uses biometrics for security. As Chandak Sengoopta has shown in *Imprint of the Raj,* fingerprinting, that most exemplary of biometrics, was born of Empire when Herschel, probably influenced by indigenous 'signatures', used fingerprints to root out fake pensioners, faked deaths, prisoners-for-hire and, more generally, to control a huge, seemingly undifferentiated population.

While Galton was the first to establish the possibilities of fingerprint classification, the traffic of Empire did not stop with him. Fingerprint classification was further refined in India by Edward Henry, Azizul Haque and Chandra Bose and was institutionalized as an indispensable weapon in the police arsenal following the recommendations of the Home Office Belper Committee in 1900. Fingerprinting also spread to Argentina where Juan Vucetich developed another classification system that jockeyed for position with *Bertillonage* before eventually winning out in South America. Indeed, by the turn-of-the-century in Argentina a register of immigrants was augmented by data on prostitutes, conscripts, bank employees and voters, eventually leading one utopian thinker to dream of an international book of persons (Ruggiero, 2001). Even earlier, in San Francisco fingerprinting was used to exclude returning Chinese labourers. Thus, by the turn-of-the-century the standardized photograph, fixed personal names represented by signatures, fingerprints and techniques for coding and data management combined to form a functioning biometric security assemblage.

Liberty and security

The emergence of fingerprinting is well documented as a chapter in the story of modern policing. However, scholars such as Cole (2001) and Sengoopta (2004) go further by situating fingerprinting amid the widespread Victorian-era problematizations of identity. Cole rescues fingerprinting from a Whiggish history that begins with its contested 'discovery' and ends with institutional acceptance as a colour, class and gender-blind science routinely accepted as proof positive. He notes the relationships between fingerprinting and other recognition technologies, its emergence from the shadow of anthropometry, and the questionable representation of its scientific basis. Sengoopta extends our understanding of its origins by uncovering its tangled roots in the broad landscape of Empire. Fingerprinting, he argues, was canalized in modern policing, forensic identification and immigration control in Britain but by the early twentieth century was a routine feature of imperial civil administration. Accordingly, he poses the following question: 'Was what was good for the Empire good for England?' His question calls attention to the important historical relationships between the securitization of identity and civil liberties:

> [*The*] individual's right to live and die unobserved by bureaucracy was a sacred principle of English liberty. Consequently, it was only in the Empire – to the heathen, subject population of which such a sacred principle did not, of course, apply – that fingerprinting came into relatively general use (Cole, 2004: 119).

Should one conclude that early on civil liberties arrested the advancement of biometrics, liberties that were available to rulers but not to the ruled? And, are civil liberties capable of preventing contemporary biometric security from crossing the 'threshold' identified by Giorgio Agamben, a Rubicon beyond which lies the full animalization of man? Sengoopta argues that clear differences separated liberal England from its Empire, namely, authoritarian government in India, the perception that personation rather than crime was the core problem, and the consequences for administration of illiteracy.

India certainly experienced authoritarian government but it is questionable whether this translated into radically different inscriptions of rule onto the body. During the era of the East India Company an individual's name and offence was tattooed onto the forehead upon being handed down a life sentence. However, as noted herein, pillory and branding remained common in many European countries until the early to mid-nineteenth century. What of the colonial preoccupation with personation? Statements on this by figures such as Francis Galton were numerous; but Galton was clear that personation was a manifestation of a broader problem of government: the inability to distinguish, on the basis of well established documentary evidence, fixed names and signatures. Empire required definite, concrete and routine bureaucratic interactions with populations in order to be (Anderson, 1991: 163–85). In this context, biometrics enabled the efficient and legible administration of populations via biometric 'signatures', ostensibly free of caste, gender, 'racc', rank or religion. Literacy was needed as a bridge to the population, and biometrics narrowed the administrative gap. But that bridge rested on broader, uneven and discontinuous processes through which colonial societies were governed, processes that unfolded simultaneously 'back home' (Comaroff, 1993).

It is also possible to argue that crime and mobile identities were thoroughly blended together in metropolitan Europe and imperial contexts. For a European example one may look no further than Charles Booth's turn-of-the-century maps of London in which the colour-coded key ranged from the stony-yellow that denoted the 'Upper-middle and Upper classes; Wealthy' to black for 'Lowest class; vicious, semi-criminal'. Booth argued that London was home to a class composed of those who 'are incapable of better work' (Booth, 1887: 375). This population could, he argued, exchange their 'half-fed and half-idle and wholly unregulated life for a *disciplined existence*' (ibid. my emphasis). Statements such as this bring into view the problematization of mobility and identity and the associated political mentalities within which improvement and reform were always predicated by discipline and security. Consequently, here I am arguing that nineteenth-century biometrics such as fingerprinting were canalized in criminal and forensic detection in the western world not because of the bulwark of civil liberties *per se* but rather because such societies were already 'disciplined' in a multitude of other ways – a *disciplined existence* that was available to the rulers but not (as yet) to the ruled.

Discipline and security

Michel Foucault begins his celebrated discussion of panopticism in *Discipline and Punish* with an example of seventeenth century plague measures, which included strict partitioning and control of persons and things. Each day, according to Foucault, syndics enumerated the living and the dead, commanding individuals to appear at their windows and acknowledge their 'true' names. This was a terrifyingly 'enclosed, segmented space, observed at every point, in which the individuals are inserted in a fixed place [*and*] all events are recorded, [...] a compact model of the disciplinary mechanism' (1995 [1977]: 197). Foucault, however, argues that this form of government was too heavy and closed to circulation to last. In contrast, panopticism, as exemplified by Jeremy Bentham's prison plan, is neither an architectural form nor a specific technology but rather a diagram of discipline. But neither Bentham nor Foucault restricted their concerns to the powerful image of the bowed inmate, subject to and instrument of the ocular tyranny of surveillance.

Discipline and Punish is concerned with the striking family resemblance between a whole series of institutions, such as the army, school, hospital, factory and prison. These seemingly enclosed institutions shared a similar microphysics of power in which the individual was a necessary unit, assigned a place, ranked according to others, and generalized in whole bodies of knowledge. As Foucault argues, 'discipline fixes; it arrests or regulates movements; it clears up confusion; it dissipates compact groupings of individuals wandering about the country in unpredictable ways; it establishes calculated distributions' (1995 [1977]: 219). In this sense, a 'disciplined existence' also denotes production and amplification:

> Discipline is basically the mechanism of power through which we come to control the social body in its finest elements, [...] which is to say individuals. [...] How to oversee someone, how to control their conduct, their behaviour, their aptitudes, how to intensify their performance, multiply their capacities, how to put them in the place where they will be most useful: this is what discipline is, in my sense (Foucault, 2007: 159).

In nineteenth and early twentieth-century European societies, biometric identification was set to the tasks of making legible certain populations, fixing their identities and movements. But the target populations for biometric identification were not just the domestic equivalents of overseas imperial subjects. Some New York banks, for example, demanded fingerprint-based identification as early as 1912, from a population unlikely to be illiterate, alien, stigmatized or exceptions to liberties and rights (see *Irish Times,* 1912: 24).

The fundamental difficulty in many discussions of biometric security is the manner in which historical evidence is weighed on scales which have liberty on one side and security on the other. Outside the specific area of biometrics, Semple (1993) has noted that scholars have often found it difficult to square the image of Bentham the social reformer with his vision of the panoptic, privatized, iron and glass prison, or his proposal for universal tattooing for the purposes of identification. But Bentham did not see a philosophical contradiction between liberty and security. For example, in 'Indirect Methods of Preventing Crimes' Bentham notes the importance to 'general security' of recognizing and locating individuals: '[*L*]ess is to be feared from those who have a fixed habitation, property, or a family. The danger arises from those who, from their indigence or their independence of all ties, can easily conceal their movements from the eye of justice' (1843 Chpt. XII: Problem ix). His infamous proposal was to combine policing through census-style knowledge with the tattooing of 'proper names' onto individuals' wrists, especially to combat fraud. Bentham does not offer a simplistic discussion of the 'balance' between security and liberty; rather, he formulates the argument that liberties would in fact be enhanced by the *universal* measures. His point was that security predicates liberty; liberty is possible only where security has been achieved. Bentham was clear that instruments of government such as identification or the panopticon would fix, identify and locate individuals in order to maximize efficiency, increase productivity – and enhance liberty.

Perhaps, then, one of the early checks to the universal roll out of biometric technologies was not liberties *per se* but a fear of excessive regulation at the expense of circulation, a fear of what one might find. Universal identification was debated in many countries and those debates frequently indicate careful consideration of the productive aspects of security. In 1909, for example, an *Irish Times* editorial demanded that if fingerprint registration were to take place for civil purposes then it should be adopted universally:

> [*People*] are gradually beginning to realize, in the words of the *Roi Soleil,* that 'L'État c'est Moi'; [...] the state thus reinforced from the multitude recognises its responsibility to the individual in a more direct and benign manner than it used to do. It looks after his health, his safety, and the quality of his food; last of all, it has stepped in to mitigate the rigours of old age. But the state cannot do these things efficiently unless it has knowledge of the individuals of whom it is composed (*The Irish Times*, 1909: 6).

Foucault explains how the institutional landscape of the disciplinary society was tending, even in the nineteenth century, towards dispersal into 'flexible methods of control' (Foucault, 1995 [1977]: 211). What, then, of the contemporary moment and the rollout of biometric security in regions such as Europe and the Middle East? It is striking that the future envisioned for the Middle East by security consultants so closely resembles Gilles Deleuze's 'postscript' to *Discipline and Punish*:

> We don't have to stray into science fiction to find a control mechanism that can fix the position of any element at any given moment – [...] a town where anyone can leave their flat, their street, their neighbourhood, using their (individual) electronic card that opens this or that barrier; but the card may also be rejected on a particular day, or between certain times of day; it doesn't depend on the

barrier but on the computer that is making sure that everyone is in a permissible place, and effecting a universal modulation (Deleuze, 1997: 182).

How close we are to this vision, and the particular and accented ways in which security is articulated today is the subject of the following discussion of biometrics in Ireland and Qatar.

From the British Isles to the Middle East

Nation-states around the world are attempting to promote the mobility necessary for participation in the global economy while at the same time controlling that same mobility. Mobility in this sense denotes international travel and access through other e-gates. Several commentators argue that biometric security is both marking and enabling an unbundling of the more traditional relationships between the nation-state and citizenship in favour of the kinds of graduated access to spaces and privileges afforded by the securitization of identity (Lyon, 2009).

For much of the twentieth century Irish immigration policy was transposed from British policy, and an unusual agreement emerged called the Common Travel Area (CTA) (Maguire and Cassidy, 2009). The CTA ensured a steady supply of Irish labour to Britain. The British Government also provided their Irish counterparts with lists of suspects and aliens to keep under surveillance or deport. Both countries also had a tacit agreement to keep their immigration legislation in step. Ireland, in common with other Member States, is now moving into a future in which the EU will be the main driver in immigration and asylum policy, a future in which biometrics will play an important role. In essence, the UK government is attempting to roll out an e-borders system that prioritizes individual biometric security over and above the security inherent in the older nation-state and citizenship-based Common Travel Area.

Since 2000 there has been a growing push to transform the UK/Irish border into a 'smart border' in the style of Canada and the USA. The UK Border Agency envisaged an advance passenger information system and identity clearance for all land and sea crossings by 2014 (Cross, 2009; British-Irish Assembly, 2009). The Irish government, propelled by the legal requirements that followed the Good Friday Agreement in Northern Ireland, moved quickly to introduce legislation on biometric security under the *Immigration, Residency and Protection Bill.* While there was no requirement *per se* to show a passport upon, for example, departing Belfast for the UK, Northern politicians rallied against the proposal, in part because of the symbolic nature of the changes, and the UK's proposal to end the CTA was defeated in the House of Lords, though the proposal was successfully resurrected thereafter. While the moves are temporarily stalled, it is nonetheless interesting to mine into UK and Irish thinking on biometric security.

The UK's position on smart borders is most clearly articulated in the 2005 Home Office publication, *Controlling our Borders.* Therein, it is Prime Minister Tony Blair's preface that remains with the reader:

> We will finger print visitors who need visas, and those planning longer stays, before they arrive. We will, where necessary, use our powers to demand financial bonds from migrants [...] to guarantee their return home. [...] We will replace out-dated and confusing rules with a clear and modern points system so we only allow into Britain the people and skills our economy needs. Those who want to settle permanently in the UK will have to show they bring long-term benefit to our country (Home Office, 2005: 6).

The stalled Irish immigration bill also indexes new thinking on mobility as a security problem, especially in its provisions for biometric data. When it is finally transposed, foreign nationals will be required to provide such data to an immigration officer or the police and it will be an offence to refuse. Biometric security will also involve a register of foreign nationals in line with broader EU systems and the Eurodac database. Responses have been given by a variety of NGOs that have expressed concern over the lack of public debate, methods of data collection, and the possible implications for data protection. No attention has been given to the evidence indicating that many of the decisions currently being made about policy, legislation and appropriate technologies are predicated by the need for interoperability with civil uses (see State Watch, 2005).

A brief survey of biometric security in Ireland reveals uses well beyond areas such as border control and asylum. Ireland is one of the world's biometrics 'research hubs' and it offers an interesting case study of the current markets for biometrics and the concerns they engender. The more controversial cases include the TruancyGuard fingerprint identification system currently operating in a South Dublin school to combat unwanted persons on school grounds, absenteeism and the faking of attendance (the system is also used in factories in Malaysia). The company and school management have been careful to note that encryption renders the data secure, and it is issues of data protection rather than liberties that have dominated debates. This case prompted the Irish Data Protection Commissioner (DPC) to issue guidelines calling for proportionality and cautioning against excessive measures. Furthermore, in January 2010, a major Irish retailer appealed to the circuit court against a DPC enforcement order over the use of palm scanners to monitor staff attendance. There are, however, many sites where biometrics are used without any public attention whatsoever, from public bodies such as county councils to the celebrated national Abbey Theatre.

Cases such as the use of biometrics in schools prompted the Irish Council for Bioethics to issue an important opinion report in 2009. The report is careful to avoid challenging biometric technologies *per se* and instead focuses on the proportionality of their uses, the implications for data privacy and, to a much lesser degree, civil liberties. Furthermore, this report begins from the position that individual rights may be overridden in exceptional circumstances, for the common good. Accordingly, the position of asylum seekers as legal exceptions is noticeably absent and only raised *vis-à-vis* the European common asylum context (2009: 23).

It is not surprising that this report has at its centre the 'balance' between security and privacy. Ireland has relatively strong data protection acts (1988, 2003). All EU Member States have similar instruments in place, and EU directives have placed privacy in the position of being a fundamental right. But the speed of the current drive towards biometric security is outpacing legislation and informed public debate. Towards the end of their comprehensive report, the Irish Council for Bioethics, citing the concerns raised by Agamben, turn to the 'balance' between security and privacy. As the individual's right to privacy does not subsist where the common good is at stake, it is recognized that we are in fact discussing the balancing of security and *liberty*. In a passage that adumbrates John Stuart Mill's *On Liberty* (1863), the report argues that 'rights may legitimately be limited or overridden where a given biometric application is deemed to be necessary to uphold some common good' (2009: 89). However, a closer reading of the work of John Stuart Mill, or indeed that of Jeremy Bentham before him, renders discussions of the 'balance' between liberty and security problematic at best. In *Utilitarianism* and *Considerations on Representative Government*, Mill identifies security as 'the most vital of interests', the basic need of all societies and the fundamental priority of government (see chapter by Karyotis in this volume). For Mill, good government implied a good disposition of persons and things: the security of persons and private property. Discussing the fallacious 'balance' between liberty and security ostensibly derived from liberal philosophy, Mark Neocleous concludes:

> If security is the supreme concept of bourgeois society it is equally the supreme concept of liberal ideology. On this view we should read liberalism less as a philosophy of liberty and more as a *technique of security*, [...] a strategy of governance in which security is deployed as liberty (Neocleous, 2008: 31).

Accordingly, biometric security must not be understood as constituting new forms of state-led surveillance threatening our civil liberties but rather as an extension of much older ways of governing individuals and populations that condition the possibility of discourses on liberties. These discourses turn on a configuration of security that suggests gates, bans, segmentation, but also security as a maximizing and productive force. Security enhances certain kinds of mobility and liberties for some, while denying others access to spaces and liberties.

The ramifications of the biometric securitization of identity in Europe may clearly be seen in the area of migration management. Take for example the recent Commission of the European Communities report on the next steps in EU border management:

> Persons could be granted 'Registered Traveller' status after appropriate screening [...]. These could as a minimum include [...] holding a biometric passport. [*And*] at the border itself, the introduction of automated gates could enable the automated verification of travellers' identity without the intervention of border guards. A machine would read the biometric data [...], one border

> guard should be able to oversee up to ten automated border gates in operation. Automated border controls for bona fide travellers would provide major benefits in time savings on crossing the external border and allow border authorities to focus their resources on those groups of third country nationals that require more attention (Commission of the European Communities, 2008: 3–6 passim).

What is envisaged here is policies, legislation and technologies being used to enhance the mobility and degree of access for 'high value' migrants – a securitized elite. Indeed, the position of elites demands particular attention not just as beneficiaries of securitization but also as a key grouping driving these processes (see chapter by Karyotis in this volume). On the other side, however, e-border gates will be closed for 'nationals that require more attention'. And even within Europe, especially for those within or falling out of the asylum determinations processes, biometric securitization promises electronic segregation and hidden populations continually shadowed by their data selves (see chapter by Schuster in this volume).

The biometric securitization of identity is therefore central to the construction of a Europe of 'Justice, Freedom and Security'. And the 'next steps' in border management are clearly understood to involve leveraging off the possibilities of new security technologies in order to imagine a world of graduated e-gates. Importantly however, those next steps have already been taken elsewhere.

'They're much more receptive there – the government, I mean'

The above epigraph/header is taken from a discussion with a biometrics consultant at the 2008 Biometrics Convention in London. He refers to is the Gulf State of Qatar. In 2005, the Ministry of Interior of Qatar moved to replace existing ID cards with high-end cards, combining biometrics – fingerprints and iris data – with contact and contactless technologies. The vision is to combine individual identity verification with a full range of techno-social activity. The ID cards will perform the role of a traditional card but also grant access to Hukoomi, the celebrated Qatar e-government portal, function as a driving license and be acceptable as access tokens for healthcare. Moreover, the e-government system is to extend into the home. The smart card will allow access to civil records and enable data transfer for household employees. And, if one leaves home, the new ID card will enable, for an additional fee, e-gate recognition to help speed through airport security. The Qatar Government intends to build a *universal* National Identification System and National Biometric Repository for all citizens and residents.

Much of the technology in operation in Qatar has been delivered under contract by Daon Biometric Systems, an Irish-based company. Daon operates in locations from the UK e-border to Reno Airport, but Qatar continues to be one of the leading 'test beds' in the Gulf – a market some consultants have estimated to be worth hundreds of millions of dollars. According to Daon's local partner Gemalto – their advertising strap is 'the security to be free' – the potential uses are matched to Qatar's exemplary e-government portal. What is envisaged in this case is something that could be termed biometric citizenship. Thus, if the Commission of the European Communities 'next steps' report (2008), with its systems for bona fide 'registered' travellers, is suggestive of an EU that will use biometrics to segregate mobility according to value, then the example of Qatar indicates that security technologies have an even greater potential to redraw the lines of societal difference.

In 2008 the *Gulf Times* reported that, due to the advent of bio-IDs, passports for travel between countries such as Saudi Arabia and Qatar would be permissible using 'smart cards' (*Gulf Times*, 17 June 2008). Regional solidarity and ease of movement were cited as the main benefits by users; while the *Gulf Times* also reported that emigration officers at borders were satisfied they could acquire 'all data about the passenger by just swiping the card on the electronic machines at borders' (ibid.). Such a comment would act as a magnet for criticism about data protection and civil liberty in Europe, yet in Qatar the mode of reception is conditioned by very specific variables.

Since independence from Britain in 1972, Qatar is an absolute monarchy in which the Emir from the ruling Al Thani family is head of government and central to the combination of civil codes and Islamic law that obtains. Less than thirty percent of the population is permitted to vote or hold office. Furthermore, while the constitution guarantees freedom of expression these rights are limited in practice. Privacy is also inviolable under article 37 of the constitution but is 'limited' by other laws, especially those laws that refer to security.

The most strikingly sociological feature of Qatar, however, is that most of the workforce is composed of foreign-born immigrants. Protests over living and working conditions are relatively frequent, arising from abuses especially in the area of domestic labour. Accusations of false imprisonment, arbitrary deportation and even torture have been made. During a fieldwork trip to Doha last year I noted gossip to the effect that employers were posting wanted posters appealing for information on run-away domestic workers. One of the key approaches taken by government is to sign bi-lateral agreements with sending countries, such as the on-again-off-again agreement for a securitized labour migration corridor with Vietnam. Since 2008 there has also been an ongoing plan for a separate workers' city in Doha in part to ameliorate conditions in existing workers' compounds, but perhaps more accurately described as an effort to ensure a better disposition of persons and things. It would be easy, therefore, to conclude that Qatar shows an excess of government. In such a view, the exceptional measures of securitization normally confined to exceptional populations such as foreign workers are extended to all. In short, should we see Doha as a future city wherein mobility is governed by the universal modulation enabled by reading the human body?

In his 1978 lectures to the Collège de France, Michel Foucault asks: can we speak of a society of security in which power is dominated by the technologies of security? He argues for attention to the relationships between security and circulation: 'a matter of maximizing the positive elements […] and of minimizing the risky and inconvenient, like theft and disease, while knowing that they will never be suppressed' (2004: 19). While the disciplinary societies of old, in which biometric security was born, allowed little to escape, 'the apparatus of security [...] lets things happen; security "laissez-faire" [*in inverted commas in the manuscript*], in the positive sense of the expression' (ibid.: 45). Theoretically and empirically, the securitization of Qatar offers an example of contemporary discourses of security in which liberty is understood *as* security. A society composed of e-gates has the advantage of maximizing the security and liberties of some, allowing things to happen, while minimizing the risks associated with the mobility and identity of suspect populations.

Conclusions

Today, biometric security is a vast global industry without borders that promises solutions to some of the most pressing governmental problems. As governments, corporations and public bodies throughout the world attempt to manage mobility, new security technologies are finding an increasing role. Recent efforts to theorize biometrics have tended, however, to over-state the degree to which both these technological solutions are new. Rather than examining biometric security as a part of state-led attempts to lead societies across a Rubicon beyond which lies the full animalization of man, here I have been concerned to unpick the cultural conditions for the possibility of biometric security. Technologies such as fingerprinting were born in nineteenth-century colonial contexts in which individual identities and mobility were a site of problematization; so too in nineteenth-century European cities, when anthropometric science, fingerprinting, photography, coding and filing were assembled as a means to fix populations imagined to be unruly and potentially set against society.

A discussion of the historical conditions for the possibility of biometric security also calls attention to the ways in which security emerged within what Foucault has termed 'disciplinary' societies. It is important to recognize that for key figures, from Bentham to Galton, the universal securitization of identity was not considered to be exceptional, inhumane or a strategy of government only befitting the poor or distant colonial subjects. It is clear from any reading of their work that attention must be given to security as intrinsic to the foundation of an orderly and disciplined society. Security in this sense meant the removal of confusions, unwanted mobility and shifting identities and, instead, the government of secure identities to increase efficiency, and maximize positive elements.

While nineteenth century developments have important lessons to teach, the contemporary world of biometric security is also marked off as different. The speed of developments in statistics, imaging and database technologies is outpacing legal protections of privacy and public debates over liberties. A history of the present reveals the need to deepen and extend critiques of security in order to develop a theoretical approach to new security technologies, and a new approach is necessary to apprehend not just the exclusionary capacities of biometrics but also the manner in which they offer 'the security to be free'.

Finally, by appreciating that current developments in biometric security in European countries are deeply enmeshed with developments elsewhere in the world it becomes possible to apprehend the scale and direction of those developments. One consequence of this is that research that imagines biometric security to be primarily concerned with border control and asylum shows its shortcomings. It is doubtless the case that Europe is facing a future of increasing xenophobia, Islamophobia and racism, and efforts to securitize migration are likely to become more intense and more applied to asylum-seeking migrants – Galton's (1910: 127) nineteenth-century term for those who would be targeted by biometrics, 'refugee criminals', has never seemed so apt. However, as this chapter has shown, there is much more to biometric security than simply border control. Today (as it did during the nineteenth-century) biometric security offers the possibility to imagine and configure society as a series of e-gates with differentiated access, a society in which technology maps onto the old problem identified in North American migration studies – 'one can be in but not of America'. Alongside more informed discussions of vanishing borders, examples such as the Gulf State of Qatar also point to mobility and securitized identities as central to emergent governmental discourses of societal maximization, efficiency and productivity. In the 'test beds' of the Middle East one can properly speak of biometric citizens for whom border crossings, secure building access, healthcare, online access, financial transactions, and perhaps even parking, are facilitated by the characteristics of their bodies and their shadow, coded selves. The anthropology of the contemporary, therefore, calls for attention to the under-explored question of biometric rights.

References

Agamben, G. (2004) 'Bodies Without Words: Against the Biopolitical Tatoo', in *German Law Journal*, Vol. 50, No.2, pp. 168–9

Amoore, L. (2006) 'Biometric Borders: Governing Mobilities in the War on Terror', in *Political Geography*, Vol. 25, pp. 336–51

Anderson, B. (1991) *Imagined Communities: Reflections on the Origins and Spread of Nationalism*, 2nd edn, London: Verso

Bentham, J. (2001) *The Works of Jeremy Bentham: Published under the Superintendence of His Executor, John Bowring. Volume 1 (Principles of Morals and Legislation, Fragment on Government, Civil Code, Penal Law)*, London: Adamant Media Corporation [first published in 1843]

Bertillon, A. (1893) *Instructions signalétiques pour l'identification anthropométrique*, Paris

Booth, C. (1887) 'The Inhabitants of the Tower Hamlets (School Board Division), their Condition and Occupations', in *Journal of the Royal Statistical Society*, Vol. 50, pp. 326–401

British-Irish Assembly (2009) 'Minutes of the Thirty-Ninth Plenary Conference', Swansea Marriott, 19 and 20 October

Browne, S. (2009) 'Digital Epidermalization: Race Identity and Biometrics', in *Critical Sociology,* Vol. 36, No. 1, pp. 131–50

Chevalier, L. (1973) *Labouring Classes and Dangerous Classes in Paris during the First Half of the Nineteenth Century*, London: Routledge & Kegan Paul

Cole, S. (2001) *Suspect Identities: A History of Fingerprinting and Criminal Identification*, Boston: Harvard University Press

Comaroff, J. (1993) 'Homemade Hegemony' in Comaroff, J. and Comaroff, J. *Ethnography and the Historical Imagination* by Jean and John Comaroff, pp. 265–95, Boulder: Westview

Commission of the European Communities, COM(2008) 69 final. (2008) *Communication from the Commission to the European Parliament, the Council, the European Economic and Social Committee and the Committee of the Regions Preparing the Next Steps in Border Management in the European Union*, Brussels: EU Commission

Cross, M. (2009) 'The Illegal e-Borders Disaster', in *The Guardian*, 18 December 2009

Darwin, C. (2009) *On the Origin of Species: By Means of Natural Selection or The Preservation of Favoured Races in the Struggle for Life*, London and New York: Penguin Classics [first published in 1859]

Deleuze, G. (1997) *Negotiations, 1972–1990*, New York: Columbia University Press

De Marsangy, A. (1844) *De la Récidive, ou des moyens les plus efficaces pour constater rechercher et réprimer les rechutes dans toute infraction à la loi pénale*, Paris: Cosse

Dillon, M. (2007) 'Governing Terror: The State of Emergency of Biopolitical Emergence', in *International Political Sociology*, Vol. 1, No. 1, pp. 7–28

EC-DG JRC-IPTS (2005) *Biometrics at the Frontiers: Assessing the Impact on Society for the European Parliament Committee on Citizens' Freedoms and Rights, Justice and Home Affairs (LIBE)*, Brussels: European Commission

Engles, F. (1952) *The Condition of the Working-Class in England in 1844*, London: Allen & Unwin [first published in 1892]

Epstein, C. (2007) 'Guilty Bodies, Productive Bodies, Destructive Bodies: Crossing the Biometric Borders', in *International Political Sociology*, Vol. 1, No. 1, pp. 149–64

Foucault, M. (2007) 'The Meshes of Power', Gerald Moore, trans. in *Space, Knowledge and Power: Foucault and Geography*, Elden, S. and Crampton, J. W. (eds) pp. 153–63, London: Ashgate

Foucault, M. (2004) *Security, Territory, Population: Lectures at the Collège de France, 1977–1978*, trans. Michel Senellart, London: Picador

Foucault, M. (1995) *Discipline and Punish: The Birth of the Prison*, London: Penguin Press [first published in 1977]

Fuchs, R. and Moch, L. (1990) 'Pregnant, Single, and Far from Home: Migrant Women in Nineteenth Century Paris', in *The American Historical Review*, Vol. 95, No. 4, pp. 1007–31

Galton, F. (1877) 'Address to Section D-Biology, British Association', in *Nature*, Vol. 16, pp. 344–7

Galton, F. (1890) *Anthropometric Laboratory, Notes and Memories no. 1*, London: Richard Clay

Galton, F. (1910) 'Numeralized Profiles for Classification and Recognition', in *Nature,* Vol. 83, pp. 127–30

Gillis, A.R. (1989) 'Crime and State Surveillance in Nineteenth-Century France', in *The American Journal of Sociology*, Vol. 95, No. 2, pp. 307–41

Gulf Times (2008) 'Gulf Cooperation Council Agrees Passports for Saudi Arabia', June 17, p. 1

Heidegger, M. (1982) *Nietzsche Vol. 4*, New York: Pfullington

Horn, D. G. (2003) *The Criminal Body: Lombroso and the Anatomy of Deviance*, London and New York: Routledge

Irish Council for Bioethics (2009) *Biometrics: Enhancing Security or Invading Privacy? Opinion*, Dublin: ICB

Irish Times (1912) 'Finger-Print Cheques', June 8, p. 24

Irish Times (1909) 'Editorial', January 15, p. 6

Jain, A. K., Bolle, R. and Pankanti, S. (eds) (1999) *Biometrics: Personal Identification in a Networked Society*, Norwell, MA: Kluwer Academic Publishers

Jain, A. K., Flynn, P. J., and Ross, A. (eds) (2008) *Handbook of Biometrics*, London and New York: Springer

Lyon, D. (2009) *Identifying Citizens: ID Cards as Surveillance*, London: Polity

Lyon, D. (2007) *Surveillance Studies: An Overview*, London: Polity

Magnet, S. (2009) 'Using Biometrics to Re-Visualize the Canada-US Border' in Kerr, I., Steeves, V. and Lucock, C. (eds), pp. 359–77, *On the Identity Trail: Anonymity and Authentication in a Networked Society,* Oxford: Oxford University Press

Maguire, M. (2009) 'The Birth of Biometric Security', in *Anthropology Today*, Vol. 25, No. 2, pp. 9–14

Maguire, M. and Cassidy, T. (2009) 'The New Irish Question: Citizenship, Motherhood and the Politics of Life Itself', in *Irish Journal of Anthropology*, Vol. 12, No. 2, pp. 18–29

Maguire, M. (2010) '"Race", Biopower and New Security Technology', paper presented at the European Cooperation in Science and Technology (COST) conference, *A Global Surveillance Society?*, City University London, April 13–15

Mayhew, H. (1967) *London Labour and the London Poor: A Cyclopaedia of the Condition and Earnings of Those That Will Work, Those That Cannot Work, and Those That Will Not Work*, New York: A. M. Kelley [first published in 1861]

Mill, J. S. (1978) *On Liberty*, London: Hackett Pub Co [first published in 1863]
Mill, J. S. (1957) *Utilitarianism,* London: Bobbs-Merrill
Mill, J. S. (2007) *Considerations on Representative Government*, New York: BiblioBazaar
Neocleous, M. (2008) *Critique of Security*, Edinburgh: Edinburgh University Press
Nye, R. (1984) *Crime, Madness and Politics in Modern France*, Princeton, NJ: Princeton University Press
Ong, A. (2006) *Neoliberalism as Exception: Mutations in Citizenship and Sovereignty*, Durham: Duke University Press
Pinkney, D. H. (1953) 'Migrations to Paris during the Second Empire', in *The Journal of Modern History*, Vol. xxv, No. 1, pp. 1–12
Rabinow, P. (2007) *Marking Time: On the Anthropology of the Contemporary*, Princeton, NJ: Princeton University Press
Ruggiero, K. (2004) *Modernity in the Flesh: Medicine, Law, and Society in Turn-of-the-Century Argentina*, Stanford, Palo Alto: Stanford University Press
Semple, J. (1993) *Bentham's Prison: A Study of the Panopticon Penitentiary*, Oxford: Clarendon Press
Sengoopta, C. (2004) *Imprint of the Raj: The Colonial Origin of Fingerprinting and its Voyage to Britain*, London: Pan
Stedman Jones, G. (1991) *Outcast London: Study in the Relationship between Classes in Victorian Society*, London: Penguin Books
Thompson, V. E. (1997) 'Urban Renovation, Moral Regeneration: Domesticating the "Halles" in Second-Empire Paris', in *French Historical Studies*, Vol. 20, No. 1, pp. 87–109
United Kingdom. Home Office (2005) *Controlling Our Borders: Making Migration Work for Britain: Five Year Strategy for Asylum and Immigration*, UK: The Stationery Office
Walid, T. (2007) 'Too Much Information?' in *Arabian Business*, July 8, pp. 1–4

Internet Sources

www.statewatch.org/news/2005/jul/11092-05.pdf

Chapter 3

Security and Insecurity at the Border: Notes from the Mexican–US Border when Considering the Mediterranean

Natalia Ribas-Mateos

Introduction

This work briefly outlines the dimensions of regional border issues in the Mexico–US region and in the Mediterranean in the context of a border security–insecurity debate. The complex Mediterranean context, characterized by sea borders, and the US–Mexico region with its extended land border, may well represent the two regions of the world where the processes of integration-interaction between asymmetric countries are currently most in evidence.

My hypothesis is that questions related to the evolution of new forms of mobility, issues of border security and conditions of insecurity (specifically vulnerability) can be examined by the study of human (in)security in highly specific places, namely borders. There, the gap between the protection migrants formally enjoy under international law, national laws and the actual experiences of individuals reveal a severe absence of human rights. It is also at such strategic sites as these that we encounter many of the main issues in this book: securitization, irregular migration, racism, gender issues, human trafficking, globally networked surveillance etc.

I will essentially use the US–Mexico region as a model of reference for (or contrast with) the Mediterranean setting. Particular reference is made to the way in which the securitization of the border has been exported. The question to be addressed is to what extent has this securitization pattern been exported to other areas? With regard to the US–Mexico side, we are mainly talking about the consequences of four so-called concentrated border enforcement operations, beginning with 'Operation Hold the Line' in the El Paso metropolitan area in 1993, followed by 'Operation Gatekeeper' in the San Diego area in 1994, 'Operation Safeguard' in the central Arizona area beginning in 1995, and 'Operation Rio Grande' in the south Rio Grande Valley of Texas beginning in 1998. So, each one of these operations was activated in a different state.

Operation Gatekeeper, an enforcement strategy established in 1994 in the Tijuana–San Diego border area, made it 'the world's most policed international divide between two non-belligerent countries' (Nevins, 2002). Indeed, many authors refer to this area as a laboratory of the postmodern security–insecurity contradictions of the world. Can such a model be easily exported to Mediterranean Europe? Following researchers such as Staudt (2008) I will consider border areas as examples of extreme security yet at the same time also of extreme insecurity for migrants (e.g. feminicides in Ciudad Juárez, deaths of migrants crossing the Arizona desert, deportation of family household heads from Texas to Mexico etc).

This chapter aims to identify security–insecurity processes related to the restriction of borders in the age of globalization. Within the scope of such processes one has also to reconsider the conception of Fortress Europe, which is based on the one hand on elimination of internal borders (focusing on free mobility and transborder interaction)[1] and, on the other, on restrictive external borders (focusing on exclusion and border management). For example, Moldova and Ukraine had a visa-free regime with the Czech Republic, Hungary, Poland, Slovakia, Bulgaria and Romania. Curiously enough, when such ex-communist countries joined the EU, they become borders to their ancient allies.

This conception of Fortress Europe has been with us since the beginning of the nineties, when critical NGOs in the UK started highlighting the exclusion of non-EU citizens and the impact on the various ethnic minorities in the construction of Europe. From then onwards it also became a *sine qua non* condition in the study of migration policies in Europe and the criminalization of migrants, and of course, a key condition in the study of the metamorphosis of contemporary borders.

Borders as key strategic sites

In general terms, the classic concept of the frontier implies a geographical notion which incorporates the administrative limits of the political and strategic functions of the state (Foucher, 1991: 9). Within the context of the new settings for the world's borders, the end of the Cold War and the fall of the Iron Curtain,[2] the clash between East and West has often been replaced by a simplistic North–South division.

This is clearly a contemporary issue. However, in contrast to classic geopolitical studies in which borders are portrayed as inert frameworks, my ethnographic attempt to visualize these borders reveals them as a form of expression for intricate socio-economic global processes, in which borders act as frameworks rather than objects, from which we can observe current socio-economic processes, linked for example to deterritorialization, delocalization and relocalization.

---

1 Even though I will not consider such issues in this chapter, my on-going research is now addressing such hidden processes in European borders. I concentrate on two research sites: La Raia Humeda ('the wet line' marked by the River Minho) between North Portugal and Galicia in Spain, and the Mediterranean Pyrenees between Rousillon and Catalonia in Spain.

2 The Mexico–US region has been popularly nicknamed 'The Tortilla Curtain'.

Indeed, such research sites are privileged spaces in which social interaction takes place (Foucher, 1991: 10), where the space represents a means (rather than being an historical actor), a framework, an *enjeu*, which contains resources of use and interest to many. This highlights the idea that borders are constructed socially, and defined through an interrelation of social groups, as stressed by Barth (1969) when discussing ethnic groups and their possible social relations between ethnic borders. Thus, borders are not limited to being a mere 'social product' or social process, but are instead alive and dynamic.

Increasingly, this concept of borders must be contextualized with the visions of other authors who are focusing their efforts on the spatial and scale reflections related to globalization. One such author is Brenner (1999), who, rather than analysing globalization within a space, refers instead to the production of space in the age of globalization by means of a number of processes.

Nevertheless, the problem with this type of contemporary approach is how to achieve an empirical vision of globalization. Authors such as Sassen have attempted to establish strategic research units in order to interpret global changes. This particular author focused her attention initially on London and New York, capturing the essence of these cities as the hubs of a world economy that extended beyond the Atlantic region, indicating a new phase in capitalist accumulation and urban structuring. Whilst Sassen (2000) centred her attention on global cities, I have opted to focus on marginal spaces, situated on the borders of the so-much discussed global capitalism. Yet I am not referring to just any type of border; instead I have targeted common global processes in a range of areas and their borders, and it is here that I have also centred my analysis in this chapter of the complex relation between security and insecurity.

This persistent focus on borders, and border cities in particular, further highlights the interest in recovering those physical places that form part of economic globalization, albeit from a spatial heterogeneousness in which multiple trajectories and processes under ongoing construction are to be found. This vision of these processes as a 'permanent construction site' is also attributed to other authors such as Nevins, who mentioned that not only walls and fences but national identities and exclusivities are frenetic works in progress. Here also politicians and bureaucrats manufacture the self-serving myths that advance the interest of the 'border control industry' (see the introduction to the work on Operation Gatekeeper, Nevins 2002). Such a dynamic view of my research site can also be seen in Staudt and Spener (1998), borders are continually made and remade, rebordered and debordered, in concert with larger circulations of migration, the projects of states, the implementation of trade agreements, and the political responses of those experiencing these processes at first hand.

In order to locate the security–insecurity debate at the border as a key strategic site I take three conditions into consideration. First the cartography of global processes, a cartography based on economic triadization and global mobility. However, this cartography seems to be more evident in some places than others, which is why I link it afterwards with the idea of the 'universal laboratory'.

The third condition necessary to understand my research site is that of border reinforcement in the time of globalization.

A cartography based on economic triadization and global mobility

In order to obtain an accurate picture of the real situation, I have identified at least two major forces: triadization,[3] which offers us an insight into global capitalism on a planetary scale, and reveals an unequal division of wealth in America, the Mediterranean and Asia, and regionalization.

First, three major differences should be highlighted, based on a triadization process that recalls the creation of new geographic centres of accumulation (the US South and West, Western Europe and Japan, plus a number of newly industrialized countries). As Fordist production systems matured, these became new and often highly competitive centres of over-accumulation. Spatial competition intensified between geographically distinct Fordist systems, including the most efficient regimes (such as the Japanese) and lower labour-cost regimes (including those found in less developed countries). In the configuration of such centres of accumulation, notions of social contract with labour were either lacking or weakly enforced (Harvey, 1990: 184).

Secondly, we should consider the importance of regionalization in the study of global migratory flows (see Castles and Miller, 2003), as this provides a pattern that enables us to verify economic processes and migratory restrictions in the light of the deterritorialization of borders. As in many other academic and militant debates, I am referring to the way in which these borders are projected outwards and extended beyond their actual physical space via specific forms of control and differentiation between populations. An example of this is the southern Andalusian border which stretches nowadays well into the Sahara. It is what is called the externalization of borders, moving south from their geographical check-points.

When referring to the importance of global mobility, I give special consideration to the new patterns of international migration (the feminization of migratory flows, the consolidation of remittance-dependent economies and the diversification of migrant types). Generally speaking, capital, products and ideas have become more mobile, whilst certain labour categories, constrained by immigration laws, continue to be controlled, with mobility penalized. The closing of borders (militarization etc.) results in types of borderland which can be thought of as sites of resistance to globalization constituting a resistance to the closed face of a territorial logic. Such a consideration confirms the status of borders as places in which to reinforce restrictive policies, physical fences that mark political and administrative borders. This closed face explains why they are a waiting place for internal migrants.

3 The new geographical centres of accumulation form the basis of this triadization: (East and South) USA, Western Europe and Japan, in addition to the new industrial countries.

However, these borders are constantly questioned by ordinary people and by those who consider that their lives have a continuum in the transnational processes, including circular migration (Ribas-Mateos, 2005: 179).

'The universal laboratory'

Consider, then, the ways in which the Mexico-US region has become embedded as a model within European border politics, as a key universal laboratory for border studies, and for security–insecurity processes in particular. Many authors reference this overall picture, including Álvarez (1995), Bustamante (1989) and Valenzuela (2006) among others.

Among them, there is one project of particular significance. In the 1990s, Martínez and his team created the world's largest oral archive on the history of border people (Martínez, 1994: xvi). Martínez believed that by using this region as his laboratory he was also achieving a degree of applicability to border life in other parts of the world, a possible universal model.

In my view, this focus on the study of border life is also very contemporary, focussing on everyday life, organizations and institutions that make the border a strategic site in which to study social practices and social identities in relatively dense and durable configurations. The security and insecurity tension was also characteristic of this border life.

Martínez's much-quoted work stresses that the universal nature of border life can be applied in accordance with four models of interaction, or predominant types of relation, namely alienation, coexistence, interdependence and integration (ibid.: 6).

The first model is characterized by a state of tension, with several clear examples in the Middle East, whilst the second is defined by an irregular stability, as in the border region separating Ecuador and Peru. In the case of the fourth model, stability is both strong and permanent, as in the case of the European Union. The third model, of interdependence (the one characterized by an essential sense of stability), is represented by the US–Mexico region as a prime example of strong asymmetrical interdependence. The importance of this case is highlighted in this quote from the back of Martinez's (1994) book: 'while the US–Mexico borderlands resemble border regions in other parts of the world, nowhere else do many millions of people from two dissimilar nations live in such close proximity and interact with each other so intensely'. But the case of Morocco and Spain is an even clearer example: even though the border is not as long, what really matters is the magnitude of the phenomenon in terms of the number of people affected.

Many other authors, such as Nevins, have given similarly unique accounts of such regions. Since Operation Gatekeeper, the Tijuana–San Diego border area has become 'the world's most policed international divide between two non-belligerent countries' (Nevins, 2002).

Border reinforcement in the context of economic globalization

The meaning of boundary zones is determined by the workings of the world economy, and this process has important territorial dimensions (Herzog, 1991: 13).

Borders have become far more multidimensional, not only in a geographical sense, but also in their broad social nature. In this chapter, the border-type analysis takes into account the various fragmentation, deterritorialization, and externalization processes which affect the definition of 'border', which in turn impacts on the type of circumstances in which the periphery takes over the centre, and where the control of margins defines state sovereignty.

The various types of enclosure materials (fences, wire etc.) are testimony to the materialization of border closures around the world. These conventional fences can be overcome, can be scaled, and indeed this happens in various ways. They are being replaced by more technically sophisticated versions, such as concrete bollard fences. A long list of technical gadgets can be added to the idea of control: televised video surveillance systems operating 24 hours a day along these heavily fortified segments, which are linked to Border Patrol dispatching centres and which can dispatch Border Patrol vehicles to the exact spots where illegal entries are detected. There are also infrared night scopes, both fixed and mobile units which detect the presence of 'illegal entrants' by their body heat. The new forms of control arising from this are outlined in this book in the chapter by Maguire, who reviews the major developments in biometric security during the past decade, from 'smart' borders to the securitization of access to spaces and privileges, demonstrating a progressive increase in the number of new ways of governing through securitization.

When discussing the current global border framework, researchers tend to limit their discourse to the relation between open and closed gateways, movement and barriers, crossings and barriers. They fail to venture into the complex task of analysing the differentiated effects of mobility/enclosure as the unequal terrain of the global capitalist system itself, as indicated by Heyman and Campbell (2004), for example when writing about their own research findings focused on the so-called ports of entry at the United States border with Mexico, as key nodes in the world system. Heyman and Campbell (2004) claim that the 'globalized' world consists not of an open terrain, but of 'a jigsaw puzzle of socially unequal spaces (…) the complex pattern of manoeuvres by border crossers and attempted state controls provides us with vital clues about the making and remaking of "globalisation"' (303, 304).

Border case studies have shown that such processes do exist in border cities, thanks particularly to the specific characteristics of such cities. In cities such as Tijuana, Tangiers, Ceuta and Melilla, a series of contradictions arise, such as the differences between *border crossing* (e.g. women who work in domestic service and who cross the border into Ceuta and Melilla on a daily basis, networks of transborder families between Tijuana and San Diego, smugglers of basic items that embark on long routes from Ceuta and Melilla) and *border reinforcing* (Schengen 1991 and the construction of Fortress Europe, the abuse of the human rights of Sub-Saharans, the reinforcing of the US–Mexican border following September 11, etc.).

Security–insecurity processes

In this chapter, my analysis of the border context aims to go further than the open-close dichotomy chosen by most analysts. The interpretation of social processes reflects more than just one basic contradiction. I therefore venture beyond the open-closed complexity, or the open-closed mobility that are normally discussed. My intention is to delve deeper into the descriptive and consequential analysis of globalization. Indeed, the object is to leave behind the dichotomy analysis, and to consider the spaces from 'the in-between' as well as other less apparent elements that play a key role in understanding border spaces.

These other elements include the idea of the 'magnet', whereby the actual borders act as poles of attraction, in keeping with authors such as Staudt, who in his latest work (2008) refers to the 'magnetic function' of borders. Another element is the idea of these areas as 'weak legislation spaces', and reflections on the vision of borders as places characterized by high levels of security for the borders themselves, but also by the insecurity that threatens the individual.

Historically, the US–Mexican border region has been a legendary and lawless periphery that outlaws have crossed in search of safe haven. This is still true nowadays, although these criminals are now mainly men charged with domestic violence and murder. Occasional consumers cross with wads of cash but no visible means of support, paying for cars or houses without paper trails. Thus, the border is not only a place of transnational activism but also an escape valve (Staudt, 2008: 14). Here we are again adding extra elements, in this case that of the escape valve, which make the border a strategic place for the study of security–insecurity processes, those elements that today qualify the notion of 'border' beyond the open-closed notion of the free movement-closed movement of the global era.

The Mediterranean as an example: from the racism of the neighbour to border cities

The search for unity amongst Mediterranean societies has been a constant in historical and anthropological analyses. The idea of convergence is quite well understood by historians, because the Mediterranean has not been considered to be a region but rather a series of divergent seas. Much has already been written about the Mediterranean as 'common space' where economic and cultural exchanges, as well as violent confrontation, take place. Nowadays, the Mediterranean represents one of the most active friction-planes when considering North-South imbalances in the globalized world; it is the setting for sharp socio-economic contrasts and for several kinds of migratory phenomena which derive from global inequality and instability.

In principle, reference to a Mediterranean area of circulation and exchange provides unity. Differences, in the form of conflict, can be attributed today to the global space and the reinforcement of borders. It is only in this sense that the Mediterranean can be thought of as a paradigmatic case of international migration in the age of globalization (Ribas-Mateos, 2005).

Changes related to globalization and scale (especially the regional and sub-national, notably cities and regions) made me consider theoretical and methodological questions in order to identify the empirical way in which strategic economic and political processes territorialize in peripheral Europe and in particular at its southern borders, zones separated by semi-closed borders with a lightly functional porosity.

The often mentioned Andalzúa′s border is 'una herida abierta' (an open wound) where the Third World grates on the first, and bleeds. Before a scab forms it haemorrhages again, the lifeblood of two worlds merging to form …a border culture (Andalzúa, 1987). This bleeding can also be said to apply to Mediterranean population contrasts in the region, which seem for some to be a curse on Southern Europe. In this border-conflict scenario, two geographical points appear to me to be perfectly clear, the 14km separating the Moroccan coast away from Spain, and the 70km separating Albania from Italy.

Throughout my work I have given international migration a key interpretative role in a proposed itinerary of the Mediterranean and its borders, playing with change and continuities together. The closing of European borders and the will of people to move (taking account of all strategic agencies involved) can be represented as two contradictory faces of the global economy. The obstacles that migrants face in the countries from which they arrive are partially in evident contradiction with the principles of a free market and a global capitalist economy. On the one hand, the free global market limits the freedom of workers through restrictive entry, and European nation-states have a sovereign right to apply such measures.

It is in this broad regional space that Southern Europe faces two major migration routes in the contemporary global geography of international migration: from the South (from North Africa, particularly Morocco, and beyond) and from the East (from Eastern Europe, particularly Albania, and beyond). Nevertheless, many borders of Europe could have been selected, such as the German–Polish frontier, or that between the Adriatic Coast and The Aegean Coast. Abandoned borders such as the Portbou/Cerbère border (between Spain and France) could be included, or borders that are starting to play a new role in refugee issues, such as Sangatte, especially when Calais is described as the European Tijuana. However, not all those borders are silent. 1700 goods vehicles heading for Morocco passed through the border crossing in El Perthus in 1991, 17,000 in 1993 and 42,000 in

1995 (Tarrius, 2002: 55). If these borders have become more open, allowing a Mediterranean circuit around the countries of Southern Europe and France, on the other hand they tend to reproduce ethnic discrimination.

How do I select specific Mediterranean cases? Morocco and Albania are seen through European eyes as being the key doors for Southern European migration. They are at the front-line, the first to suffer from the closure of Fortress Europe. Furthermore, Albanians and Moroccans are the most criminalized groups in Southern Europe. Morocco and Albania seem to conform to the characteristic role of the European periphery. We can speak of a division of groups based on the degree of stigmatization they face from the receiving society. The Albanian and Moroccan communities appear to suffer the effects of a heavily stereotyped image in all Southern European countries apart from Portugal, although this depends to a considerable extent on the level of their presence in the country. In addition to the heavy criminalization of the Albanians in Italy, their image in Greece is linked to the stereotyped connotation of 'backwardness', as they are associated with the figure of the peasant (see research conducted in Greece and Spain, in Ribas-Mateos, 2005).

In the case of southern European countries, constructing the enemy involves a peculiar relationship with their southern and eastern neighbours. In accordance with the Baumanian concept of the *ante portas* alien (Quassoli, 2004), it refers to all those who threaten the security of home, corrupting the space for semiotic and social habitation. In these cases, the insecurity and uncertainty that comes from globalization are expressed through the defence of the territory: our district, our city, our apartment block, and complex paradigms are constructed regarding the figure of the alien, neighbour and enemy. Therefore this concept also indicates how the construction of the foreigner as an enemy, and the issue of protection of space or territory, can be analysed as the typical response of contemporary society to the negative consequences of globalization processes.

Analysing previous data on our neighbouring migrants led me to conduct research in the border sites of such realities (for example, which were the most numerous groups, the incidence of unattended minors by nationality). The borders to Southern Europe were to be exemplified by two selected cities, Tangiers in Morocco and Dürres in Albania, where financial movements, capital investment and transnational companies were highlighted, as well as through the actors′ social agency (their expressed creativity against closed doors). This meant looking at how household members generate alternative production and consumption patterns to ensure survival at a time of closed borders; it is through them that we learn how barriers are circumvented. It is a way of showing how repressive border control encourages, paradoxically, the reconstruction and strengthening of reciprocity.

Consequently, changes in international migration related to feminization, the construction of new routes, higher forms of circulation (both elite circulation and the increase of circulation among lower classes) have been viewed in the context of the relation of those places with Southern Europe. Through interviews and ethnographic work I have been able to analyse how migrants often live and plan their lives in transnational circuits (woven by peer-groups, relatives, friends and neighbours, Ribas-Mateos, 2005). Thus, border cities emerge in relation to the limits of Southern Europe, as being among the main closing doors of the intensified Fortress Europe[4], the borders of the Schengen construction. I also established through the notion of the border city how I could focus on nodal points, on a very specific type of place, that serve a border function for Fortress Southern Europe. First, in relation to circulation (the example of commodities), transit, waiting place (the example of the Subsaharian migrant in Tangiers) and restriction (the examples of the so-called 'mafias' working on such restrictions). Secondly, in relation to a socio-historical context, communism, colonialism, Islam. Thirdly, in relation to a global commodity chain framework (border economy, clothing manufacture and international subcontracting, telecom circuits, commodification of the body), which is deeply articulated with processes which are present in the global economy.

I focus here on Mediterranean borderlands as research sites, encompassing the very specific friction, the liminal sites of globalization processes. These sites offer the possibility of exploring the globalization process in great detail, within a bounded setting and with its contradictory features. Simply put, global economic activities such as the mobility peculiar to the export processing zones (the example of the textile and car industry) can also be deeply embedded in place. It is where global processes become structured by the local constraints of border cities (especially those in which mobilities are restricted to some nationals by states and their policing powers) as well as by their own characteristics which define the heterogeneous forces. However, it is important to recognize that borderlands are not the rule for all socio-economic processes in the countries under study. Globalization tendencies such as the development of export industries are not as marked elsewhere in the countries concerned, but the picture is clearer when one focuses on these border cities, which are the nodes of the privatization processes in those countries.

How do I focus on border cities? I consider here the effects of the global (the mobility of labour, capital, information), the role of the state (in how the notion of the state has become a critical issue in the context of its neoliberal restructuring and in the forms of transgression of the nation state), and the role of international migration (e.g. in circularities as a post-Fordist expression and as an explicit form of the contradiction of globalization). So I identify on one hand the open door, the metaphor of mixing communities and the free circulation of goods, and on the other, borders as physical fences of the Europeanized Nation State. Finally, I

---

4 The EU financial resources programme for external aid 2000–2006 linked to the migration issue provides an indication of the key importance attached to the management of migration flows. We can deduce this through the following budget breakdown: management of borders, €321,971,760 combating illegal immigration, €65,042,526 and management of flows, €51,367,336 while to development (sources of emigration, €71,569,477) (Commission of the European Communities, 2002: 51).

identify the core notion of the border, as enforcement or as a form of resistance to its process of insecurity and blockade.

There, we can locate the feedback from the global economy towards the most intimate worlds of everyday life. Firstly, from the outside, families inherit a context of labour inequality, a context closely tied to international migration, and, in interesting cases, to a patchwork economy for lone women. In the case of Tangiers, this development may probably be attributed to low salaries and the need to meet the demands of the family, in a border context defined by sharp inequality. It is there where the patchwork economy emerges as a crucial concept not only through theoretical categorization of the data but also through empirical descriptive detail. The place-ness of a border zone creates peculiar conditions for the examination of the informal economy as a pooling tool as well as of acts of resistance against a globally determined context. Apparently submissive to socio-economic restrictions, families adopt strategies including innovative forms of marriage strategies and interesting forms of balance between family and work life, as we have seen through the development of the factory-emigration hypothesis: this involved the identification of a way in which families balance extended kin and family income support from factories through remittances.

Hence, the family reacts towards the outside by using mobility as a resource, knitting the family in a global and European extended reciprocity network, not only in shaping ways of departure but also of settlement. Therefore, families try to move beyond the structural conditions of the global economy and sealed borders, stimulating the family to become a main actor for social agency in the border city and in migrant transnational strategies, using the family net as a form of transnational security.

How do I focus on the security–insecurity debate in diverse border-cities?

Increasingly, studies, either by groups working in civil rights or authors such as Bustamante (2001), emphasize the importance of the analysis of human rights in border sites. Worthy of note here is the analysis of human rights as a basis for an approach to border migration. This analysis began with the Zapata Canyon Project in Tijuana, Mexico, and the early surveys into the crossing points, when the majority of undocumented migrants crossed the border at Tijuana, Mexicali, Ciudad Juárez, Nuevo Laredo and Matamoros. As for the Mediterranean, the shipwrecks of undocumented migrants in Algeria (e.g. on the coast of Oran in the direction of Spain), Western Sahara, Greece, Italy, Libya, Tunis, Turkey, Malta, Spain (Straits of Gibraltar, the Canary Islands) etc. have also led to acts of resistance in such areas.

The examples given here reveal the complexities of the security–insecurity processes and the way in which the Internet has provided a space from which to fight for human rights in the two border regions under discussion. For example, with regard to the condition of the Sub Saharians in Morocco, or the repatriation of Moroccan minors from Spain who have crossed through the border-city of Tangiers.

By 'vulnerability of human rights' I refer here to the experiences of vulnerable populations in border regions, which include relocalization of industry and its impact on the local labour market, gender and class inequality, and also forms of resistance against human rights abuses which specially occur in border regions. I refer also to the problems of death in border regions, where no alternative apart from missionary charity and activist global action has been found.

A particular example: Spanish prisoners in the city of Tangiers

I will review here a small case study.

My intention was not to work empirically, but rather to critically reflect on border sites (in the US and Europe) as key places for the security–insecurity debate. However, I have worked on many border cases, and full accounts of these can be found elsewhere (Tangiers as a border city, Dürres, El Paso, Tijuana, Thrace in Greece etc.), most cases being described in *The Mediterranean in the Age of Globalization: Migration, Welfare and Borders* (Ribas-Mateos, 2005). Many other examples could be provided, but this one is little known, and no literature exists.

So I acknowledge that borders and movement of people can take complex forms, and can be illustrated by numerous examples, such as the situation of Spanish prisoners in a border city such as Tangiers. According to the latest data from Europa Press by the Spanish Ministry of Foreign Affairs and Cooperation, on 31 October 2008, 219 Spaniards, including 26 women, are currently serving prison sentences in Moroccan jails, the majority for crimes related to drugs-trafficking, and with sentences of less than five years. Morocco has the largest number of Spaniards in prison for drugs-related offences, followed by Peru and Brazil, which each have more than 150 Spanish nationals sentenced for this type of crime. More than half of these 219 Spaniards are housed in the Tangiers Prison, a jail renowned for its cramped conditions, lack of hygiene and hot water, according to reports from Amnesty International and Asochofer (the association of drivers and truck drivers, one of the sectors with the largest number of workers in Moroccan jails accused of drugs trafficking).

One of the particular border contexts in relation to the security–insecurity processes is the unknown situation of foreigners in border-cities. I refer in particular to my own research in Tangiers (ethnographic visits during June, July and August 2009 to the city prison). Even though poor Moroccan families constitute the majority of visitors, there are always Spanish visitors among them. In such situations we can easily recall the border configuration of such places; in this case how the prison symbolizes the border of the Strait of Gibraltar, the tough side of the line in both senses, and also the under-explored question of foreign prisoners in a border place such as this, one of the busiest Straits in the World.

Many others Spaniards are dispersed in other Moroccan prisons, but Tangiers is the hot spot in drug trafficking (hashish), as well as the node for communications with Spain, which makes the circuit denser not only in terms of tourists but also in terms of the possibility of visits by family members coming from Spain.

Getting information on the life of Spanish prisoners in Tangiers is an additional aspect of the community life of the most numerous foreign population group in the city. The Spanish community is involved in all aspects of city life, but transparent information about it is difficult to find. To focus on the Spanish community in prison is to do the reverse of what researchers tend to do, focusing on other more obvious aspects and circuits of the border city.

The living conditions of prisoners in the city which contains the largest number of Spanish prisoners outside Spain have been severely criticized over many years by their family members. Conflicts and protests over the conditions of foreigners in Tangiers have pushed Spanish law towards improving the welfare of Spanish detained citizens abroad. Spanish prisoners abroad now enjoy a 100 euro monthly allowance, unknown for other foreigners held as prisoners in Morocco.

However, the hidden miseries of border cities do not end here. We can think for example of the 40 Subsaharian Africans (who were here in transit to Europe) buried in the Christian cemetery of Boubana in Tangiers. Again and again, we must underline the border experiences of human insecurity in a situation in which human realities are not visible to the world at large.

To conclude

Carlos González thought of the title 'La Frontera que vino del Norte', The Border which came from the North, a phrase which is also suitable for me to use in completing my discussion, which has brought to light two main issues, one concerning the use of the hegemonic Mexico-US region in order to study borders, and secondly how well can we locate the security–insecurity debate using the border as a research site.

I have tried to determine global-border processes in the context of specific closure impacts. This chapter has made us think over the presumed universality of social processes; I have considered the US–Mexico border region in principle as a point of reference, and also reflected on it in light of other authors' thinking as a type of laboratory, which can even imply a form of ethnocentrism in the analysis of social processes.

The key role of this chapter is to help clarify and determine complex mobility forms as an axis within the security–insecurity context. A discussion of the conditions for the possibility of border security in specific regions also calls attention to the type of actors involved. In such contexts, we can refer to the repeated attention by current scholarship to vulnerable groups, normally involving the various vulnerable actors in global processes in border areas such as minors[5] and

5 In 2002, a total of 52,532 children under eighteen were returned from the US to Mexico and in 2003, 39,690 (Palmer, 2008: 60).

women. The consolidation of this ideology has implications on the border stage, where minors and women have repeatedly become pinpointed as the main actors in global processes which impact on border spaces, whether in the Mediterranean or the US–Mexico border region.

Using the border as a research site enables the acknowledgment of the power of certain social processes that are independent of each other but which can be brought together in a dynamic field of interaction. Such interaction occurs in the constant movement affected by global changes, a constant transformation of the environment which also transforms our selves, the way we think about border regions, the way we think about security–insecurity.

In this security–insecurity debate, the border appears as an escape valve containing many elements, common to both regions, related to the externalization of borders, the extreme racism expressed in such areas, and so on, but there are radical differences in the nature of the fight for human rights at the border. For example, such activism at the US–Mexico border is much more religiously oriented than Mediterranean activism. Even though both regions have lived through a period of restrictions on particular populations as opposed to global commodities, the impact and the reactions associated to it have been different. Finally, by appreciating current developments in restrictions and border security in the United States and European countries, the future of activism at the border will emerge as an important research field and will has the potential to give us new keys for action.

References

Álvarez, R. (1995) 'The Mexico-U.S. Border: the Making of an Anthropology of Borderlands', in *Annual Review of Anthropology*, Vol. 24, pp. 447–70

Andalzúa, G. (1987) *Borderlands: La Frontera*, San Francisco: Aunt Lute Books

Barth, F. (1969) *Ethnic Groups and Boundaries: The Social Organisation of Culture Difference*, Amsterdam: University of Amsterdam

Brenner, N. (1999) 'Global Cities, Local States: State Re-scaling and the Remaking of Urban Governance in the European Union', Phd Dissertation. The University of Chicago, August 1999

Bustamante, J. (1989) 'Frontera México-Estados Unidos: reflexiones para un marco teórico', in *Revista Frontera Norte*, Vol. 1. No. 1. pp. 7–24

Bustamante, J. (2001) 'Proposition 187 and Operation Gatekeeper: Cases for the Sociology of International Migration and Human Rights', in *Migraciones Internacionales*, Vol. 1, No. 1, pp. 7–34

Castles, S. and Miller, M. (2003) *The Age of Migration: International Population Movements in the Modern World*, 3rd edn, New York: Guilford Press

Commission of the European Communities (2002) 'Communication from the Commission to the Council and the European Parliament. Integrating Migration

Issues in the European Union´s Relation with Third Countries', Brussels, 3.12.2002 703 final

Coubès, M. L. (2008) 'Maquiladora or Cross-Border Commute: The Employment of Members of Households in Five Mexican Border Cities' in Márquez, Raquel R. and Romo, Harriett D. (eds) *Transformation of La Familia on the U.S.–Mexico Border*, Notre Dame: Indiana

Foucher, M. (1994) *Fronts et frontières: un tour du monde géopolitique*, Rev.ed, Paris: Fayard

González Herrera, C. (2008) *La frontera que vino del norte*, México D. F.: Taurus

Harvey, D. (1990) *The Condition of Postmodernity: An Enquiry into the Origins of Cultural Change*, Oxford: Blackwell

Harvey, D. (2001) *Spaces of Capital: Towards a Critical Geography*, New York: Routledge

Herzog, L. (1991) 'Cross-National Urban Structures in the Era of Global Cities: The US–Mexico Transfrontier Metropolis', in *Urban Studies*, 28: 519–33

Heyman, J. and Campbell, H. (2004) 'Recent Work on the U.S. Mexico Border', in *Latin American Research Review*, October 2004, Vol. 39. No. 3, pp. 205–20

Márquez, R. and Romo, H. (2008) 'Introduction' in Márquez, Raquel R. and Romo, Harriett D. (eds) *Transformation of La Familia on the U.S.–Mexico Border*, Notre Dame: Indiana

Martínez, O. (1994) *Border People: Life and Society in the US–Mexico Borderlands*, Tucson and London: University of Arizona Press

Nevins, J. (2002) *Operation Gatekeeper: The Rise of the 'Illegal Alien' and the Making of the U.S-Mexico Boundary*, New York and London: Routledge

Palmer, C. (2008) 'An Overview of Children and Youth on the Northern Mexican Border' in Márquez, Raquel R. and Romo, Harriett D. (eds) *Transformation of La Familia on the U.S.–Mexico Border*, Notre Dame: Indiana

Quassoli, F. (2004) 'Making the Neighbourhood Safer: Social Alarm, Police Practices and Immigrant Exclusion in Italy', *Journal of Ethnic and Migration Studies*, Vol. 30, No. 2, pp. 1163–81.

Ribas-Mateos, N. (2005) *The Mediterranean in the Age of Globalisation: Migration, Welfare and Borders*, New Brunswick: Transaction

Sassen, S. (2000) *Cities in a World Economy*, Thousand Oaks: Pine Forge Press

Staudt, K. and Spener, D. (1998) 'The View from the Frontier: Theoretical Perspectives Undisciplined' in Spener, David and Staudt, Kathleen (eds) *The U.S.–Mexico Border: Transcending Divisions, Contesting Identities*, Boulder: Lynne Rienner

Staudt, K. (2008) *Violence and Activism at the Border: Gender, Fear and Everyday Life in Ciudad Juárez*, Austin: University of Texas Press

Tarrius, A. (2002) 'Au-delà des Etats-Nations: réseaux et sociétés de migrants en Méditerranée Occidentale'. Colloque Economie de bazar dans les métropoles euroméditerranéenes. Aix-en-Provence, 29–31 May 2002. Maison Méditerranéenne des Sciences de l´Homme

Taylor, J. T. and Hicky, M. (2001) *Tunnel Kids*, Tucson: University of Arizona Press

Valenzuela, M. (2006) *Por las fronteras del norte: una aproximación cultural a la frontera México-Estados Unidos*, México: Fondo de Cultura Económica

Vila, P. (2002) *Crossing Borders, Reinforcing Borders: Social Categories, Metaphors, and Narrative Identities on the US–Mexico Frontier*, Austin: University of Texas Press

Vila, P. (2003) 'Conclusion: The Limits of American Border Theory' in Vila, P. (ed.) *Ethnography at the Border*, Minneapolis and London: University of Minnesota Press

Chapter 4
The Impacts of the Global Economic and Financial Crisis

Khalid Koser

Introduction

In 2010, the global economy is showing tentative signs of recovery from the worst economic downturn since the Great Depression in the 1930s. What began as a financial crisis in the USA in the second half of 2008 quickly became a global economic crisis. In October 2008 the International Monetary Fund (IMF) predicted that global growth, measured in terms of purchasing power parity, would reduce to three per cent in 2009, compared with 5.2 per cent in 2007. This forecast was revised downwards in April 2009 when the IMF forecast that the world's economy would shrink by 1.3 per cent in 2009: advanced economies were predicted to shrink by 3.8 per cent, while the emerging and developing economies were predicted to continue to grow but only by 1.6 per cent, compared with six per cent in 2008. In July 2009 the IMF lowered its forecast for the global growth rate in 2009 by a further 0.1 per cent. But by October 2009, the IMF had raised its global forecast for 2010 to a positive growth rate of 3.1 per cent, with growth once again predicted in the advanced economies too (IMF, 2009).

Against this overall trend of gradual recovery from a short, sharp, shock, the Euro area (comprising the Economic and Monetary Union of 16 European Union member states) has lagged behind the other advanced economies. France and Germany, the largest economies in the Euro area, reported a growth of only 0.3 per cent in the second quarter of 2009, after five consecutive quarters of negative growth; the British economy shrank by 0.3 per cent; and economic growth in the Euro areas as a whole declined by 0.1 per cent. The Greek economy has also undergone a major new crisis, resulting from the national budget deficit and debt burden. For 2010, forecasts released by the European Commission in October 2009 predicted a modest increase of 0.7 per cent in the GDP of the 27-member European Union (EU). Meanwhile Eastern Europe is still being affected by an economic slowdown. In October 2009 a forecast by the European Bank for Reconstruction and Development (EBRD) suggested that the contraction of that region's economy would be worse than it had foreseen six months before – with a 6.3 per cent decline compared with the initial prediction of 5.2 per cent. Growth in 2010 was however foreseen by EBRD to be higher than previously forecast, at 2.5

per cent, up from 1.5 per cent, although the upturn, it added, would remain fragile (EBRD, 2009).

In the Euro area unemployment had reached 9.7 per cent in October 2009, and was expected to peak at 11 per cent in 2010. In Spain, it reached 19.3 per cent, the highest for more than twelve years, and youth unemployment was as high as 35 per cent. Germany saw a rise in its unemployment rate from 8.1 per cent in March to 8.3 per cent in April 2009. In the United Kingdom (UK), the number of unemployment benefit claimants per vacancy had more than doubled between 2008 and 2009 to an average of ten according to the British Trades Union Congress. Sweden was expected to have an unemployment rate of 11.7 per cent by 2011. In December 2008, Turkey had an unemployment rate of 13 per cent; Estonia 11 per cent; Latvia 15.7 per cent; and Lithuania and Ireland over 13 per cent (Ghosh, 2010).

Evidence on the impact of this global economic and financial crisis on international migration patterns, processes, and policies, in Europe and more widely, is gradually emerging, but it remains limited. For many European countries data on migration trends in 2009 are not yet available, and even for those countries where data are available, they are incomplete. There are some estimates for example by the International Organization for Migration (IOM) on migrant flows and stocks of foreign workers, but virtually no evidence on other important issues such as integration and social protection. There is some disaggregation of the existing data according to economic sector, but more detailed disaggregation for example on the basis of gender, age or nationality is largely missing (IOM, 2009). Analysis is also hampered by the fact that the impacts are complex and vary across economic sectors, migrant types, countries and regions, and in any case it can be hard to establish causal relationships. What is more there is always a time-lag for data on migration and remittances. As a result much of the existing evidence is in the form of policy and press reports.

Bearing in mind such reservations, the main purpose of this chapter is to assess the impacts of the recent global economic and financial crisis on international migration patterns, processes, and policies, primarily in Europe, but also referring where relevant to wider global impacts. Although its impacts appear not to have been as significant in Europe as the 1973 Oil Crisis, which resulted in severe restrictions on labour migration and a concomitant growth in asylum applications and irregular migration in Europe, they have still been important, may have lasting implications, and have attracted policy responses. The second, shorter part of the chapter considers the security implications of these impacts – in the short and longer term, and for migrants and host societies, as well as for states. The conclusion makes a number of policy recommendations in foreseeing future shocks to the global economic system.

The impacts of the global financial crisis on migration in Europe

The global financial crisis has had an observable impact on international migration patterns, processes, and policies in Europe, across the following areas:

A reduction in labour migration

Almost every member state of the Council of Europe has reported a slowdown in labour migration inflows since 2008 (Council of Europe, 2010). One reason has been a significant decline in international recruitment by employers. In Spain, for example, new entries under the employer-nominated system (*régimen general*) fell from more than 200,000 in 2007 to 137,000 in 2008 (OECD, 2009). In the United Kingdom (UK), the number of approved applicants to the Workers Registration Scheme was down from 46,600 in the first quarter of 2008 to 21,300 in the corresponding quarter in 2009 (a reduction of 54 per cent). In Ireland, fewer than 9,000 people from the 12 new European Union (EU) member states were granted Personal Public Service numbers in the first quarter of 2009, a 57 per cent fall from the same period the year before (OECD, 2009). Second, as unemployment has risen in Western Europe, labour markets there have become less attractive to foreign workers. In the UK, for example, there was a 60 per cent decrease in quarterly inflows from Eastern Europe between 2006 and the second quarter of 2009. Similarly, Spain experienced a 25 per cent decrease in new arrivals in 2008, especially from Romania and Bulgaria. A third reason for the slowdown in labour migration, expanded below, has been a general trend in government policy especially in Western Europe towards restricting access to the labour market for foreign workers.

The extent to which this slowdown is likely to persist largely depends on wage and employment differentials between traditional countries of destination for labour migrants in Europe, and the countries where they originate. If owing to higher rates of growth origin countries make a faster recovery than the destination countries, for example, labour migrants may be less inclined to move to these destination countries. An econometric study published by the UK government in April 2009 argues that a higher rate of growth and employment in countries that send migrants to the UK would lead to a reduction in net immigration there of 50,000 annually by 2015, as compared with the non-recession scenario (Department of Communities and Local Government, 2009). This is turn may reduce the stock of foreign workers in the UK, and the same study projected that the migrant population in the UK would be smaller by some 360,000 by 2015, as compared with a pre-recession projection in July 2008, and would reduce the UK labour force by 200,000.

Another important determinant of whether labour migration resumes to pre-crisis levels will be political reactions to public sentiment, which has hardened against foreign workers during the recession. An April 2009 Harris/Financial Times poll, for example, revealed that the majority of those polled in the UK and

Germany opposed citizens of other EU countries being able to work there. Another survey conducted in Autumn 2009 by the Pew Research Center showed that the majority polled in five EU countries – Italy, Spain, the UK, Germany and France – were in favour of more restrictions on immigration; in the case of the first three of these countries, the percentage was above 75 per cent (Ghosh, 2010).

Rising unemployment

Disproportionate job losses for migrant workers have been recorded around the world, especially in employment sectors that are most sensitive to economic cycles, such as construction, manufacturing, financial services, retail, travel and tourism. Unemployment rates for foreign workers are reported to have increased more quickly than those among the local population across most EU countries, as well as in the Russian Federation. In several countries unemployment among migrant workers is now more than double that of the local population – in Spain 28 per cent compared to 15.2 per cent, in France 18.5 per cent compared to 9 per cent, and in Portugal 17.2 per cent compared to 8.5 per cent (IOM, 2009).

Unemployment rates among migrants are reported to have varied considerably from one sector to another. In the construction sector, which employs many short-term labour migrants, job losses have been particularly high. For example, in Spain, the unemployment rate of African immigrants, who work mostly in construction, reached almost 33 per cent in the fourth quarter of 2008, almost double the recorded average for immigrants as a whole (Council of Europe, 2010). Given their sectoral distribution, it is likely that the crisis has had a greater impact on men than women migrant workers. The International Labour Organization (ILO) has reported that in some European countries redundancy rates for male migrant workers, who are concentrated in construction and manufacturing sectors, is twice as high as for female migrant workers who are mostly engaged in health, education and domestic services (ILO, 2009). Similarly the Council of Europe has concluded that in Europe low-skilled migrant workers have tended to be more affected than the highly-skilled, as the latter group often has skills in demand. In particular, the persistence of labour shortages in engineering and healthcare has positively affected work opportunities for highly skilled migrants (Council of Europe, 2010). An additional concern is reports that migrant workers who have lost their jobs have faced difficulties in accessing back-pay and accrued social welfare benefits (Koser, 2010).

Preliminary evidence from Spain suggests that unemployment may also be rising among irregular migrant workers (IOM, 2009). Competition for jobs between the local population, legal migrants and irregular migrants has intensified in the agriculture, building and service sectors, which are unable to absorb the demand. This has resulted in a situation where most male irregular migrants in urban areas in Spain are now unemployed. Although domestic service continues to be an alternative for female migrants in an irregular situation in urban areas in Spain, the economic difficulties facing Spanish families are having a considerable impact on demand for their services too (Council of Europe, 2010). Furthermore, the conditions pertaining to the renewal of residence and work permits in Spain may mean that some legal migrants who have become unemployed may find themselves in an irregular situation there.

Deteriorating working and living conditions

Deteriorating working and living conditions have been reported for those migrant workers who have kept their jobs. There have been press reports from the Russian Federation, for example, of non-payment of wages for foreign workers; and reductions in wages, working days, and the availability of overtime. There has also been a significant drop in wages and excessive working hours for domestic workers in Spain, who are increasingly dependent on the families employing them (and thus increasingly vulnerable to exploitation), and confront reduced opportunities to move into the formal sector of the economy.

Coupled with worsening conditions in the workplace, sporadic instances of discrimination against migrant workers and a rise in xenophobia have also been recorded, including in the Russian Federation, Spain, and the UK. In Spain the apprehension of irregular male migrants in urban areas has fuelled xenophobia and worsened resentment in an increasingly competitive labour market. Similarly in the Russian Federation industrial unrest and rising resentment against unemployment has coincided with growing violence against immigrants. The Sova Centre, a Moscow-based organization that monitors hate crimes, reported 428 racially motivated attacks in Russia in 2008, in which 97 people died, and cautioned that these figures almost certainly underestimate the true scale of the problem (Ghosh, 2010). The Moscow Bureau for Human Rights, a Russian NGO, reported that 113 migrants were murdered between January and October 2008 (The Economist, 2009), double the rate of the previous year. In Italy, there have been reports of attacks on Romanian workers in Rome, and on Roma Gypsies in Hungary; and a number of xenophobic protests have also taken place in the UK.

More returns

Although both globally and in Europe returns have taken place at less of a scale than was predicted early during the crisis, there have nevertheless been some significant return movements. A survey in Ireland in late 2008 indicated that one-third of 200,000 Polish migrant workers there expected to return home within a year (IOM, 2009). In this particular case the relative strength of the Polish economy compared to many Western European economies was an important factor, as was the fact that these migrants could return to Ireland and other Western European destinations after the crisis as a result of EU labour mobility agreements, and also encouraged by the relatively short distance and low travel costs. Other return flows have been reported from Spain and the UK to Poland as well as other new EU countries such as Bulgaria and Romania. In addition it has been reported that up to one million migrant workers returned from the Russian Federation back to some Commonwealth of Independent States (CIS) countries, although many have apparently subsequently gone back to the Russian Federation as a result of deteriorating conditions in their own countries, especially in Central Asia (Council of Europe, 2010).

In response to the growing rates of unemployment among migrant workers and the prospect of some of them subsequently losing their legal status, some countries, such as Spain and the Czech Republic, have offered incentive packages to encourage the voluntary return of unemployed migrants. The Spanish Ministry of Labour adopted a new regulation at the end of 2008 to support the voluntary return of unemployed non-EU nationals. Eligible persons could receive their unemployment benefits in two lump sums on the condition that they returned home and did not come back to Spain for at least three years. According to data from the Organization for Economic Cooperation and Development (OECD), however, only 10,000 unemployed immigrants and 3,600 family members had signed up for this programme by the end of January 2010 of an estimated 137,000 unemployed immigrants eligible for the programme in June 2009 (Council of Europe, 2010).

The 'Voluntary Returns Programme' introduced by the Czech Republic, and aimed at returning a proportion of 12,000 foreign workers predicted to lose their jobs in 2009 (a large share of whom were Vietnamese, Ukrainian and Mongolian migrants), offered potential returnees an incentive allowance of 500 EUR in the first wave (during the first half of 2009) and 300 EUR in the second wave (during the second half of 2009) to cover living costs on arriving home. Some 1,871 foreign workers with legal status returned within the first phase of the project (mostly Mongolians), but as of 23 October 2009, only 154 persons had left under, and another 167 had applied to, the second phase (Council of Europe, 2010). The Czech Ministry of Interior has also extended the return programme to migrants with irregular status, but between 15 September and 15 December 2009 only 53 persons applied for return.

A reduction in irregular migration flows; an increase in irregular migrant stocks

The experience of most European countries as regards irregular migration during the global economic and financial crisis has been similar to that found in destination countries around the world; namely a decrease in flows but an increase in stocks. Frontex (the European Agency for the Management of Operational Cooperation at the External Borders of the Member States of the European Union) reported a 16 per cent decrease in illegal border crossings in the first quarter of 2009 compared to the year before (IOM, 2009). According to the Office of the United Nations High Commissioner for Refugees (UNHCR) there was also a 20 per cent reduction of apprehensions for illegal arrivals by sea in Greece in 2009 (Sumption, 2009). The number of people arriving without authorization in Spain and intercepted by the police fell from 18,057 in 2007 to 13,424 in 2008 (in other words by 25 per cent) and is estimated to have fallen even further in 2009 (Council of Europe,

2010). It may well be that part of the explanation for these reductions is more effective border control, but most experts also believe that the lack of employment prospects has made irregular migration less attractive, especially given the high upfront costs often paid by irregular migrants.

At the same time, the existing evidence is that the overall share of irregular migrants among the total immigrant population in most European countries is increasing (Kuptsch, 2009). The main explanation appears to be that legal (as well as irregular) migrants who have lost their jobs would often prefer to remain in the destination country in an irregular situation, rather than leave and risk not being able to re-enter after the economic upturn (CLANDESTINO, 2009). This effect is exacerbated because in many cases the recession has been even more severe in their countries of origin, and thus there is little likelihood of finding work if they do return. It has been estimated that hundreds of thousands of Tajiks are in this situation in the Russian Federation (IOM, 2009). According to the Council of Europe one of the dangers of the current crisis is that it appears to be leading to the 'normalization of irregularity' as informal work is likely to become more commonplace among migrant workers – both regular and irregular. Migrants, especially those without legal status, may be willing to accept the lowest of wages and the most oppressive of working conditions for fear of remaining completely unemployed and with no means at all for subsistence.

A reduction in remittances

For the first time in at least twenty years, the World Bank has predicted a decline in the growth of remittance flows for 2009, at best by 0.9 per cent and at worst by six per cent (as compared with a growth rate of 6.7 per cent in 2008). In addition, changes in exchange rates affecting currencies in significant destination countries (for example the US dollar, Pound Sterling, and the Russian Ruble) are decreasing the value of remittances for families at home even where their volume remains constant. Significant declines in remittances have been reported in Morocco, the Philippines, Indonesia, and Sri Lanka. A similar trend has also been found among major remittance-receiving countries in Europe. In Turkey, for instance, official remittances have fallen by 43 per cent since 2008 (Fix and Papademetriou, 2009). In Moldova remittances have fallen by 37 per cent, and the impact is expected to be especially severe because remittances are equivalent to one-third of the country's national income (Ghosh, 2010). Remittances to Armenia have declined by more than 30 per cent; while Poland and Romania have also experienced a sharp slowdown in remittance flows.

Central Asia is one of the regions most affected globally, notably in countries like Tajikistan, Kyrgyzstan, Moldova and Uzbekistan, where reduced labour migration to the Russian Federation, combined with growing unemployment for migrant workers already there, has been exacerbated by the sharp depreciation of the Russian Ruble alongside falling oil prices. A decline in remittances to sub-Saharan Africa has largely been due to the negative effects of the recession on migrant jobs and wages and on migration flows in host countries both within Africa and outside the continent, notably in the European Union (Ghosh, 2010).

Mixed effects on other migration flows

While global statistics are not yet available for overseas student enrolment in 2009, the Higher Education Council in the UK, as well as equivalent bodies in other EU states, has made public statements predicting a reduction in enrolments. Because family reunification is subject to income limits, entries under this regime may have been delayed as potential sponsors have become unemployed. Alongside the closing down of restrictions in labour migration opportunities, many countries have simultaneously tightened family unification opportunities. For example, in Portugal, where presently people with a two-year legal stay can bring in their families including their parents, a reform of the Law on Foreigners is envisaging raising the requirement to a five-year legal stay for the parents to join (Council of Europe, 2010). The impact of the economic crisis on humanitarian migration is even more difficult to assess since it depends primarily on changes in the economic, social and political situation in less developed countries and the most fragile states. The Council of Europe has predicted that Europe may be affected by an increasing risk of abuse of the asylum channel within mixed migratory flows.

Restrictions on new admissions of migrant workers and non-renewal of work permits

With some notable exceptions – for example Sweden where the government decided to liberalize its migration policies in December 2008 and follow a demand-driven labour migration model – the predominant policy response to the potential migration impacts of the global economic and financial crisis in Europe has been to implement more restrictive policies towards migrant workers in order to reduce the inflow of new migrants and protect national labour markets. This has been achieved through various means, including cutting the numbers of work permits for foreigners, mainly in low-skilled sectors (Greece, Spain, Italy, Russia, UK); offering incentives for migrants to return including one-way tickets and lump-sum payments (Spain and the Czech Republic); reducing shortage occupation lists and reinforcing labour market tests (France, Spain, UK); introducing changes in visa levels and entry requirements (Italy, UK); limiting possibilities to change status and/or renew work permits (Russia); and tightened controls on family inflows (Portugal, Spain).

Spain, for example, sharply reduced its annual regional ceilings for non-seasonal labour migrants to a total of 901 for 2009, compared to 1,571 in 2008. Italy also lowered its ceiling for 2008, and announced its intention to set a quota of zero for non-seasonal work in 2009. In November 2008 the United Kingdom decided to cut back its originally-planned target of one million skilled migrants to meet special shortages in specific occupations to 800,000. The government suspended recruitment of non-EU workers for low-skilled occupations and made changes in the points-based system for recruiting foreign workers, requiring a higher score than previously for new entrants. It also tightened labour market tests for certain skilled jobs and raised the standards for admission of foreign workers for some others (UKBA, 2009).

Security implications

Given the overall theme of this volume, this shorter second part of the chapter considers the implications for security and insecurity of the impacts analysed above. Three preliminary comments are appropriate. First, as already demonstrated, one of the impacts of the global economic and financial crisis has been to elevate migration still higher on political agendas and in some cases to fuel anti-immigration sentiment and even violence. Such a policy setting makes a discussion of the links between migration and security very sensitive, and careful and objective analysis is required, for example to avoid imputing migrants with tainted intentions without any substantiation (Uehling, 2004; also see chapters by Karyotis and Togral in this volume). Second, and in part in response to this concern, this section considers the implications for 'human security' as well as for more traditional national security. Human security is a concept that focuses not on security as states normally understand it, but on security from the perspective of human beings. It can be thought of as the difference between a top-down and bottom-up approach to security. Human security addresses a far wider human condition than conventional security, and the human security debate is designed to elevate these matters to a higher priority in the international system (CHS, 2003). Third, given that the full impacts of the crisis on migration are still not clear, it is too early to make an accurate assessment of whether the crisis has exacerbated or intensified the 'migration–security nexus' as it was configured in Europe before the crisis began.

Irregular migration is often described as constituting a threat to state sovereignty (Waever *et al.*, 1993). Simply stated, it is argued that states have a sovereign right to control who crosses their borders and remains on their territory, and that irregular migrants threaten sovereignty by undermining this control. It thus follows that a fundamental way to reassert full sovereignty is to stop irregular migration. In this context the preliminary evidence on the impacts of the global economic and financial crisis on irregular migration have multiple implications. On one hand, it appears that fewer migrants are crossing borders into Europe in an irregular manner. The reason may however be as much to do with the decision of the migrants themselves not to move until after the recession, as with the effectiveness of border controls; and it has been predicted that the scale of irregular migration in Europe will resume pre-crisis levels soon (IOM, 2010). On the other hand, it appears that the stock of irregular migrants in many European countries has increased, as unemployed migrants opt to stay on even if this means they will no longer have legal status.

While again reiterating that direct evidence is still lacking in the recent European context, it is possible at least to conceive of implications for national security of growing stocks of irregular migrants, especially where their numbers are substantial and have increased rapidly (as appears to be the case for example in the Russian Federation). Their willingness – even desperation – to work may expand the scope of the informal labour market. In certain circumstances, the irregular migrants may compete with legal migrants and even local populations for work, especially in low-skilled sectors, and especially during a period of recession and rising unemployment. Where they cannot find work some may turn to petty crime in order to subsist. Irregular migrants can also represent a strain on scarce resources, especially where they are barred from using the full range of services available to citizens and legal migrants, meaning that already hard-pressed non-governmental organizations (NGOs), religious bodies, and other civil society institutions are obliged to provide assistance, at times compromising their own legality.

Another concern is that, as they become more obvious and visible, irregular migrants may become a catalyst for anti-immigration violence and racist attacks, the victims of which may extend to including legal migrants and even settled ethnic communities. While clearly a gross exaggeration, it is interesting in this context to note that in response to the rising incidence of street protests in countries as varied as Greece, Iceland, and Latvia, in February 2009 the new US Director of National Intelligence, Denis Blair, told the United States Congress that instability caused by the global economic crisis had become the biggest security threat facing the United States, outpacing terrorism (Ghosh, 2010).

A final issue is that the growth in irregular migrant stocks has the potential further to undermine public confidence in government policy on migration and integration. This is particularly salient as most experts have advised that European (and other industrialized) countries should lift temporary restrictions on labour migration in order to accelerate economic growth after the end of the recession, and those governments will need public support to do so.

Indeed another possible link between the impacts of the global economic and financial crisis on international migration and national security is the possible knock-on effect on future migration. Several studies have shown the links between immigration and economic growth (for example OECD, 2009). The April 2009 UK government study cited above suggested that an estimated fall by around 360,000 in the (legal) migrant stock by 2015, and a consequent reduction in the labour force by 200,000, would lead to a 0.1–0.125 per cent decline in the country's economic growth (Department of Communities and Local Government, 2009). Recent research presented to the EU Parliament has also showed that migrants from within the EU had boosted the region's aggregate Gross Domestic Product (GDP) by 0.28 per cent since 2004 (Ghosh, 2010). In the longer term a sustained decline in migration might restrain technological and economic progress and inhibit cultural innovation; and also further exacerbate Europe's problems of ageing and labour supply, social security funding and demographic decline.

The overview of impacts provided in the first part of this chapter highlighted what might be considered some of the human security implications for migrants themselves, including growing unemployment, deteriorating working conditions, the transition from legal to irregular status, the rising incidence of violent attacks, and for some the prospect of returning home to even greater poverty. Of course it is not just the migrants themselves who may be affected in this way: their families in destination countries will also suffer the consequences of a reduced income; in some cases changing regulations have made it more difficult for migrants to exercise their right to family reunion thus extending periods of separation; and as alluded to above even settled ethnic communities may become the target for growing anti-immigration sentiment and action.

The reduction in remittances also has a wider impact, on families and sometimes entire communities, in countries of origin. In Tajikistan, for example, remittances have provided a lifeline for many families, adding on average USD 250 to the annual income of every household in 2008 (and also representing over 45 per cent of national GDP). A heavy reliance on remittances, a poorly diversified national economy, and inadequate access to other sources of external revenue, make countries such as Tajikistan and the people living there extremely vulnerable to the effects of the global economic and financial crisis. And in the absence of adequate social safety nets, the poor are likely to be the hardest hit.

Conclusions

The global economic and financial crisis of 2008–2010 may have been the worst global recession since the Great Depression, but it is unlikely to be the last. Some experts are predicting a 'double-dip' effect whereby at least in some parts of the world national economies will begin to recover temporarily then re-enter a second recession; while most agree that shocks to the global economic system are cyclical and can be expected again. This chapter therefore concludes by considering some of the policy implications arising from the preceding analysis of the impacts of the global economic and financial crisis on international migration in Europe, and their implications both for national and human security.

One conclusion is that policy in destination states for migration in Europe and elsewhere should be based on sound economic evidence and be consistent (Koser, 2010). Just as during previous economic and financial crises, governments in Europe today have been under political and popular pressure to respond to the impacts of the crisis through migration policies, especially to reduce competition with local workers in the labour market. But as has been explained in this chapter, the impact of such crises on labour markets is complex and unpredictable, and it can take some time for a clear picture of labour market requirements to emerge. Governments, in contrast, are under pressure to respond immediately, and risk reducing the supply of labour migrants when it in fact may be required to support economic recovery. This is especially important as migration policies can have unintended consequences. Restrictions on renewing work permits for migrant workers, for example, are one reason why the stocks of irregular migrants have increased in some countries, in turn countervailing policies targeted on deporting migrants with irregular status. It has also been suggested that certain governments may find it difficult to reverse restrictions and open up the labour market to foreign workers once the crisis recedes, potentially with long-term knock-on effects for economic competitiveness.

A second conclusion is that migrant workers require protection during economic and financial crises. Those who have lost their jobs have been found to face difficulties in gaining back pay and other social benefits owed them by their employers; and have also experienced pressure to leave the country either formally through government policy or informally as a result of public resentment. For those who retain their jobs there is evidence of deteriorating working conditions and exploitation. More generally migrants have experienced increasing surveillance and the effects of rising xenophobia. In these ways the securitization of migration has been exacerbated by the effects of the crisis. In all these situations attention needs to be paid to the protection of the rights of migrant workers. Policies and procedures are required to guard against the arbitrary use of power to expel migrants, there is a need to protect the right to accrued wages, employment agencies need monitoring, and migrant workers' rights should be reinforced. These and other rights are enshrined in a series of ILO Conventions and protocols, the ratification and implementation of which however remains problematic. National laws and policies to reinforce the protection of migrant workers may be more effective.

A final conclusion is that international cooperation on international migration is even more important during periods of recession than periods of economic growth. One of the immediate responses to the global economic and financial crisis in Europe was the reinvigoration of Euro-scepticism with particular concerns about the freedom of the movement of labour in the EU zone and calls for the strengthening of the EU's 1999 'posted workers directive' to uphold local workers' rights (Koser, 2010). In contrast, one of the lessons from earlier crises is the importance of cooperation between pairs of migrant-sending and migrant-receiving countries, and also more generally at the regional level, for example to match labour supply and demand. Cooperation is also required to cope with growing stocks of irregular migrants; to support the return and reintegration of migrant workers; and to help developing countries cope with the implications of reduced remittance levels. World Bank President Robert Zoellick, for example, has called for the creation of a new 'vulnerability fund' to help poor countries cope with the impact of the global economic and financial crisis.

References

CLANDESTINO (2009) 'Report on the Workshop on Irregular Migration and Informal Economy', Athens, 27 April 2009, accessed at: http://clandestino.eliamep.gr

Commission on Human Security (CHS) (2003) *Human Security Now*, New York: Commission on Human Security

Council of Europe (2010) 'The Impact of the Global Economic Crisis on Migration in Europe', Report to the Parliamentary Assembly Committee on Migration, Refugees, and Population, AS/ Mig (2010) 06

The Economist (2009) 'The People Crunch', 15 January

European Bank for Reconstruction and Development (2009), Economic Statistics and Forecasts, accessed at: http://www.ebrd.com/country/sector/econo/stats/index.htm

Fix, M. and Papademetriou, D. (2009) 'Migration and the Global Recession: a Report Commissioned by the BBC World Service', Migration Policy Institute, Washington DC, September

Ghosh, B. (2010) *The Global Economic Crisis and Migration*, Geneva: IOM (forthcoming)

International Labour Organization (ILO) (2009) *The Global Economic Crisis and Migrant Workers: Impact and Response*, Geneva: ILO

International Monetary Fund (IMF) (2009) *World Economic Outlook*, Washington DC

International Organization for Migration (IOM) (2009) *IOM Policy Brief*, March 2009

International Organization for Migration (IOM) (2010) *World Migration Report 2010* (forthcoming)

Koser, K. (2010) 'Responding to the Impact of the Global Financial Crisis on International Migration: Lessons Learned', Migration Research Series 37, Geneva: IOM

Kuptsch, C. (2009) 'Impact on the Migrant: Special Repercussions on Migrants and their Families arising from the Economic Crisis, presentation at the PACE forum 'Remain, Migrate or Return: What to do in a Global Recession?' Antalya, 12–13 October

Organization for Economic Cooperation and Development (OECD) (2009) 'International Migration: Charting a Course through the Crisis', in *OECD Policy Brief*, June

Sumption, M. (2009) 'Migration and the Economic Downturn in the European Union', Report for the Migration Policy Institute, October 12

Uehling, G. (2004) 'Unwanted Migration: Combating and Unwittingly Creating Irregular Migration in Ukraine', in *New Issues in Refugee Research* 109; Geneva: UNHCR

UK Border Agency (2009) 'Immigration and Asylum Statistics', 24 February

United Kingdom. Department of Communities and Local Government (2009) 'Projections of Migration Inflows under Alternative Scenarios for the UK and World Economies', Economics Paper 3, London, April

Waever, O., Buzan B., Kelstrup, M., Lemaitre, P. (1993) *Identity, Migration and the New Security Agenda in Europe*, Pinter: London

World Bank (2009) *Migration and Development Brief 8*, 11 November

Chapter 5

Globalization, EU Enlargement and the Challenge of Financial Crisis: East–West Migration and the Search for EU Solidarities

Branka Likic-Brboric

Introduction

The latest phases of eastwards EU enlargement, bringing the final accession of ten Central and Eastern European (CEE) countries, are seen as a successful EU-driven post-communist institutional transformation towards democracy and a functioning market economy, as well as the realization of a regional peace and security project. Furthermore, the regional model of European integration has been attributed with a potential to satisfactorily address the migration related 'liberal paradox' (Hollifield, 2004), a conflict between the logic of transnational economic opening and free movement on the one hand, and the political and economic 'irrationalities' of national closure in the face of rising migration, on the other. However, the underlying dynamics of and forces behind the processes of globalization-*cum*-EU-enlargement have, until now, primarily been infused with a neo-liberal impetus underpinning the 'incontestable superiority' of the logic of capital.

Following the launch of the 2000 Lisbon Strategy, a ten-year EU programme for sustainable development and economic growth which confirmed the pre-eminence of the EU's commitment to global competitiveness, the promotion of the European Social Model (ESM) across the enlarged EU has undergone successive repositioning in the discourses of European solidarity.[1]

Thus, after the 2004/05 Lisbon Review, EU enlargement eastwards was explicitly directed towards global competition, unrestrained mobility of capital, and deregulation of the labour market with a redefinition of competitive edge in terms of an ever cheaper labour force. 'Conditionality', with respect to the social dimension of Europe, was largely restricted to formal transposition requirements of the EU social *acquis*. Substantive change has remained more problematic and contested.

The persistent implementation of radical structural reforms in the post-communist space has resulted in regional configurations of political economies of inequality, marked by rising unemployment, poverty and increasing social

1 A gradual shift from explicit EU commitment to 'social policy' towards an open-ended 'social policy agenda'.

exclusion. In the area of migration, contrary to market optimists' expectations concerning positive migration effects of EU enlargement for both the old and the new member states, we have seen the realities of de-industrialization, East-West migration pressures, related informalization/precarization of employment, 'wage squeeze' and 'social dumping', and a challenge to national welfare states. Furthermore, migration pressures created by the latest enlargement incorporating Romania and Bulgaria, in the context of the global financial crisis and economic contraction of several sectors that depend on cheap informal migrant labour, have led to mainstreaming of rising anti-immigrant and racist populism, but also to legitimate trade union grievances. EU member states' political elites' responses to the challenges brought about by the new 'geo-political and conceptual widening' of migration involved a spatial reconfiguration of migration regimes and their dualization (Geddes, 2005: 791). An encouraging business-friendly approach to high-skilled labour migrants is thus only a benign facet of the new EU-level migration management that is pervaded by the logic of securitization of asylum policies (Karyotis in this volume). It is claimed in this chapter that circular migration policies informed by the combined logic of neo-liberal economic policy and securitized asylum *acquis* not only undermine migrants' rights as human rights and create a vicious circle of their enduring 'statelessness' (Pajnik in this volume), but also emasculate social citizenship and labour rights of the citizens in the enlarged EU.

In conjunction with a presentation of the main features of the ESM and a broad overview of the processes of post-communist transformation as well as its socio-economic consequences, this chapter addresses contingencies for the realization of EU strategies for the configuration of a transnational social space that places at its core a European social citizenship and a rights-based labour mobility regime. After the critical analysis of the asymmetric policies of EU enlargement, as well as their impact on the configuration of the EU-wide political economy of inequality and new migratory landscapes, the final discussion scrutinizes the EU's current responses to the global crisis. It concludes that while the unfolding process of EU enlargement has not successfully addressed the inherent liberal paradox(es) of the EU project, the EU's responses to the current financial crisis reveal regionally differentiated reaffirmation of economic austerity measures. Coupled with the securitization of migration, this reinforced neo-liberal policy trend discredits social dialogue and generates a decay of democratic politics and European social citizenship. These mutually reinforcing frames of security and economic austerity breed multiple paradoxes that can only be challenged by an institutional rebalancing of EU policies in terms of an alternative that promotes European solidarities, social protection, 'decent work' and a rights-based mobility/migration regime.

European Social Model and a 'Liberal Paradox'

The European 'Social Model', a term coined by Jacques Delors to differentiate European from American market integration, is understood here as a set of discourses, supported by hard and soft regulations. Together, these seek to establish a common institutional framework for the realization of European social solidarity among the workers and citizens of the EU. The ideas of the ESM have been advanced since the 1960s, expressing the political understanding of the necessity to underpin European market integration by European social solidarities, translated into workers rights, an employment model that promotes workforce participation and a co-operative industrial relations system that acknowledges the legitimacy of key actors in the labour market. For migrant workers, different treaties have since forwarded different facets of social considerations such as freedom of movement and social security (the Treaty of Rome) and various directives support equal opportunities, non-discrimination and health and safety provisions.[2] In addition, non-binding but overarching policy orientations are provided by the 1989 Community Charter of the Fundamental Social Rights of Workers and the social protocol of the Maastricht Treaty. Thus, considerations of social issues such as unemployment, poverty, income inequalities and different forms of ethnic and gender exclusion and labour market segmentation, and even issues of 'undocumented labour' and migration, resulted in the inclusion of the employment objectives in the 1997 Amsterdam Treaty. The treaty endorsed the European Employment Strategy (EES) and the launch of the 1997 Luxemburg process for its implementation, leading to the 1999 European Employment Pact. However, the Employment Pact undermined the original aspirations and enabled a flexible employment policy supporting the macroeconomic stability policies of the 1997 Stability and Growth Pact. The latter spelled out fiscal constraints, most notably the threshold of a 3 per cent public deficit/GDP ratio and a 60 per cent public debt/GDP ratio, as convergence criteria that all member states have to fulfil for a final union within the European Monetary Union (EMU) and for inclusion in the euro.

In the twenty-first century we have seen the subsequent subordination of the EES to the imperatives of macroeconomic stability, exemplifying the contradictory character of the processes of EU formation and enlargement, burdened, on the one hand, by the conflicting demands for stabilization, economic efficiency and

2 The accession negotiations with the new member states (NMS) involved a transposition of 61 social policy directives, ranging from labour law directives (the directives on employment contract information (91/533/EEC), protection of pregnant workers (92/85/EEC), working time (93/104/EC), protection of young workers (94/34/EC), parental leave (96/34/EC), posting of workers (96/71/EC) and part-time work (97/81/EC) to more recent anti-discrimination directives (The Racial Equality Directive 2000/43/EC and The Employment Equality Directive 2000/78/EC). Yet, the problems of uneven implementation and compliance across both new and old member states pertain (Cf. Keune, 2009).

competitiveness and, on the other, by the objectives of social cohesion and political legitimacy (cf. Likic-Brboric, 2004; Schierup *et al.* 2006). Thus, the discourse of 'global competitiveness' framed the 2000 Lisbon Strategy and was further endorsed by the EU heads of state at the Nice Summit, adopting the Social Policy Agenda (CEC, 2000) as an essential component in support of competitiveness. The aim of the agenda was 'creating more and better jobs' and development of a 'knowledge-based economy' through greater mobility, 'modernizing' social protection and promoting social inclusion, gender and ethnic equality, as well as human rights.

The Lisbon Strategy can be seen as reflecting a growing awareness of the social downside of globalization, seeking to promote compensatory European solidarities. Anti-globalization movements, starting with Seattle 1999, had by the beginning of the twenty-first century managed to place global poverty and inequalities on the political agenda. Indeed, something of a global consensus had emerged in support of the UN Millennium Goals and Global Poverty Reduction Strategy. Between 2000 and 2004, the EU had rhetorically endorsed claims towards global solidarity and a European social agenda which embraced solidarity discourses on the promotion of labour rights, migrant rights and necessity for a social governance of the forces of globalization (CEC, 2000; CEC, 2001; CEC, 2004). Deacon (2005) enthusiastically proclaimed that there existed a broad global consensus for the promotion of global social policy, at a time when the EU was still feeling its way towards an understanding of the social implications of its own policies in the transfigured context of successive enlargements.

Thus, with respect to the new EU member states from the CEE countries and their emerging systems of social welfare, Deacon identified '(a) tension between the aspiration towards a European-style social-market economy (or conservative corporatism) and a budget-induced and IMF-World Bank-backed residualism' (Deacon, 2000: 156). Warning of the latter's negative influence on social policy, especially of its promotion of 'political globalization', Deacon identified the ideologically driven imposition of minimalist and commercial approach to social reforms and an uncritical bias towards financial globalization and its speculative character. At the same time, the EU was seen as an important actor able to promote the European social model eastwards through the accession instruments and an asymmetrical relation of 'conditionality' between applicants and the Union that provided strong compliance incentives, at least in formal terms. Pro-European ideologists saw this period of pre-accession as providing a window of opportunity to export the social model eastwards. Thus, Vaughan-Whitehead (2003) strongly advocated the decisive promotion of Social Europe in an enlarged EU, as a unique opportunity to counter the undermining trends of free market liberalism in prospective member states and the neo-liberal 'Europeanization' of the welfare models in the old member states (ibid.: 530).

Similarly, Hollifield (2004: 903–4) characterized EU regional integration as a project being able to move forward both 'the trading' and 'the migration state', by promoting a rights-based regional migration regime and basic citizenship rights across member states. This offered resolution of the 'liberal paradox' in reconciling a 'rights-based perspective' with the 'economic perspective' on migration. Free migration was expected to have positive effects on the economic growth of both the receiving and sending countries, thus reducing global inequalities in the long run. While the adherents of the former (a rights-based perspective) advocate promotion of transnational citizenship rights, subscribers to the latter (an economic perspective) tend to promote the principle of 'necessary' differentiation of the freedom of movement in terms of skills and competitive cost-advantage (Solimano, 2001: 21). As yet, this debate remains unresolved both at policy and practical levels, although significant challenges to social solidarity have arisen both in the realm of labour relations (protests against the exploitation of labour by foreign contracting companies) and in the realm of European jurisprudence (Laval, Viking, Ruffert cases).[3]

Moreover, rising irregular migration and informalization of the global economy present profound social and political challenges. The downward pressure that irregular migration has placed on wages and the deterioration of working conditions clearly demands a critical exploration of the consequences of international migration (Solimano, 2001), especially considering the on-going securitization of existing migration regimes. Hollifield sums up qualms concerning the actual contingencies of enabling escape from the inherent dilemmas of liberalism:

> But I am not sanguine about this possibility because the asymmetry of interests, particularly between the developed and the developing world, is too great to permit states to overcome problems of coordination and cooperation. Even as states become more dependent on trade and migration, they are likely to remain trapped in a liberal paradox for decades to come (Hollifield, 2004: 905).

The following sections suggest that the unfolding process of EU enlargement has been informed by market fundamentalism, a narrow neoclassical economics approach to free movement within the EU and the securitization of the migration

3 These cases are situated at the heart of the neo-liberal Europeanization of national labour regimes by the judicial interpretation of the posted workers directive and the freedom of establishment and service provision directives (Article 49 EC Treaty) that enable employers to substitute protected domestic labour force of the host countries by far cheaper workers from the NMS. The Latvian construction company *Laval un Partneri* that used cheaper and less organized Latvian construction workers on Swedish construction sites was opposed by the strong Swedish trade unions in the construction sector (Byggnads), which started an industrial blockade in order to protect the 'Swedish model' and prevent 'social dumping'. *The Rüffert* case concerns the freedom of public authorities to contract private enterprises that disrespect local collective agreements on wages and working conditions. The Viking case concerns the Finnish shipping company, which tried to re-register its ships in Estonia in order to substitute Finnish crews by much cheaper Estonian ones. While the ruling in the Viking case still entertained trade unions rights, the recent ECJ rulings in support of Laval and Rüffert's claims have clearly undermined the rights of trade unions' industrial action, and strengthened the employers' strategies of 'wage squeeze' http://www.etuc.org/r/846 (accessed: 21 May 2010).

regime that has arguably undermined migrants' human rights. The inherent liberal paradox(es) of the EU project seem to be further 'trapped' by the ECJ rulings that reaffirm imbalances between the fundamental freedoms, favouring freedom of capital over labour rights.

EU enlargement, TINA and the political economy of inequality

Following the 1989 fall of communism, the former communist countries pursued a neoliberal policy approach to a transition to the market, involving simultaneous and rapid liberalization of trade, macroeconomic stabilization and a massive privatization of state assets. Informed by the so-called Washington Consensus, and vigorously advocated in the Thatcherite spirit of TINA ('There Is No Alternative'), the 'shock therapy' approach copied structural adjustment programmes (SAP) previously utilized in South American and other indebted transition economies in need of IMF/World Bank credit lines (Likic-Brboric, 2003). However, the sweeping economic restructuring that ensued has been accompanied by economic dislocation, rising unemployment, falling living standards, and new forms of social exclusion in which an informal economy has begun to flourish. Problems of deindustrialization and a dwindling formal economy were interpreted as a healthy and necessary 'creative destruction' in order to make space for a market economy to thrive. Hence, the typical responses to the post-socialist transformation crisis: a growing number of informal businesses, the rising importance of the survival economy and related trends of informalization in the labour market (Neef, 2002). The 'decapitalization of the state' as a method for 'private capital accumulation', proved to be effective in configuring both a new political system in terms of 'state capture', and an economic transformation in terms of illegal privatization and hidden economy. Economic polarization has further intensified informalization: private businesses grow 'at the expense of the state, the state transfers its losses to the population, and the population at large does its best to minimize them' (Chavdarova, 2002: 678). Similarly, Bettina Musiolek illustrates the linkage of privatization, the absence of investment capital, ill-directed financing of small micro-business and self-employment engendering illegal financial flows, to what is described as the '*demarketization and barterization of the economic activities*'. Furthermore, both the privatization of state companies and the deterioration of public services have re-positioned female workers to contingent, low-paid employment in low-protected sectors such as the domestic services and the clothing industry, with substandard working conditions and a lack of recognition of workers' rights (Musiolek, 2002: 5–7).

The 1997 Luxemburg summit of European leaders endorsed the fifth EU enlargement towards CEE countries, while devising accession partnerships as instruments to support applicant states (Poland, Czech Republic, Hungary, Slovakia, Latvia, Estonia, Lithuania, Slovenia, Bulgaria and Romania) in the process of membership negotiations. However, the EU trade regime and attendant accession negotiations have entailed persistent flexibilization and deregulation of the labour markets in order to attract foreign direct investments (FDIs) (Musiolek, 2002). The post-socialist transformation has thus been accompanied by rising unemployment, and a considerable increase in income inequalities and both relative and absolute poverty in comparison with Soviet-era standards. These phenomena have persisted even after the economic growth resumed following the chaotic early years of transition. Inequality of disposable income measured by the *Gini* coefficient in ten new member states increased in the years preceding enlargement between 2001 and 2004 from 0.29 to 0.33, while the 'at risk of poverty' rate rose from 14 to 16 per cent (Heidenreich, 2007). Following EU enlargement there has been a trend towards a slight reduction or stabilization of inequalities between states, however within-state regional inequalities, especially in the former CEE member states, have increased considerably.

There is also a rather significant difference *between* new member states, depending on their previous level of development and chosen transformation policies. In spite of the initial prevalence of neoliberal transition policies, we have also seen a differentiation of institutional choices across Central and Eastern Europe, depending on domestic political circumstances, historically shaped socioeconomic preferences and 'national identities'. Dorothee Bohle and Béla Greskovits (2007) for example distinguish different institutional configurations in the CEE political economies as reflecting different emerging transnational capitalist regimes, which have been shaped through the translations of internationally formulated institutional arrangements into different national contexts each with different historical legacies. So, the Baltic states are subscribed to a 'neoliberal type', the Visegrád states Poland, Hungary, Czech Republic and Slovakia to an 'embedded neoliberal type', while Slovenia stands out as the only example of a 'neocorporatist' regime unfolding through post-communist transformation. The 'new' member states (NMS) differ in their level of economic development. Slovenia had reached a level of 88.6 per cent of an average European GDP in 2007, and scores 0.929 HDI (Human Development Index), the highest of all the NMS.[4] At the lower end we have Bulgaria and Romania reaching only 37.7 and 41.6 per cent respectively of EU average GDP (Online database Eurostat).[5] Although the CEE countries differ in their levels of social expenditures, the largest disparities are recorded in the area of employment and the labour market, best exemplified by minimum wages ranging from 77 euro in Bulgaria to 491 euro in Slovenia in 2005 (Peña-Casas, 2007: 101–2). No doubt, the policy of EU enlargement, without an adequate promotion of equalization of wages, working conditions and social expenditures has created the political economy of inequality, and labour market segmentation and dualism.

4 http://hdrstats.undp.org/en/countries/country_fact_sheets/cty_fs_SVN.html (accessed: 21 May 2010).

5 http://epp.eurostat.ec.europa.eu/tgm/table.do?tab=table&init=1&plugin=1&language=en&pcode=tsieb010 (accessed: 21 May 2010).

East–West migration, transitional arrangements and the landscapes of precarious work

Transformation crisis, unemployment, income inequalities and protracted poverty in the region have also created structural conditions that have provoked an East–West migration (Ardittis, 1994; Iglicka, 2002). In a UNESCO study on labour migration in Eastern Europe and Central Asia, Katja Patzwaldt (2004: 4) depicts how workers migrate between the countries, 'cascading from poorer to richer countries'. Thus, people from the least developed countries in the region such as the Caucasus Republics and Central Asian states in the south are attracted to Kazakhstan or Russia. While labour migrants from eastern Ukraine and Moldova choose Russia as their destination, migrants from Russia, western Ukraine and Moldova move to Central Europe to continue from there, together with workers from new EU member and accession countries, towards the more advanced economies of the older EU member states. Furthermore, migration related to the ethnic wars in former Yugoslavia in the early 1990s and post-socialist national state-building projects in the fragmented post-Soviet space (Brubaker, 1992) generated a new awareness in Western Europe of 'the flood' of refugees and asylum seekers.

The challenges of new East–West labour migration, and of asylum seekers from crisis-driven and war-torn transition countries, not only prompted discussions on a common European migration regime, but also problematized the vision of an ever-expanding European Union and its enlargement eastwards. In the course of the 1990s, the EU member states worked eagerly to devise the contours of an EU migration regime, erecting the walls of 'Fortress Europe'. The bricks of this edifice have been painstakingly laid in the Maastricht Treaty that further expanded the Treaty of Rome, the Single European Act, the Dublin Convention and the Schengen Agreement.

Nevertheless, Fortress Europe was challenged by the new wave of market-driven East-West migrations, enabled by proliferating bilateral agreements between old and prospective EU member states, candidate countries and neighbourhoods. By the end of the 1990s, even Britain and German policy elites accepted that migration would be the best solution to problems of demographic deficit and the increasing demand for both high-skilled and low-skilled labour forces (Favell and Hansen, 2002). Following the Amsterdam Treaty of 1997 this pragmatic policy shift led to a reformulation of the European migration regime, away from 'zero migration' towards 'efficient' and 'flexible' migration management, and control of undocumented migration through 'intensive trans-governmentalism' (Samers, 2004: 30).

This 'intensive trans-governmentalism' involves the formulation of migration policy by 'governmental actors below the levels of chiefs of government such as ministerial officials, law enforcement agencies, and other bureaucratic actors' and 'is represented in part by the proliferation of inter-governmental cooperative groups such as the Ad Hoc Immigration Group, the Schengen Group, and the JHA EU working groups' (ibid.). Being dominated by interior ministries' personnel and agenda, they have consistently pursued 'co-operative' securitization, especially concerning illegal migration, in a hardly transparent and democratically unaccountable manner. Ahead of the 2003 Thessaloniki Council, the Commission presented a communication on illegal immigration, based on the preparatory work of the so-called Santiago Action Plan.

Samers (2004: 42–3) identifies three main characteristics of this emerging EU migration policy regime. The first one is an upward re-scaling and 'gradual communautarization' of policy-making that involves a continuing configuration of the EU-level migration regime. The second is a horizontal re-scaling of securitization to the third countries that have been made responsible for migration control though new development aid conditionality (ibid.). Indeed, after the latest waves of EU enlargement we have seen the erection of an intricate multi-level institutional architecture for the securitization of migration, erecting the EU's frontiers both beyond and within its actual territory at the expense of freedom and human rights (Karyotis this volume). Finally, fear of 'terrorism' and a related public sentiment of insecurity have not only led to new identity boundaries and discursive framing of the irregular migrants as a social threat, but also to policy responses that actually create illegal immigration by the very securitization techniques (Samers, 2004: 43).

No doubt, the new migration approach has also become the instrument of further flexibilization and commodification of labour markets in the Western economies as an important component of the European strategy to enhance its global competitiveness. According to Overbeek, the commodification of labour is central to the shaping of the global capitalist system, and the EU has been constructive in its delivery. It involves the search for cheap labour through the integration of formerly detached areas (former socialist and developing economies) into the capitalist world market through 'semi-proletarization' and even the 'commodification' of non-market activities in advanced capitalist societies through restructuring policies of privatization and liberalization (Overbeek, 2003: 17–19). The latter processes have resulted both in the growth of informal sectors and a 're-commodification' of labour, the reverse outcome of the de-commodifying impact of the welfare state (Esping-Andersen 1990). The dynamics of the re-commodification of labour are, according to Carlos Frade and Isabelle Darmon (2005), clearly related to the rise of new flexible forms of business organization relying on the inception of external work providers in the new service economy (call centres, performing arts and care of the elderly). Their common denominator has been the increase in precarious employment, deterioration of working conditions, and transfer of business risks onto workers with non-standard employment contracts, such as temporary, part-time and the self- employed.

This downward flexibilization of labour and the 'normalization' of informal and precarious employment would not have been possible without tapping into available resources of female and migrant labour. The latter was made available through persistent pursuit of the neo-liberal development policies that have generated inequality structures and configured a global differentiated labour market. At the same time, contradictory regional migration regimes have increasingly involved both 'survival' and 'opportunity' migration. The emerging hierarchy of migrant statuses corresponds to the reconfigured landscapes of precarious work, marked by the vertical chains of risk-sharing and with irregular (undocumented) migrants bearing the unequal burden of 'hyper-precarity' (Woolfson and Likic-Brboric, 2008).

It is arguable that the informalization in advanced European economies, especially a migration-related rise of new informal economic sectors in the contexts of the European metropolises, is interlinked with increasing ethnic and gender labour market segmentation, considerably involving socially marginal immigrant communities and irregular migrants (Schierup *et al.* 2006). Here these phenomena are understood as inherent to global capitalist strategies of liberalization and de-regularization, and involve corporate actors' pursuit of competitiveness through vertical chains of subcontracting. These linkages of states and regions blur the distinction between formal and informal activities and reintroduce pre-industrial forms of labour exploitation and sweatshop organization into the First World economies. However, as Schierup, Hansen and Castles (2006) point out, such phenomena have been more extensive in Britain as a corollary to its 'neo-American' economic trajectory. Informal employment of irregular migrants in the clothing industry, the domestic sector, hospitality, agriculture and construction has also become an important business strategy in southern European economies, Germany and the Netherlands, in the context of state retrenchment, social polarization, inadequate welfare provision and aging populations (Jordan and Düvell, 2002: 59–64). Zoran Slavnic (2010) rightly identifies significant state and economic actors' strategies of adjustment to post-Fordist restructuring and re-commodification of welfare states that generate dualization and vertical segmentation of the labour market as 'informalization from above'.

But informalization and precarious work are not only enhanced by the new geo-political mapping and global positioning of the EU economy. They are developments intimately bound up with the concomitant internalization of the adverse outcomes of the post-socialist transformation and the resultant political and economic integration that now interconnects East and West. These developments have also generated a contradictory introduction of the free movement of labour in the EU-27 across the 'old' EU-15 member states through restrictive transitional arrangements. Following 2004, the majority of migrants from the NMS chose the UK and Ireland as their destinations; by 2007 the UK registered 680,000 immigrants and Ireland 380,000. Germany and Austria also attracted a considerable number of Eastern European migrants. In spite of transitional restriction, 350,000 work permits were issued to Polish workers in Germany, in line with previously signed bilateral agreements (Krings, 2009). The increasing numbers in this new flexible workforce have given rise to trade union concerns over 'social dumping' and downward pressure on wages and working conditions, although they did not initially compete with domestic workers (Krings, 2009: 55). Considering the increasing demand for workers from the NMS following enlargement, these transitional restrictions have not curtailed the already existing incidence of demand-driven irregular migration from supplementary sources such as new neighbourhood and third countries nationals (TCNs).

Although, there have been some improvements in employment contracts following legalization of residence of NMS citizens in countries without transitional restrictions, the restrictive approaches to Bulgarian and Romanian workers, even in otherwise liberal Ireland and Great Britain, illustrate political fears of nationalistic responses to the pressures of labour market competition from cheap foreign labour. Finally, following the emergence of the global financial crisis and the mass shedding of workers' jobs across both new and old member states, we have seen trade unions in the UK and Ireland organize nationally in order to preferentially secure jobs for nationals, thus departing from the strategies of inclusion of migrants and trans-nationalism.

Following enlargement, we have also seen an increasing demand for labour in the NMS, often consequential on labour shortages resulting from out-migration of the domestic labour forces. Yet these pressures for replacement labour have not eliminated important push factors for East–West migration in terms of low wages and bad working conditions. Thus, the NMS became an attractive destination for immigrants from the new neighbouring Eastern European and CIS countries, whose irregular condition is created by new border controls. Poland, for example, has tried to attract immigrant workers from Ukraine, Romania and Bulgaria, but could not match the wage levels and working conditions offered in the 'old' member states (OMS) (Iglicka, 2006). The NMS have also become immigrant destinations and transit countries for a new kind of undocumented immigrant from China, Vietnam, India, Sri Lanka and Bangladesh. They thus have to cope with the double challenges of regional and global migratory trends. In this connection Krystyna Iglicka (2002) maintains that the enlargement and implementation of the Schengen regime causes the precarious position of the most vulnerable irregular immigrants in the CEE countries to deteriorate, and may lead to the explosion of both xenophobia and nationalism. These attitudes towards distant 'others', in conjunction with the implementation of EU-driven policies of 'circular condition' (Pajnik in this volume) undermine migrants' rights and their prospects for attaining social citizenship.

This brings us to the 'paradox of post-communist desolidarization' in the NMS: norms and attitudes of solidarity that show greater awareness of social injustices than in the OMS, but without adequate social mobilization in response. This reflects the problem of the re-articulation of social citizenship in the context of post-communism, the simultaneous failure of regeneration of solidarity relationships between individuals and 'the excluded other', on the one hand, and between the individual and the state, on the other (Reiter 2007). In the absence of proper institutions, strong trade unions and civil society that preclude violation of labour and migrants' rights and bad working conditions, it is difficult to challenge the attitudes 'inside and towards' the CEE region that tolerate an informal economy (Musiolek, 2002: 12).

Institutional framework for a social Europe and a 'liberal paradox' revisited

Considering the substantial income and wage inequalities between the new and the old member states, it is not surprising that trade unions and various social movements in the OMS made support for EU enlargement and globalization conditional upon the promotion of the ESM and the ILO's Decent Work Agenda, in an attempt to pursue 'fair globalization' (Likic-Brboric, 2007). However a constellation of factors has frustrated this aspiration, including the manner of implementation of the Lisbon Strategy itself under the influence of Britain's New Labour government, as well as by international factors including a political shift towards conservatism in the US post-September 11, and the re-election of Bush in 2004. These have undermined social aspirations by political and economic individualization, the policy advocacy of Anglo-Saxon financial capitalism and the concomitant flexibilization of labour markets and welfare regimes (Gowan, 2004: 15). By the same token, EU-level intervention in the Europeanization of multi-level industrial relations systems reflects an asymmetric settlement between competing interest groups that supports a stronger role for the Commission, capital accumulation and economic growth (Nieminen, 2005).

Following EU enlargement on 1 May 2004, we have seen the intensification and geographically differentiated expansion of the EU project towards the creation of a global capitalist production system, and its rearrangement in terms of 'buyer-driven' flexible global production chains (Gereffi and Kaplinsky, 2001). The Lisbon Review in 2004/05 reflects a renewed commitment to a neoliberal position in support of business-friendly approaches to competitiveness and to the creation of a global labour market. New neoliberal Baltic states were praised for their exemplary radical reforms, double-digit growth rates and decreasing unemployment, although these were stimulated by an enormous inflow of financial capital supplied by predominantly Swedish banks and an outflow of workers made possible by the free mobility regime. Even Slovenia turned to the Right for the first time in the post-communist era, and its neoliberal government looked to the 'Baltic miracle' as a leading star.

By the same token, the transformation of the ESM within these parameters clearly implied a programmatic decline of the welfare state as the institutional basis of social citizenship. In the context of the EU, social and migration policies are becoming post-national, involving the reassignment of the promotion of universal social justice to multi-level governance, involving global, European, regional and local level articulations. The 2004 Hague Programme (CEU, 2004) formulated the EU's priorities for Justice and Home Affairs for the period 2005–2010. It emphasized security as 'a new urgency' and formulated a strategy to create a European Area of Freedom, Security and Justice and move forward the global approach to migration management. Arguably, this has resulted in a dual migration regime, supporting employer-driven labour migration and creating a security-driven regime to fight 'illegal' migration from 'third countries' (Hansen, 2005). Hansen claims that these contradictory articulations of the EU migration regime, intended to address populist anti-immigrant sentiments, have undermined prospects for anti-discrimination policies and in the long run, clearly erode the social justice claim to a universal right of asylum. Undoubtedly, the Hague Programme underscores the essential 'ambiguity' of the EU's stance on issues of broad social and labour rights and undermines the role of the EU Charter of Fundamental Rights in promoting a wider European integration.

The most recent phases of EU enlargement have intensified the conflict between market- and socially-minded politics, translated into multiple strategic dilemmas:

- between global competitiveness and a social model;
- between EU workfare models and national welfare state provisions;
- between wide and deep integration;
- between migration management *cum* 'citizenship squeeze' as against a migration state based on free movement, together with migrant rights and multiculturalism;
- between diluted human rights and full-fledged social, economic and workers rights;
- between the 'logic of closure' and 'the logic of opening'.

The solution to these dilemmas has been the reassignment of social policies to public–private networks, soft regulation and meta-governance interchanges between state and non-state actors on the EU and global level (Jessop, 2003). European solidarities were to be 'nested' (Ferrera, 2009) into an essentially contradictory institutional and ideological framework consisting of (Barbier *et al.*, 2007):

- employment policy framed by the 2006 *Green Paper on modernizing labour law* promoting flexibilization and the removal of key regulatory protections, but at the same time, the promotion of an idealistic Decent Work Agenda and establishment of the European Global Adjustment Fund;
- citizenship defined and reaffirmed by the voluntary EU Charter of Fundamental Rights and associated with a comprehensive migration policy;
- gender equality and non-discrimination to assist labour market integration;
- promotion of decent working conditions to guard against producers seeking unfair competitive advantage.

There is an ongoing argument concerning the question of whether EU level frameworks have a potential to resolve the enlarged EU's inner tensions. Maurizio Ferrera (2009: 224) argues that they do, and sees the possibility for bridging the distinct 'EU economic space' of the EMU and the 'EU social space' by the strategic promotion of the EU social model, institutionally 'nesting' the welfare state within the European Union. This involves the generation of transnational, cross-regional and supranational sharing schemes that simultaneously protect national welfare states, but allow multilevel institutional innovation. A similar standpoint has been adopted by Monica Threlfall (2007) claiming that the EU's institutional framework and set of policy instruments have consistently furthered the European social dimension, strengthened European citizenship rights, and represent the best global example of flexible and broad methods to promote regional social integration.

However, critical reviews of the socio-economic consequences of EU enlargement point to a rather perverse impact of EU hard and soft regulations on industrial relations in the new member states. Guglielmo Meardi (2007) contends that the EU has failed to 'export' various social directives to the new member states which would preclude social dumping, in spite of favourable economic developments that have lifted these economies following enlargement in 2004. He points to the generally weak enabling effects of the European social dialogue, and to the low impact of the Working Time Directive. Similarly, Charles Woolfson (2006, 2010) questions the possibility of the emergence of the common ESM in the promotion of working conditions, as well as in the implementation of the Racial Equality Directive.

Drawing on Hirschman's (1970) 'voice/exit' model, Meardi (2007) claims that workers' response to deteriorating working conditions does not involve social action. They rather follow various strategies of 'exit' from the labour market, such as informal labour, social benefits and migration. However, he argues that migration, in the context of rising labour demands across the enlarged EU, has created opportunities for the expression of workers' 'voice' (i.e. organized protests) both in the sending and the receiving countries, and for revival of grassroots level workers' mobilization. We have also seen the political mobilization of civil society actors and trade unions that, following the unsuccessful ratification of the Lisbon Treaty, raised demands for social solidarity, concrete policy measures and financial pledges to promote ESM in the NMS. Nevertheless, the EU level formative institutional practices, such as previously mentioned ECJ rulings in the employment and social area, have enhanced the re-commodification of labour as the main momentum behind the formation of the European labour market, systematically undermining the power of trade unions and local governments.

The recently communicated review of the EU Global Approach to Migration (CEC, 2008) and the simultaneous adoption of the directive on return of illegal migrants hardly tamper with the general trend in the EU's primary concern with economic growth and low inflation. Although the issue of migration is framed in terms of 'partnership and solidarity', the policy approach to 'illegal migration' emphasizes 'effective' return and security-oriented border control. This orientation locks in the 'liberal paradox' and constrains the contingencies for development and emulation of a rights-based mobility/migration regime and the European social model. At the same time, the low level of wages and social protection in the NMS, with new forms of labour casualization–*cum*-informalization enable a continued 'wage squeeze', recommodification and vertical segmentation of labour in the OMS.

The configuration of the EU-level institutional framework for Freedom, Security and Justice has reinforced a new EU-wide landscape of precarization and insecurity that has been shaped by market fundamentalism. Moreover, the EU's target to combat irregular migration through stricter border control, neighbourhood and development policies, can be seen as a regional strategy of neo-liberalization of citizenship and recommodification of labour that creates a 'useful migrant' (Geddes, 2005: 788) and an 'employable' citizen. These uneven processes of liberalization of movement within the enlarged EU have maintained the existing 'wage squeeze' (cf. Peterson, 1994) and promoted the reality of a 'citizenship squeeze' that has undermined the basis of a 'European citizenship' in the making.

Concluding remarks: global financial crisis, TINA revisited, and the future of social Europe

The financial crisis, discussed by Kosner in this volume, has undoubtedly undermined the legitimacy of the unregulated global financial system, bringing about at least a temporary momentum towards 'regulatory renewal' (Woolfson and Likic-Brboric, 2008). The social unrest, actual and feared, accompanying the crisis has provoked the return of a European social model onto the EU's agenda. However, while the political and public discourses reveal the prominence of the issues of social solidarities, the political practices have so far been limited to bailing out of the banks and tentative steps towards re-regulation of the global financial system. Furthermore, we have seen regionally differentiated policy responses to the crisis, clearly reflecting the division of the enlarged Europe between the old and the new member states. The almost bankrupted Baltic States and many of the aspiring candidate countries in the Eastern and South Eastern neighbourhoods have not received any tangible acts of 'European' social solidarity. On the contrary, the EC has consigned their economic recovery to the austerity measures of the IMF that essentially do not differ from the 'old' and now-discredited Washington Consensus. The futility of mass popular and workers' protests against budget cuts in crisis-driven Latvia and Lithuania, but also in Greece, best exemplify reaffirmation of the TINA principle, coupled with a discredited social dialogue and a decay of democratic politics. Woolfson (2010a) conceptualizes the resulting political apathy as 'a double failure of voice'. Without any possibility of political deliberation of the prescribed remedial policies, a new downward cycle of radical restructuring, further budget cuts and real 'internal devaluations' is underway that reassigns the burden of the crisis to workers and their families. Solidarities are sacrificed to class interests and this, in turn, reinforces both 'external' and 'internal' exit strategies of migration and xenophobia to the ultimate detriment of social sustainability, especially in the (failing) post-communist new market economies.

But, the current economic decline and rising unemployment have also led to mounting populist mobilization against migrant workers and an enlargement 'fatigue' across the EU old member states. The imperative to implement a 'decent work agenda' and preclude a spreading of hyper-precarity in the face of a global economic recession has provoked the question: 'what are the appropriate and accessible forms of new accountability that might enable migrants as carriers of a double burden of risk exposure to secure their basic rights in the global labour market?' (Woolfson and Likic-Brboric, 2008: 306).

The institutionalization of the global and EU discourses on social justice, globalization and Europeanization is contingent on translation into national policy contexts. Yet it is precisely here that the limitations imposed by developed states' unconditional devotion to the World Trade Organization (WTO)-defined parameters for global competitiveness, and the new political 'closure' brought about by the financial crisis, are most keenly felt. In order to address the problems brought about by EU policies in favour of free flows of capital and (a less) free flow of people, the EU needs to promote a rights-based migration regime and a European citizenship that is based on the understanding of post-communist transformation at the core of a 'Europeanization' that 'equally affects both East and West' (Likic-Brboric, 2007: 182).

Acknowledgement

Charles Woolfson and Gabriella Lazaridis contributed comments and suggestions to previous version of this chapter. All errors remain those of the author.

References

Arditti s, S. (ed.) (1994) *The Politics of East-West Migration*, London: Macmillan Press

Barbier, C., Ghailani, D. and Pochet, P. (2007) 'Digest', in *Journal of European Social Policy*, Vol.17, No.2, pp. 169–82

Bohle, D. and Greskovits, B. (2007) 'Neoliberalism, Embedded Neoliberalism and Neocorporatism: Towards Transnational Capitalism in Central-Eastern Europe', in *West European Politics*, Vol. 30, No. 3, pp. 443–66

Brubaker, R. W. (1992) 'Citizenship Struggles in Soviet Successor States', in *International Migration Review*, Vol. 26, No. 2, pp. 269–91

CEC (2000) 'Social Policy Agenda', COM (2000) 379 final. Brussels

CEC (2001) 'Promoting Core Labour Standards and Improving Social Governance in the Context of Globalisation', COM (2001) 416 final. Brussels

CEC (2004) 'The Social Dimension of Globalisation: The EU's Policy Contribution on Extending the Benefits to All', COM (2004) 383 final. Brussels

CEC (2008) 'A Common Immigration Policy for Europe: Principles, Action and Tools', COM (2008) 359 final. Brussels

CEU (2004) 'The Haag Programme: Strengthening Freedom, Security and Justice in the European Union', 16054/04, JAI 559. Brussels. Accessed at: http://ec.europa.eu/justice_home/doc_centre/doc/hague_programme_en.pdf

Chavdarova, T. (2002) 'The Informal Economy in Bulgaria: Historical Background and Present Situation' in Neef, R. and Stânculescu, M. (eds) *The Social Impact of Informal Economies in Eastern Europe*, Burlington: Ashgate

Deacon, B. (2000) 'Eastern European Welfare States: the Impact of the Politics of Globalization', in *Journal of European Social Policy*, Vol. 10, No. 2, pp. 146–61

Deacon, B. (2005) 'From "Safety Nets" back to "Universal Social Provision": Is the Global Tide Turning?', in *Global Social Policy*, Vol. 5, No. 1, pp.19–28

Esping-Andersen, G. (1990) *The Three Worlds of Welfare Capitalism*, Oxford: Polity Press

Ferrera, Maurizio (2009) 'The JCMS Annual Lecture: National Welfare States and European Integration: In Search of "Virtuous Nesting"', in *Journal of Common Market Studies*, Vol. 47, No. 2, pp. 219–33

Favell, A. and Hansen, R. (2002) 'Markets against Politics: Migration, EU Enlargement and the Idea of Europe', in *Journal of Ethnic and Migration Studies*, Vol. 28, No. 4, pp. 581–601

Frade, C. and Darmon I. (2005) 'New Modes of Business Organization and Precarious Employment: Towards the Recommodification of Labour', in *Journal of European Social Policy,* Vol. 15, No. 2, pp. 107–21

Geddes, A. (2005) 'Europe's Border Relationships and International Migration Relations', in *Journal of Common Market Studies*, Vol. 43, No. 4, pp. 787–806

Gereffi, G. and Kaplinsky, R. (eds) (2001) 'The Value of Value Chains. Spreading the Gains from Globalisation', in *IDS Bulletin* 32 (3), Special Issue

Gowan, P. (2004) 'The Politics of the Regime Shaping in the EU', in *Research Seminar Series DESTIN*

Hansen, P. (2005) 'A Superabundance of Contradictions: The European Union's Post-Amsterdam Policies on Migrant "Integration", Labour Immigration, Asylum and Illegal Emigration', CEUS, Linköping University

Heidenreich, M. (2007) 'The Development of Social Inequalities in Europe' in Magnusson, L. and Stråth, B. (eds) *European Solidarities: Tensions and Contentions of a Concept*, Brussels: P. I. E. Lang

Hirschman, A. O. (1970) *Exit, Voice and Loyalty: Response to Decline in Firms, Organizations and States*, Cambridge, MA: Harvard University Press

Hollifield, J. F. (2004) 'The Emerging Migration State', in *International Migration Review* Vol. 38, No. 3, pp. 885–912

Iglicka, K. (2002) 'Towards Converging Migratory Trends in Europe', in *The Brown Journal of World Affairs,* Vol. 8, No. 2, pp. 203–10

Iglicka, K. (2006) 'Free Movement of Workers Two Years after the Enlargement: Myths and Reality', Reports and Analysis 11/06, Warszawa: Center for International Relations. Accessed: 25 May 2010 at: http://csm.org.pl/fileadmin/files/Biblioteka_CSM/Raporty_i_analizy/2006/Krystyna%20Iglicka_Free%20Movement%20of%20Workers%20Two%20Years%20Afte.pdf

Jessop, B. (2003). 'Changes in Welfare Regimes and the Search for Flexibility and Employability' in Overbek, H. (ed.) *The Political Economy of European Employment: European Integration and Transnationalization of the (Un)employment Question*, London: Routledge

Jordan, B. and Düvell, F. (2002) *Irregular Migration: The Dilemmas of Transnational Mobility*, Cheltenham: Edward Elgar

Keune M. (2009) 'EU Enlargement and Social Standards: Exporting the European Social Model' in Orbie, J. and Tortell, L. (eds) *The European Union and the Social Dimension of Globalization: How the EU Influences the World*, London: Routledge

Krings, T. (2009) 'A Race to the Bottom? Trade Unions, EU Enlargement and the Free Movement of Labour', in *European Journal of Industrial Relations*, Vol. 15, No. 1, pp. 49–69

Likic-Brboric, B. (2003) *Democratic Governance in the Transition from Yugoslav Self-Management to a Market Economy: The Case of the Slovenian Privatization Debates 1990–1992*, Uppsala: Uppsala University

Likic-Brboric, B. (2004) 'Creating a Stable Flexibility in an Enlarged Europe' in Likic-Brboric, B. and Murhem, S. (eds) *The Challenge of Industrial Restructuring*, Brussels: Work Life and EU Enlargement, Uppsala University

Likic-Brboric, B. (2007) 'Globalisation, EU Enlargement and New Migratory Landscapes: The Challenge of the Informal Economy and Contingencies for "Decent Work"' in Berggren, E., Likic-Brboric, B., Toksöz, G., Trimikliniotis, N. (eds) *Irregular Migration, Informal Labour and Community: A Challenge for Europe*, Maastricht: Shaker Publishing

Meardi, G. (2007) 'More Voice after More Exit? Unstable Industrial Relations in Central Eastern Europe', in *Journal of Industrial Relations*, Vol. 38, No. 6, pp. 503–23

Musiolek, B. (2002) 'Decent Work in the Informal Sector CEE/CIS Region', Geneva: International Labour Office. Accessed: 25 May 2010 at: http://www.ilo.org/dyn/infoecon/docs/41/F260146966/Employment%20sector%20cee%20cis.pdf

Neef, R. (2002) 'Observation on the Concept and Forms of the Informal Economy in Eastern Europe' in Neef, R. and Stanculescu, M. (eds) *The Social Impact of Informal Economies in Eastern Europe*, Burlington: Ashgate

Nieminen, A. (2005) *Towards a European Society?: Integration and Regulation of Capitalism*. Academic Dissertation, Helsinki: University of Helsinki

Overbeek, H. (2003) 'Globalization, Neo-liberalism and the Employment Question' in Overbeek H. (ed.) *The Political Economy of European Employment.*

European Integration and Transnationalization of the (Un)employment Question, London and New York: Routledge

Patzwaldt, K. (2004) 'Labour Migration in Eastern Europe and Central Asia and Next Political Steps', *UNESCO Series of Country Reports on the Ratification of the UN Convention on Migrants*: UNESCO. Accessed: 25 May 2010 at: http://unesdoc.unesco.org/images/0013/001395/139530e.pdf

Peña-Casas, R. P. (2007) 'European Integration and Social Inequalities from an Empirical Perspective' in Magnusson, L. and Stråth, B. (eds) *European Solidarities: Tensions and Contentions of a Concept*, Brussels: P. I. E. Peter Lang

Peterson, S. W. (1994) *Silent Depression: Twenty-five Years of Wage Squeeze and Middle-Class Decline*, New York and London: W. W. Norton & Company

Reiter, H. (2007) 'The Post-Communist Triangle of (Non-)solidarity and Beyond: Youth and the Perception of Unemployment in the "New West"' in Magnusson, L. and Stråth, B. (eds), *European Solidarities: Tensions and Contentions of a Concept*, Brussels: P. I. E. Peter Lang

Threlfall, M. (2007) 'The Social Dimension of the European Union: Innovative Methods for Advancing Integration', in *Global Social Policy*, Vol. 7, No. 3, pp. 271–93

Samers, M. (2004) 'An Emerging Geopolitics of "Illegal" Immigration in the European Union', in *European Journal of Migration and Law*, Vol. 6, No. 1, pp. 27–45

Schierup, C-U., Hansen, P. and Castles, S. (2006) *Migration, Citizenship and the European Welfare State: A European Dilemma*, Oxford: Oxford University Press

Slavnic, Z. (2010) 'Political Economy of Informalization', in *European Societies,* Vol. 12, No. 1, pp. 3–23

Solimano, A. (2001) 'International Migration and the Global Economic Order: An Overview', Policy Research Working Paper 2720. Washington DC: World Bank

Vaughan-Whitehead, D.C. (2003) *EU Enlargement versus Social Europe?: The Uncertain Future of the European Social Model*, Cheltenham: Edward Elgar

Woolfson, C. (2006) 'Working Environment and "Soft Law" in the Post-Communist New Member States', in *Journal of Common Market Studies*, Vol. 44, No. 1, pp. 195–215

Woolfson, C. (2010) 'The Race Equality Directive: "Differentiated" or "Differential" Europeanisation in the New EU Member States?', in *European Societies*, Vol. 12, No. 4, pp. 543–66.

Woolfson, C. (2010a) '"Hard Times" in Lithuania: Crisis and "Discourses of Discontent" in Post-communist Society', in *Ethnography*, Vol. 11, No. 4, pp. 487–514.

Woolfson, C. and Likic-Brboric, B. (2008) 'Migrants in the Unequal Burdening of "Toxic" Risk: Towards a New Global Governance Regime', in *Debatte: Journal of Contemporary Central and Eastern Europe*, Vol. 16, No. 3, pp. 291–308

Chapter 6
Engendering Insecurity: The EU Asylum Regime

Liza Schuster

> Me tired France. Me tired Europe. Three years Europe. No life here – sleeping eating waiting nothing. Sleeping bridge, park. Eat street [mimes squatting, eating from bowl] No life waiting fighting. Say leave Iran – go Afghanistan. Me go Afghanistan – no family Afghanistan. Family dead, family Iran. Two month Afghanistan – very bad. Me leave. Iran, go Turkey. Turkey five month prison. [why did they put you in prison?] For nothing, for nothing. Go Greece – boat. Greece very bad – police very bad. Go England – six month England – four month prison [detention]. Me angry, me fighting – 'okay, deport, deport'. Deport Greece. Come back France. Me Paris one year. Me tired. Me tired. Thinking, thinking. Go back Afghanistan. Finish. Dead. Better for me (Afghan man, Paris, 27 years old, speaking in English).

The European Union, a federation of 27 states bound together by an explicit commitment to liberal democratic values manifested in the requirement to sign up to the *Acquis*,[1] has made an equally explicit commitment to offering some form of security to those in need of protection (Schuster, 2003). The EU, according to Article 3 of the Reform Treaty, 'shall offer its citizens an area of freedom, security and justice without internal frontiers, in which the free movement of persons is ensured in conjunction with appropriate measures with respect to external border controls, asylum, immigration and the prevention and combating of crime'. The promise of security is central and not just to the citizens of the EU. While it is the duty of all states to protect their citizens, all EU Member States have also, by signing up to a range of international and domestic legal instruments, undertaken to protect non-citizens who are unable or, owing to a well-founded fear of persecution, unwilling to avail themselves of the protection of their country of nationality. Security is understood as freedom from threats to one's life and liberty and the capacity to enjoy one's human rights. The state is seen as key in protecting and in securing people, but in the context of asylum in Europe, human security functions 'as a biopolitical technology … wedded to the architecture of the territorial nation-state' (Duffield, 2007: 111). This technology not only does not

1 The body of EU law already in existence (already 'acquired') that would-be Member States (MS) must put in place before accession to the Union.

protect, but maintains in a state of extreme insecurity, those humans who challenge the capacity of European nation-states to maintain their territorial integrity. This chapter explores the damage inflicted by liberal democratic states on non-citizens, painfully articulated by the young man above.

The EU Member States reserve to themselves the power to protect and secure, and to refuse to protect and secure, those seeking asylum. In this chapter, the practice of nation-states in relation to the granting or withholding of security in the shape of papers or documents, and the benefits that come with them, should be understood as an example of the functioning of biopower, of 'techniques for achieving the subjugations of bodies and the control of populations'(Foucault, 1998: 140). Based on whether or not and where biometric data, such as finger-prints, are to be found in European databases like Eurodac, young men in particular are maintained in a situation of physical need, subjugation and availability to the police in public spaces such as soup kitchens where they eat and the parks where they sleep. In this way, the state maintains control of a population that, while physically vulnerable, is also constructed as threatening. Elsewhere, I have considered the way in which these technologies of control are racialized (Schuster, 2003a), but Maguire in this volume provides a fluent and powerful account of the place of finger-printing within the history of colonization and racialization.

While the concept of human security, in which the security, rights and freedom of the individual should take precedence over the state, is becoming, at least rhetorically, the dominant paradigm (Duffield, 2007), the story told here indicates that practically, perceptions of threat and risk to the state and its citizenry trump the security needs of vulnerable 'others'. Over at least a decade now, numerous scholars have explored the concepts of biopower and biopolitics in relation to the control of mobile populations (Agamben, 1998, 2000; Balibar, 2004; Bigo, 2001), but this chapter is less a contribution to that theoretical literature and more a reflection on the consequences of subjugating one small group of people to particular technologies of control embodied in EU law and practices. Here, security, or rather its absence, is seen from the perspective of those the EU Member States are seeking to control, to manage, in the interests of *state* security.

In 1999, the Member States made a commitment to the development of a common European asylum system based on 'full and inclusive' application of the 1951 Convention relating to the Status of Refugees (Commission, 1999, see also Vedsted-Hansen, 2005). However, the asylum regime currently in place is marked by a narrow and exclusive interpretation of the Convention, one designed to keep those eligible to make a claim to the absolute minimum and to maintain those seeking the protection of the Convention in a state of *insecurity* for as long as possible, in order to facilitate their removal to another state or to their country of origin, and to deter others from seeking its protection. It does this through two instruments in particular – Dublin II and Eurodac (see below). Yet those most in need are at least as likely to be affected as those the EU believes to be 'bogus' asylum seekers. There is no recognition of the two key realities; that it is conditions in the countries of origin that drive up the numbers (Castles, Crawley & Loughna,

2002), and that so long as some continue to win the lottery that is European refugee status, others will continue to play in the hope that one day they too will find a place in which to construct a secure future for themselves and their families.

Using the experiences of a group of Afghan migrants in Paris who have travelled overland through a number of countries, this chapter argues that in order to evade legal and moral responsibility for providing protection to those in need, EU Member States endeavour to maintain asylum seekers in a state of legal insecurity, which in turn creates the kind of physical and psychological insecurity expressed in the words at the beginning of this chapter. The two instruments that do most to maintain new arrivals in this state of insecurity and anxiety are Dublin II and Eurodac (Schuster, 2010), the mechanisms used by Member States to avoid a 'full and inclusive' application of the Refugee Convention. Eurodac is a European database storing the fingerprints of asylum seekers, 'making legible certain populations, fixing their identities and their movements' (Maguire, in this volume), while Dublin II is the regulation that identifies the country responsible for examining a claim for asylum, usually the first 'safe' country entered by someone claiming protection. This regulation is used by European Member States to evade responsibility for processing claims by pushing those in their territories back to the first 'safe' countries of arrival. Dublin II keeps asylum seekers in a state of insecurity, not knowing whether their claim will be examined and depriving them of the minimal bundle of rights available to those who are allowed to make a claim (such as accommodation, and access to some minimal allowance – if not to the right to work and control one's environment). The young Afghan men[2] referred to here were at the time of writing in Paris and 'sous Dublin' – that is, the French state was seeking to return them to another Member State using Dublin II and Eurodac. Some remain in this limbo state for many months, unsure of what will become of them (Schuster, 2011). By the time they arrive in Paris, they will have already crossed a number of countries in their search for security, and this chapter traces their experiences in these countries and the insecurity that drives them onwards. One of the most significant sources of insecurity is the constant threat of deportation (Chimienti and Achermann, 2006), a threat that affects those without any resident status as well as those who have a precarious status such as asylum seeker, or, more particularly, asylum seekers 'sous Dublin'.

The testimonies and insights are drawn from the experience of the author working over a period of about 18 months as a volunteer with the *Collectif de soutien des exiles du 10eme*,[3] a group of activists that provides information and support to the Afghans who spend their time in and around the parks, canal and

2 The Afghans referred to here are exclusively men, because the majority of young Afghans sleeping rough in Paris and assisted by the *Collectif* are men. Although we occasionally encountered women as family groups, families could be found accommodation very quickly (though this was often in very cheap hotels far outside Paris on industrial estates with no cooking facilities and no shops or public transport nearby).

3 See: http://www.exiles10.org/.

soup kitchens of the 10th arrondisement of Paris, trying to provide information and answer questions on the European regime. It is informed by many dozens of conversations[4] in English, French, Greek and very broken Dari with the young men about their journeys from Afghanistan, by the summaries of the conversations with other volunteers posted online each evening and by attendance at a number of meetings attended by activists, charities and local politicians. The chapter details the experiences of the Afghans inside the EU in order to illustrate how it is a combination of legal, physical and psychological insecurity that drives them to move from one Member State to another, and the ways in which Dublin II and Eurodac compound and trap them in that insecurity. Writing on the vulnerability of the undocumented and asylum seekers, Chimienti and Acherman (2006: 67) note that while the precariousness of these groups has been documented in a general way, there is little written on specific challenges they face, such as '[p]auperization, instability, the breakdown of social structures and future prospects (Bourdieu, 2001), and social exclusion (Castel, 1995)'. This chapter is an attempt to document some of the specific challenges faced by a specific group of vulnerable people in the heart of Europe, and as such is a critique of EU asylum policy, which continues to privilege state over human security.

To start near the beginning: Iran and Pakistan

The overwhelming majority of Afghans who have left their country in the last three decades (6.4 million according to UNHCR, 2009) sought refuge first in Iran and Pakistan, and many of those currently present in Europe have spent time in those states. It is difficult to find accurate figures especially as insecurity in Afghanistan has led to repeated movements back and forward across the Afghan-Iranian and Afghan-Pakistan borders. The most recent census puts the number of Afghans in Pakistan at 1.8 million – even *after* returns totaling 3 million (UNHCR, 2009), and in Iran at approximately 970,000 (UNHCR, 2009; see also Kronenfeld, 2008). The 2005 Amayesh (census) put the total number of registered Afghans living in Iran at just over one million (1,021,323) (cited in Abassi-Shavazi, 2008). There were also estimated to be 500,000 unregistered Afghans, including single labour migrants and, according to Abassi-Shavazi *et al.* (2008), there may well have been a substantial number of Afghans in Iran who were neither documented nor labour migrants. Until 1992, these refugees or *mohajerin* who had fled to Iran would

4 These were conversations that took place in the course of 'patrols' by the collective in the evening with those queuing at Salvation Army soup kitchens, as questions were asked about conditions in France and elsewhere in Europe. They were not interviews conducted for a research project, though I explained that I was a *Ma'alem* (teacher) and a writer and would write about what I saw.

be issued with 'Blue cards' which gave them permission to remain indefinitely.[5] Although initially supportive towards the Afghan refugees, who were allowed to look for work and accommodation and permitted 'access to healthcare, basic education and subsidized food on the same terms as Iranian citizens' (Harpviken *et al.*, 2004), after the fall of Kabul in 1992, when Afghans fled again in large numbers to Iran, they were regarded as a burden who should return 'home' (IPS 2009). And instead of being treated equally, as Abassi *et al.* put it, 'policies which differentiate non-nationals from nationals shape the opportunities and experiences of second-generation Afghans in Iran' (Abassi-Shavazi *et al.*, 2008: 4)[6] – in other words, whatever papers or status Afghans have, unless they have the security provided by Iranian citizenship they remain vulnerable to changes in law and policy.

Since 1992, when a series of mass deportations was initiated, life in Iran has become much more insecure. The testimonies of Afghans in Paris who have lived in Iran bear out the earlier accounts in the briefing by Harpviken *et al.* (2004) that they were no longer welcome (in spite of continuing demand for labour in the Iranian agricultural sector), that, without papers, they felt unprotected in the face of discrimination, crime and violence. As one Afghan who had arrived in Iran aged six months explained 'in Iran, anyone can stop you, take your money, beat you, you can do nothing. The police will not help you. Afghans have no rights there. Iran people are very bad to Afghans'. While competition for scarce accommodation in Paris may explain some of the tension, it seems this hostility was also at the root of fights that broke out in Paris between Afghans and Iranians waiting for the bus that would take the homeless to emergency shelters over the winter of 2009/10 when the numbers of Iranian asylum seekers increased sharply. As one Afghan who had lived in Isfahan, Iran, explained 'the Iranians think they are better than us. They treat us badly in Iran'. Nonetheless, there were mixed feelings. Some Afghans in private conversations said they preferred the Iranians because they had lived among them for so many years, that having returned to Afghanistan recently they found they had no family there, no support network they could trust, and so their experiences of their 'Afghan home' had been very difficult, but they were not allowed to return to Iran.

Similar stories abound among those who had lived in Pakistan, with the Human Rights Commission of Pakistan reporting 'that the "push" to send refugees back home had acquired the dimensions of a "shove" by creating an environment of fear and persecution to force refugees out' (HCRP, 2009) even though many of those living in these two countries were born or spent most of their lives there and even

5 For an analysis of Iran's relationship with Afghanistan and its attempt to create greater security for itself through using the Mohajedin as an indigenous Afghan force sympathetic to Iran, see Milani (2008).

6 They go on to point out that 'Second-generation Afghans in Iran have been raised in an arguably more liberal social and religious environment, and exposed to values, attitudes and practices that are different from those of their parents'. More research is required on this aspect of exile.

though segments of Pakistan's economy has largely become reliant on Afghan labour (Kronenfeld, 2008: 45).[7] Although the years after the Allied invasion saw large numbers of Afghans crossing back to Afghanistan, many with the help of UNHCR's voluntary repatriation schemes, others had settled and founded families in Pakistan. More significantly perhaps, as Kronenfeld (2008: 49) points out '*the majority of Afghans currently residing in Pakistan were actually born there*, and may have little or no familiarity with Afghanistan. Yet these children of refugees are considered refugees themselves and do not have Pakistani citizenship'. In the words of an Afghan in Paris, 'I cannot go back to Afghanistan. I have no family there now. Without family I am dead'. This is a reference to a society that relies heavily on the extended family for security and support. Without a father, uncle or grandfather to guarantee a place in the community, to offer work and shelter, an Afghan stranger is treated with suspicion and will not find protection in Afghanistan. Kronenfeld (2008: 55) writes that many Afghans who have returned to Afghanistan from Pakistan, return again, citing lack of shelter (57%), no means of making a livelihood (18%) and insecurity (18%) as factors preventing their long term re-integration.

Given such insecurity, it is perhaps surprising that so few Afghans make their way to Europe. Nonetheless, faced with an uncertain future in Iran and Pakistan, or forced return to a country of which all they know is that they would not have either physical or economic security there, a proportion of these Afghans do seek some form of security further afield, though it is important to note that relatively speaking the number of Afghans who seek that security in Europe is tiny. According to the Afghan Ministry of Refugees and Returns the number of Afghan refugees living in all of Europe is 500,000 (cited at IRIN, 2009).

En route: Greece

Until relatively recently, the usual route for Afghans coming to Europe overland was through Iran and Turkey, with Van, Istanbul and Izmir in Turkey being important staging posts on the route (Yaghmaian, 2005). Though most of those in Paris referred to the precariousness of life in Turkey and the brutality of the police there, they insist their problems began in earnest once they entered the European Union – that is, on the Greek side of the Turkish border. The asylum regime in Greece has been the target of considerable criticism by scholars (Karamanidou, 2009, Sitaropoulos, 2000, Papadimitriou & Papageorgiou, 2005), NGOs (AI, 2008, Human Rights Watch, 2008), other EU Member States, in particular Norway (IAB, 2009, Norwegian Organization for Asylum Seekers, 2009) and the European Commission (2007). The problems in Greece are multiple and occur at different

---

7 Kronenfeld extrapolates from Pakistan government figures that at least 55 per cent of Afghan refugees under 18 were born in Pakistan (2008: 49).

levels, and can be understood in terms of different forms of insecurity: physical, legal, economic, social.

Signatories of the Geneva Convention are prohibited by Article 33 from forcibly returning or refouling people to 'a country in which the refugee has reason to fear persecution'. There is ample evidence both from NGOs (Amnesty International, 2004; Human Rights Watch, 2008; Norwegian Organization for Asylum Seekers, 2009), journalists such as Sarah Prestianni,[8] scholars (Karamanidou, 2009, Yaghmaian, 2005) and from conversations with Afghans in Paris and that the Greek authorities engage in practices which amount to the refoulement of asylum seekers, that is, the return of people to 'the frontiers of territories where his life or freedom would be threatened on account of his race, religion, nationality, membership of a particular social group or political opinion' (Art. 33, 1951 Convention). Those caught by the Greek authorities on or near the land border with Turkey are liable to be returned across the border without being offered the opportunity to claim asylum. There is also evidence that the Turkish authorities have continued the refoulement, pushing some back into Iraq (Norwegian Organization for Asylum Seekers, 2009).

Afghans in Paris told of being arrested on crossing the border between the two countries and thrown into jail near Alexandroupoli, near the Turkish border, echoing similar stories gathered by Yaghmaian (2005), including the use of dogs to harass and bite the migrants. One young man told how he was picked up and kept with others in a large concrete room. After 20 days, when about 100 men had been detained, Greek 'commandos' came and moved them in groups of 10 onto a boat on a river (the Evros). On the Turkish side, they were forced off the boats at gunpoint, before it turned round and headed back to Greece. Sara Prestianni, an Italian photojournalist working with NGOs, and campaign groups on this subject confirmed the stories told in Paris, and added that before being returned Turkey the Afghans were stripped of anything that might betray their presence in Greece – including clothes with Greek labels or markings. In spite of this, the Turkish authorities sometimes insist on pushing back to Greece those they find on their side. In one case documented by UNHCR, Turkish police tried to force 18 men, including five recognized as refugees by UNHCR, across a river separating the two countries leading to the deaths of four of them by drowning (UNHCR, 2008). Other young men in Paris confirmed reports of boats being returned to Turkish waters by the Greek coastguards, with their occupants being threatened with guns (Pro Asyl *et al.*, 2007). According to the Human Rights Watch Report (2008: 38), the Turkish General Staff was cited as claiming 'that Greece "unlawfully deposited at our borders" nearly 12,000 third-country nationals between 2002 and 2007'. At this stage, these young men had no physical or legal security and were extremely vulnerable.

---

8 Interviewed in Paris, 11 July 2009 – accessed at: http://www.flickr.com/photos/saraprestianni/page4/.

For those not returned to Turkey from Greece, there is still little hope of any form of legal security. Despite a doubling of the number of asylum applications in Greece from 12,267 in 2006 to 26,735 in 2007 (OFPRA, 2008), these numbers are very low compared to other Member States, given that Greece is the first EU state reached by many potential asylum seekers, especially those coming from Afghanistan. From the accounts of the men in Paris, and from work by Karamanidou (2009) and Karamanidou and Schuster (2010), the primary explanation for the very low figures is the difficulty of making an asylum claim in Greece, though with the election of a new government in 2009 this may change (Karamanidou and Schuster, 2010). All the Afghans spoken to in Paris on this subject (more than 20) in response to a question about claiming asylum in Greece, said without exception that they had not been allowed to do so. This was a recurring theme, brought up spontaneously by many others. In spite of the many thousands of Afghans transiting the country, only 2,287 claimed asylum in 2008. None were granted any form of status (UNHCR, 2009) so none have any form of legal security. The Greek document most often seen in Paris is a single sheet of paper written in Greek that obliges the holder to quit Greek territory within 30 days. It offers no protection, no security, no access to rights of any kind. Without papers, potential asylum seekers in Greece are liable to harassment and to arrest by the police. Two of the Parisian Afghans in July 2009 reported having spent 3 months in a prison near Alexandroupoli without charge, without trial, and without having been allowed out of their cell once – a cell they shared with 13 others. Almost all told of physical beatings, some retaining the scars. I accompanied one 18 year old to hospital to be treated for damage to his kidneys resulting from a beating by Greek police. When speaking of experiences on the way to Paris, young men would often interrupt each other to show off different injuries, sustained frequently as they struggled to resist giving their fingerprints – in one case a broken nose as the 17 year old's face was pushed onto a table. Broken fingers were a common souvenir.

However, it is not merely the physical abuse that persuades the Afghans to move deeper into Europe. Even where people have not been beaten and have managed to find work in Greece, make friends and learn the language, the lack of any secure legal status makes it difficult for people to build a life, to earn enough money to send home or to get the education and training many seek. As a result, most Afghans headed for Patras, or more recently Igoumenitsa, the Greek ports from where the ferries for Italy depart. There they joined the squatters camps which until recently housed up to 3,000 migrants. In early summer 2009, the camps in Patras were again destroyed by the police as part of a crack down on undocumented migrants. Another young Afghan told me he had been so shocked by his experiences in Greece, which he 'thought was part of Europe', he almost gave up and went back to Iran, but thought he would try a little bit further on into Europe. Nonetheless, transit through Greece, especially if one has had one's fingerprints taken, means that access to legal security, and all the other forms of security attendant on having papers, is extremely difficult. Nearly all the Afghans in Paris have come through Greece, and many have had their fingerprints taken there, and so once they come to the attention of the authorities anywhere else in the EU, if their fingerprints are found in the Eurodac database, it is to Greece that they will be returned – whether or not they have applied, or been allowed to apply, for asylum in Greece.

Some security: Italy

Some Afghans in Paris had had to make a number of attempts to get to or through Italy, having been picked up on ferries or small boats and returned immediately to Greece. Among the young men in Paris, none appeared to have been returned to Greece from Italy under Dublin II, and it has not been possible to discover how many applications are made by Italy to return asylum seekers to its Eastern neighbor using EU law. Although in a decision of the Council of State of 3 February 2009, three young Afghans had their expulsion to Greece, ordered by the Ministry of Interior under the Dublin Regulation, stopped (ordinance of 3 February 2009).[9] Most of the criticisms directed at Italy have related to the treatment of people arriving on the islands of Lampedusa and Sicily, or the mainland coast, who may be detained in crowded and unsanitary conditions, and to the rapid mass deportations of people who have not been offered the opportunity to claim asylum to North African countries (Council of Europe, 2009: 3, 16). However, in 2008, for example, 1,732 Afghans did claim asylum in Italy, and 208 were granted refugee status and 969 a subsidiary status (UNHCR, 2009), which is a very high recognition rate by comparison with Greece (see Table 6.1 below).

One might expect that this status would offer Afghan refugees a degree of security, and to a certain extent it does. Those recognized as refugees are unlikely to be refouled and those who are rejected will not be returned in most cases but left to live as *clandestine*, without accommodation, the right to work or any support. Those who are awaiting a decision, may, if they are lucky, find accommodation

Table 6.1 Recognition rates (%) of Afghan asylum seekers in selected EU MS 2008

	Aus.	Den.	Fin.	Fra.	Ger.	Gre.	Ita.	UK
Refugee status	44	4	0	36	29	0	12	8
Total positive decisions*	80	50	95	42	60	0	67	55

*Including subsidiary protection

Source: Adapted from UNHCR (2009)

9 'Il Consiglio di Stato ha dunque bloccato l'iniziativa del Ministero dell'Interno di allontanamento dall'Italia dei giovani afgani, in ragione delle note, ripetute ed ingiustificabili violazioni della normativa comunitaria e dei diritti umani perpetrate in Grecia' (accessed at: http://www.meltingpot.org/articolo14033.html).

in one of the *centri di accoglienza* (accommodation centres), which are similar to emergency shelters provided to the homeless – that is they open at 21.00, close their doors at 22.00, and 'guests' must leave before 08.00. These are distinct from the closed *Centri di identificazione ed espulsione (CIE)* (centres of identification and expulsion) where those who have been rejected, or who are in a fast-track process, are held awaiting deportation.

However, from August 2009, a steady stream of Afghans with Italian papers indicating that they had been granted refugee status in Italy began arriving in Paris – 12 on one day alone. When asked why they had left Italy, they responded because of the racism. They are frequently mistaken for Rom (perhaps the most victimized population in Europe) and share the Rom experience of violence at the hands of fascists, racists and police (Council of Europe, 2009). They could no longer bear the daily exposure to implicit and explicit hostility, such as people refusing to sit beside them on the bus or shouting at them in the street. However, they also explained that a significant factor was difficulty finding work, or accessing training or education. This they put down to a combination of discrimination and economic problems.

None had thought to apply through legal channels for permission to change residence, to move to another European Member State. Those I spoke to could not see that it was anyone's business where in Europe they lived or worked – they were not looking for anything from the state, just the chance to find secure accommodation and employment. However, while refugee status allows one to travel around Europe, it does not enable one to access the labour market, or economic or social support. As a result, those who have left Italy looking for economic security, the chance to build a life and start sending money to family in Afghanistan, often find themselves living on the streets in Paris, alongside those who have no papers, or only the most temporary residence permits.

Engendering security: France

In France, 42 per cent of Afghans who *succeeded* in making a claim for asylum in 2008 were granted either refugee status or some form of temporary, renewable status (see Table 6.1 above). However, most Afghans will have travelled through Greece and Italy and had their fingerprints taken en route. Once their fingerprints have been found in the Eurodac database, using the Dublin II regulation, France will apply to Greece to return them there without allowing them to make a claim for asylum. In 2006, France applied to return 2,647 people to other countries (all Member States) using Dublin II, succeeding in 849 cases (32%), including some Afghans (actual numbers not known, but some of those in Paris have been returned to Greece under Dublin and have returned again to Paris). Being put 'under Dublin' should in theory offer some protection from arrest by the police, as those in this condition would normally be able to produce a paper demonstrating that they have made an attempt to claim asylum (i.e. they are not undocumented and trying to hide). However, among those in Paris, many had experienced regular harassment by the police, in particular during the night if sleeping rough when they would be woken to be questioned and, if they were lucky, moved on. The Afghans themselves would only rarely comment on such everyday occurrences,[10] but occasionally emails among members of the Collective would circulate as volunteers living locally would become aware of a spike in the harassment, usually a prelude to attempts to deport.

Having applied for asylum in France, their fingerprints have been taken. Even if their fingerprints have been taken in Greece, they are not always found in Eurodac (Greece has an interest in not making these returns easy since this would only increase the pressure on their system!) and, even if found, Greece does not always respond to requests within the two months allowed. If people cannot be returned to a country they have transited, their claim must be examined in the country where they are, however precariously, resident. However, by the time France takes responsibility, those who have been through this process are frequently bitterly disappointed by their treatment in France. At least two or three times a week, members of the *Collectif* would hear of Afghans 'under Dublin' being detained in *centres de retention* (detention centres). In theory, this should only happen towards the end of the process, to facilitate the transfer. In practice, it seems far more arbitrary. Although they are entitled to some form of accommodation, dozens of Afghans 'under Dublin' are sleeping rough in Paris, and members of the C*ollectif* wage a constant battle to insist that the local authorities find accommodation for them (often in B&Bs). This insecurity about their future, about whether they will be allowed to claim in France, about whether they will be granted recognition, places a heavy psychological burden on those who effectively find themselves in a legal limbo.

Afghans arriving in Paris for the first time are shocked by the reception conditions of the other Afghans in Paris, including those who have applied for asylum, and often decide within days to continue their journey, unable to believe that they would be unable to find asylum in other EU Member States. These young men want to pursue their journeys because France's inhospitality to Afghans is evident to the new arrivals, who are disheartened to find friends and fellow Afghans sleeping in parks, under canal bridges and in the streets, or accommodated with alcoholics and drug addicts in emergency hostels that resemble Bedlam. Young Afghans in Paris can spend many months outdoors. Winter 2008/9 saw temperatures drop to −10°C, to which the City of Paris's response was the provision for minors of 25 beds in a hall. Each night there were approximately 60 Afghans, many without blankets or sleeping bags, some of whom would walk all night, afraid that the cold would kill them if they lay down. Those left outside include many who have applied for asylum, so that there is little incentive for new arrivals

10 Recently, two Afghans who are volunteers with the homeless charity *Restos du Coeur* explained that they were unable to make their regular shifts because they had been stopped on their way to the soup kitchen and held in a police station for 2.5 and 4 hours.

to follow suit. The consequences of living outdoors include lice, scabies, dental problems, depression and psychosis. On occasion, distressed young men have requested help for friends who have threatened suicide, who wander aimlessly at night, or who in two cases were found banging their heads against the stones of the bridge under which a group were sleeping. A psychologist from *Médécins Sans Frontiers* visits one of the parks frequented by the Afghans to try to offer support, but it is difficult to build relations of trust in these circumstances.

However, it is not just the lack of accommodation that pushes people to try elsewhere. Among the other push factors are the lack of training opportunities, the lack of work permits, and a growing anger and bitterness at their treatment. Of the group assisted by the *Collectif*, many wait a while before applying for asylum. In some cases they are consulting with friends in Paris, trying to work out whether they would be better off trying to stay in France or moving on. Some choose positively to apply for asylum in France, while others claim because there is nowhere else for them to go. This group includes those who have made it to Britain and been deported back to Afghanistan from where they have once again made their way west and northwards. Among this group there is an intense sense of frustration. Many are not allowed to claim asylum in France because of Dublin II, but are also not allowed to claim asylum anywhere else, and they feel a sense of abandonment, caused by the French state's refusal to take responsibility for them, while at the same time refusing to allow these young men to take responsibility for themselves, to find work and accommodation.

The French state is particularly reluctant to take responsibility for providing, or allowing NGOs to provide, more than the scarcely adequate assistance currently available because of its experience in Calais with the Red Cross shelter at Sangatte.[11] It is argued that as most of those, particularly minors, in Paris are only passing through, there is little point in investing time and effort in integrating them into the system. As the former director of France Terre d'Asile explained in conversation one evening 'our caseworkers spend hours, days trying to find a host family, a school, a social worker – and then pouf, they disappear – heading for England or Norway'. Against this, Jean Michel Centres of the *Collectif* argues that if these young people were taken in charge quickly, rather than just 'stored' in B&Bs, they would not want to continue their journeys. Local politicians are concerned that if more support is offered, this will encourage more people to come, and to stay for longer periods. Maintaining people in a situation of uncertainty, of destitution and distress, has become a weapon in the battle to deter new arrivals and encourage others to return (Schuster, 2005) or to move on in the search for security.

---

11 Sangatte has become something of a bogeyman – an article in *L'Express* (7/8/2008) on the park and the Afghans opened with the line 'Sangatte in the heart of Paris …'

The myth of elsewhere: Norway? UK?

In these circumstances, the UK and Scandinavian countries have taken on a mythical status. On the nightly patrols carried out by members of the *Collectif*, questions are regularly asked about where people should go next, about conditions elsewhere. The *Collectif* produces leaflets explaining the asylum processes in different EU states and in Norway, together with the recognition rates for Afghans and the provisions for minors (see Parusel, this volume). The goal is to ensure that whatever decision people make, it is based on information that is as accurate and up-to-date as possible. Volunteers do not advise on where to go, though they try to inform people about what to expect.

For minors, or those who hope to pass as minors, Norway in particular is seen as somewhere they will be safe. Those who have made it phone friends still en route and explain that they are quickly taken into care, placed with a family and in school. In Paris, I have met only one person who had returned from Norway, and that was to visit friends stuck in Paris. In 2008, Norway suspended returns to Greece, because of conditions there, though reinstated them for single men in 2009, but only after an individual examination of their case, when particular difficulties experienced in Greece may be taken into account. Families and those under 18 are not returned, but are cared for and integrated. This perhaps accounts for the increase in the number of applications to Norway in recent years (Table 6.2), especially among those who are under 18 (1719 in 2009, 579 in 2008). Though not a member of the EU, Norway has signed up to the Dublin Regulation and has access to Eurodac.

Table 6.2 Applications from Afghans (Norway)

Year	2000	2001	2002	2003	2004	2005	2006	2007	2008	2009
	326	603	786	2032	1059	466	224	234	1363	3871

Source: Norwegian Directorate of Immigration (www.udi.no)

Others remain determined to make it to England. Those who have families there tell of successful businesses and of educational opportunities, and believe they will be among the lucky ones. Those who are in Paris having been returned from the UK under Dublin II, or caught on the Channel ferries, or having been returned by the UK to Greece or Afghanistan, still speak of better treatment in the UK. In part this is because the Afghans compare British police favourably to French, Italian and Greek police, in part it is because they were accommodated until they were deported, so were not left to sleep in the streets. The impact of a lack of accommodation, especially in winter, should not be underestimated. The sense of vulnerability and insecurity felt among those forced to sleep in the open wears people down very quickly both physically and psychologically.

Among this group of Paris Afghans are those who understand that they have no hope of asylum in the UK, and that they risk deportation to Afghanistan, but they make the calculation that at least in the UK they will be able to find work (even if it is in the underground economy) and support from their community. Even if one does not have papers oneself, access to a support network that includes those that do can provide some security. A bed, familiar food, advice and information (however partial) in one's own language and above all work, make it seem a risk worth taking, even though there is a far greater chance of being expelled from the UK than from France to Afghanistan. These Afghans are engaged in a calculation about how to maximize a complex understanding of security – trading legal security against psychological and/or economic insecurity.

Conclusion

Having experienced difficulties in each country along their route, the Afghans continue their journey, hoping the next place will be better, believing that it cannot be worse. Above all what drives asylum seekers from one European Member State to another is lack of security, especially legal security, as it is this that furnishes people with the economic and social rights they need to be able to build their lives. But legal status alone is not sufficient. These young people, even those with papers of some kind, have frequently spent months (and occasionally years) sleeping on the streets of European cities, vulnerable to harassment and violence, their lives at constant risk. Without accommodation, attending language classes and finding work is almost impossible. Many do find security, are recognized as refugees and allowed to remain. Others find their disappointment growing, and grow more bitter. The use of EU legal instruments such as the Dublin regulation and Eurodac by EU Member States to avoid their responsibilities to a vulnerable population inflicts more damage on this group.

However, while the costs are borne most cruelly by those like the young man cited at the beginning, there are also costs to European societies. In the tradeoff between security, as a value and policy goal, and other key values, such as freedom or justice, (Karyotis, this volume) the freedom and justice of European citizens are also compromised. The link between some of the security measures and instruments discussed throughout the chapters in this volume and the attendant racism and discrimination that infects societies employing them have been explored elsewhere (Schuster, 2003; Maguire, this volume). One is reminded of the words of Hannah Arendt (1973: 269) 'those whom the persecutor had singled out as the scum of the earth – Jews, Trotskyites etc. – actually were received as the scum of the earth everywhere', such that in our pursuit of 'our' security, we Europeans are then implicated in the abuses committed far from our territories.

References

Abbasi-Shavazi, M., Glazebrook, D., Jamshidiha, G., Mahmoudian, H. and Sadeghi, R. (2008) *Second-generation Afghans in Iran: Integration, Identity and Return*, Kabul: Afghanistan Research and Evaluation Unit

Agamben, G. (2000) *Means without Ends: Notes on Politics*, Minneapolis: Minnesota University Press

Agemben, G. (1998) *Homo Sacer: Sovereign Power and Bare Life*, Stanford: Stanford University Press

Amnesty International (2008) *Amnesty International Report 2008: State of the World's Human Rights*, accessed at: http://thereport.amnesty.org/eng/regions/europe-and-central-asia/greece

Arendt, H. (1973) *The Origins of Totalitarianism*, 2nd edn, New York: Harcourt Brace Jovanovich

Balibar, E. (2004) *We, the People of Europe: Reflections on Transnational Citizenship*, Princeton: Princeton University Press

Bigo, D. (2001) 'The Möbius Ribbon of Internal and External Security' in Albert, M., Jacobson, D., and Lapid, Y. (eds) *Identities, Borders, Orders: Rethinking International Relations Theory*, Minneapolis: Minnesota University Press

Castles, S., Crawley, H. and Loughna, S. (2002) *States of Conflict: Causes and Patterns of Forced Migration to the EU and Policy Responses*, London: IPPR

Chimienti, M. and Achermann, C. (2006) 'Coping Strategies of Vulnerable Migrants: The Case of Asylum Seekers and Undocumented Migrants in Switzerland' in Cuadra, C. B. and Cattacin, S. (eds) *Migration and Health: Difference Sensitivity from an Organisational Perspective*, Malmö: Malmö University/IMER

Commission des Communautés Européennes (1999) *Conseil Européen – Tampere: conclusions de la Presidence*, SI (1999) 800; Bruxelles

Council of Europe (2009) *Report by Thomas Hammarberg, Commissioner for Human Rights of the Council of Europe, Following his Visit to Italy on 13–15 January 2009*, CommDH(2009)16 published Strasbourg, 16 April

Duffield, M. (2007) *Development, Security and Unending War: Governing the World of Peoples*, Cambridge: Polity Press

European Commission (2007) *Report from the Commission to the European Parliament and the Council on the Evaluation of the Dublin System {SEC(2007) 742}* accessed at: http://eur-lex.europa.eu/LexUriserv/LexUriserv.do?uri=CELEX:52007DC0299:EN:NOT

Foucault, M. (1998) *The History of Sexuality: Volume 1: The Will to Knowledge*, London: Penguin

Harpviken, K., Strand, A. and Suhrke, A. (2004) *Afghan Refugees in Iran: From Refugee Emergency to Migration Management*, policy briefing paper for Chr. Michelsen Institute, Norway, accessed at: http://www.cmi.no/pdf/?file=/afghanistan/doc/CMI-PRIO-AfghanRefugeesInIran.pdf

HCRP (2009) *Afghan Refugees in Pakistan: Push Comes to Shove,* Report by the Human Rights Commission of Pakistan, Lahore, accessed at: http://www.hrcp-web.org/Publications/RRefugees.pdf

Human Rights Watch (2008) *Stuck in a Revolving Door: Iraqis and Other Asylum Seekers and Migrants at the Greece/Turkey Entrance to the European Union*, accessed at: http://www.hrw.org/sites/default/files/reports/greeceturkey1108web_0.pdf

IAB (2009) *Press Release: Conditional Returns to Greece*, Norwegian Immigration Appeals Board, accessed at: http://feed.ne.cision.com/wpyfs/00/00/00/00/00/0E/CA/2C/wkr0003.pdf

IPS (May 2009) *Iran: Afghan Refugees – Pawns in Standoff with West*, International Press Service, accessed at: http://ipsnews.net/news.asp?idnews=37727

IRIN news (2009) 'Sharp rise in attempted illegal migration of Afghans to Europe', RAWA news, accessed at: http://www.rawa.org/temp/runews/2009/06/04/

Karamanidou, E. (2009) 'The Discursive Legitimation of Asylum Policies in Greece and Ireland', unpublished PhD thesis, City University

Karamanidou, E. & Schuster, L. (2010) 'Greece, Dublin II and Eurodac', paper prepared for the *Journeys and Justice: Forced Migration, Seeking Asylum and Human Rights* Conference, University of Leeds, 29 January 2010

Kronenfeld, D. (2008) 'Afghan Refugees in Pakistan: Not All Refugees, Not Always in Pakistan, Not Necessarily Afghan?', in *Journal of Refugee Studies*, Vol. 21, No. 1, pp. 43–63

Milani, M. (2008) 'Iran's Policy towards Afghanistan' in *Middle East Journal*, Vol. 60, No. 2, pp. 235–6

Norwegian Organization for Asylum Seekers (2009) *Out the Back Door: The Dublin II Regulation and Illegal Deportations from Greece*, report by the Norwegian Helsinki Committee, (NHC), the Norwegian Organization for Asylum Seekers (NOAS) and Aitima, accessed at: http://www.noas.org/?p=news&news_id=66

OFPRA (2008) *Rapport 2007*, Paris, Office Français de Protection des Réfugiés et Apatrides, accessed at: http://www.ofpra.gouv.fr/documents/Rapport_OFPRA_2007_BD.pdf

Papadimitriou, P. and Papageorgiou, I. (2005) 'The New "Dubliners": Implementation of European Council Regulation 343/2003 (Dublin-II) by the Greek Authorities', in *Journal of Refugee Studies*, Vol. 18, No. 3, pp. 299–318

Pro Asyl, Group of Lawyers for the Rights of Refugees and Migrants, Athens (eds) (2007) *The Truth may be Bitter but it Must be Told: The Situation of Refugees in the Aegean and the Practices of the Greek Coast Guard*, Frankfurt/Main: Förderverein PRO ASYL e.V

Schuster, L. (2011) 'Turning Refugees into Illegal Migrants: Eurodac and Dublin II', in *Ethnic and Racial Studies* (forthcoming)

Schuster, L. (2010) 'Dublin II and Eurodac: Examining the (Un)intended(?) Consequences', in *Gender, Place and Culture* (forthcoming)

Schuster, L. (2005) 'A Sledgehammer to Crack a Nut: Deportation, Detention and Dispersal in Europe', in *Social Policy and Administration*, Vol. 39, No. 6, pp. 606–21

Schuster, L. (2003) *The Uses and Abuses of Political Asylum in Britain and Germany*, Frank Cass: London

Schuster, L. (2003a) 'Common Sense or Racism: The Treatment of Asylum Seekers in Europe', in *Patterns of Prejudice*, Vol. 37, No. 3, pp. 233–55

Sitaropoulos, N. (2000) 'Modern Greek Asylum Policy and Practice in the Context of the Relevant European Developments', in *Journal of Refugee Studies*, Vol. 13, No. 1, pp. 105–17

UNHCR (2009) *2008 Global Trends: Refugees, Asylum Seekers, Returnees, Internally Displaced and Stateless Persons*, Report of the United Nations High Commission for Refugees, Geneva, accessed at: http://www.unhcr.org/4a375c426.html

UNHCR (2008) *UNHCR Position on the Return of Asylum Seekers to Greece under the 'Dublin Regulation'*, accessed at: http://www.unhcr.org/refworld/docid/4805bde42.html

Vedsted-Hansen, J. (2005) 'Common EU Standards on Asylum: Optional Harmonisation and Exclusive Procedures', in *European Journal of Migration and Law*, Vol. 7, No. 4, pp. 369–76

Yaghmaian, B. (2005) *Embracing the Infidel: Stories of Muslim Migrants on the Journey West*, New York: Delacorte Press

Chapter 7
Security, Insecurity and Refugees in The Netherlands

Kees Nagtegaal

In the first decade of the twenty-first century immigration policies seem to be increasingly dominated by security issues. Within the European Union (EU) and especially within the Schengen area of the EU, immigration policies and, connected with it, border control, are hotly debated topics. It seems that 9/11 and the fear of terrorism have increased the debate on this issue. Linking security issues with immigration is, however, not new. It has been a decisive factor throughout Europe during the twentieth century. European states have always been vigilant when immigration is concerned (see chapter by Maguire in this volume). What has changed in the past decades is the technical means states have available to operate in this politically delicate field. Technical development links with the ability of central governments to control societies and borders, dictating the extent of the power of the authorities. Yet it also presents governments, especially with the present rapid development of digital data collection, with extra problems such as how to safeguard against data theft. Dutch refugee policies will be used as an example of the framing of immigration policies in the twentieth century, showing that the link between security and immigration policies is an old phenomenon. Additional attention will be given to a new aspect of security and immigration policies: the risks connected with digital data collection as an integral part of these policies.

This chapter will argue that the way states deal with refugees in the first decade of the twenty-first century does not deviate from refugee policies executed since the early twentieth century. The Netherlands is usually portrayed as an example of a hospitable country for refugees. The Dutch case will be used as an example of how a 'liberal' country has dealt with refugees in the last hundred years. It has ratified all the relevant important international treaties[1] and modified its national legislation accordingly. In spite of this, Dutch refugee policies bear witness to the tension between on the one hand security and the responsibility authorities feel towards their national population and on the other hand the duty, emanating from international human rights treaties, to protect refugees. This double character of security poses a problem to many state authorities. I will, by giving examples of a number of refugee crises in the twentieth century, demonstrate that Dutch authorities have practiced, and still do, refugee policies which tend to give more

1 E.g. The 1951 Geneva Convention and the European Charter on Human Rights.

emphasis to the security of the Dutch nation state than to that of refugees, when the least one may expect is that some sort of balance may be struck in this conflict between two policy frameworks.[2]

Refugees

Refugees and asylum seekers are terms which are often used in a confusing way. The first is the general generic term, and the second is the legal term for refugees who have asked for (political) asylum and are in the process of having their claim processed by authorities. Since I will describe the way refugees have been treated by Dutch authorities since 1914, a remark should be made here. In The Netherlands the procedure to formally request asylum, have this request processed, and be permitted to appeal in case of a negative decision, is based upon legislation introduced in the final year of the Great War. Further development of this system did not take place until after the Second World War (Nagtegaal, 2005) and with the introduction of a possible appeal to the European Court of Human Rights in Strasbourg. In this chapter I will refer to both categories as refugees. Only where necessary will I make a distinction between the two.

The sovereignty or freedom of a state in formulating and executing its refugee policies is limited by national and international legislation and treaties, and by scrutiny by parliament. Execution of refugee policies is also under the scrutiny of national courts. States which have ratified the European Charter of Human Rights are also subject to scrutiny by a supranational court, the European Court of Human Rights in Strasbourg. Judicial scrutiny is not always welcomed by governments (Nagtegaal, 2005), but is a fact of life and, if formalized through legislation, refugees have a right of appeal against government decisions.

State and sovereignty

Weiss *et al.* (2001: 5) note that 'the emergence of the territorial states [...] was accompanied by the notion that the state was sovereign'. In refugee crises the state and its sovereignty are a central issue. According to Aleinikoff (1995: 257–8) 'refugees [...] present a failure of the state system, a problem to be solved. As involuntary migrants, refugees evidence a breach of the state of origin/citizen relationship' [...] The acts of the country of origin, then constitute an injury to both the refugee and the other states. The existence of refugees, and the weight of their moral claim to protection, also puts pressure on a basic premise of international law, that state sovereignty entails plenary power to determine admissions policies'. It is therefore important to take a closer look at state and sovereignty.

2 The fact that these two obligations make the framing and executing of immigration policies very difficult is elaborated upon in the chapter by Parusel in this volume.

Weber, quoted by Bendix (1977: 418), sees a state as 'having a legal order, bureaucracy, compulsory jurisdiction over a territory and monopolization of the legitimate use of force'. A state is sovereign within its territory and will defend itself against any attack that is considered to constitute a threat. Bos (2003: 57) says that 'sovereignty in its traditional sense [...] signifies independence, or the right to exercise within a portion of the globe and to the exclusion of any other state, the functions of a state, and to represent these functions externally'. In spite of all this, due to the increased formation of international laws, treaties and other regulations, states increasingly transfer aspects of their sovereignty. Russett *et al.* (2000: 474) see sovereignty as a problem, in so far as 'sovereignty does not foster a sense of international obligation'. And it is international obligations that increasingly constitute an attack on the sovereignty of the state. Yet sovereignty also gives freedom to a state, to a certain extent, to hand over parts of that sovereignty and reduce it. This was due to a trans-national demand for the effective treatment of pressing problems (Weiss *et al.*, 2001). One of the areas in which transfer of sovereignty occurred was the sphere of human rights, and as a consequence it applies to issues connected with immigration and refugees. Forsythe (2000: 3) notes that 'despite the rhetoric of universality, however, human rights remained essentially a national matter, to be accepted or not, until 1945 when they were recognized in global international law'. He notes that this has been seen as 'a radical reformulation of sovereignty' (Forsythe, 2000: 21). As a consequence 'states came to share jurisdiction over human rights issues with various international organizations and even foreign governments' (Forsythe, 2000: 23). Yet according to Hannah Arendt (1968: 269–70) modern power conditions have made a mockery of national sovereignty, with the exception of giant states. The rise of imperialism and the pan-movements undermined Europe's nation-state system from the outside. Yet Arendt also states that political refugees, of necessity few in number, still enjoy the right to asylum in many countries, and this right acts, in an informal way, as a genuine substitute for national law (Arendt, 1968: 295).

Since before the Second World War, human rights issues have been subject to national law and as a consequence national policies.[3] The importance of the Great War should however not be underestimated. Hammar (1985: 239) states that both wars have affected immigration policies and thus human rights regulations and refugee policies. Before 1914 travelers could move freely while 'during the First World War systematic immigration regulation and aliens control appear(-ed) in many European countries. Legislation was later passed, in the 1920s, to make permanent the provisional measures taken during the war' (Hammar, 1985: 240).

3 There are a few exceptions, although these have had no implications for The Netherlands. One was the attempt to aid Russian refugees through the 1933 Convention on Russian Refugees. Article 7 of this Convention stated that Russian refugees would be exempted from the increasingly restrictive immigration legislation in Europe. The only country ratifying this convention was France (1933). A similar Convention, this time concerning refugees fleeing Nazi-Germany, was ratified by Belgium in 1939 (Caestecker, 1998: 90).

Hammar is not alone in this analysis. Lucassen states that 'for the specialists in the field the "Great War" generally stands out as the major watershed' (Lucassen, 1998: 45).[4] One of the reasons was an increased nationalism; as Hammar points out, demands to identify and control foreigners increased (Hammar, 1985). After the Second World War an international system of human rights developed. This system, Forsythe (2000: 21–3) states, 'caused [...] a radical reformulation of state sovereignty'. This means that since then states have been confronted with limitations on their ability to formulate refugee policies, including legislation, since international treaties have to be taken into account.

Control of immigration

All immigration and refugee policies include an aspect of protection, in the widest possible sense, of the state and its citizens. To put it boldly: the state protects itself and its citizens against an external threat by making legislation fitting the national mood. This requires that apart from ethical considerations concerning human rights and refugee policies, a certain degree of realism is embedded into the execution of refugee policies. *In extremo* there is a choice between absolute protectionism of the state and its citizens (categorical refusal of entry to refugees) or the unconditional protection of refugees by giving them asylum. In reality many authorities develop a policy between these extremes, thus creating a frame within which refugee claims can be dealt with.

Hammar gives this definition of immigration control: 'immigration regulation is the control that a sovereign state exercises over the entry of foreign citizens and their access to residence and employment' (Hammar, 1985: 249). Brochmann refers to this in terms of barriers to movement with which prospective and actual migrants are confronted (Brochmann, 1994: 1). These barriers are part of a control policy, and are introduced to 'satisfy social, economic and security needs for immigration regulation, without violating international treaties and conventions for asylum and human rights protection, and without violating the public sense of fair treatment of human beings' (Brochmann, 1999: 1). The control system governs selection, admission and deportation of foreign citizens.[5] Another important aspect of this issue is described by Brochmann. She states that this control

> ... will always be influenced by historical precedents and traditional patterns of behaviour – the history of immigration (magnitude and length of experience) and the

4 Although Lucassen primarily writes about the connection between migration and labour laws, it should be said that many refugees were able to use the possibility of free movement from one state to another. The Dutch government restricted the free migration of labourers in order to protect the Dutch labour market from foreign labour. This meant that for refugees entrance into The Netherlands became almost impossible.

5 Admission and deportation can be found in the 1849 Dutch immigration law.

> state's 'control culture' as it were. Within this 'control culture' there are normative elements; traditions in terms of humanitarian values and cultural openness towards the outside world. Legitimacy as to using economic conditions as a premise for policy-making in relation to immigration control (Brochmann, 1999: 16).

The state is influenced by, for example, trade unions trying to influence policies since they want to protect the national labour market (Moore, 1986; Nagtegaal, 2005).

A government has to ensure its sovereignty by territorial control and 'secure its continuous formation and consolidation of the social and cultural national distinctions and secure the political integration and continuity' (Brochmann, 1999: 6). This means that a number of interests have to be considered: national security, national economy, demography, and finally social and cultural cohesion. A selection of these factors will form 'the basis for a state's legitimization of its national control policy towards immigrants' (Brochmann, 1999: 6). Three dimensions play a role in this process. First is the ethical dimension dealing with the question of whether the policy is in accordance with fair treatment of humans. Second is the dimension of positive law embedded in articles of (inter-)national regulations about immigration and human rights. Third is the threat dimension. This contains a number of sub-dimensions which consist of potential threats to the state which have to be considered and dealt with. Table 7.1 gives insight into the three dimensions in refugee policies, and the underlying factors involved.

Table 7.1 Dimensions of refugee policies

Ethical dimension	**Legal dimension**	**Threat dimension**		
Humanitarian	Legal	Political	Financial/ Economic	Law and order
To give aid to (homeless) refugees	International treaties	Sharing burden of refugees between states	Refugees are a financial burden	Prevent (procedural) chaos in the host state and protect security
Uphold human rights	National legislation	Political stability. Threat to the political elite / coalition	Labour + unemployment	Protect against crime by the 'strangers'
Quality of aid	Scrutiny by national and international courts (jurisprudence)	Speedy repatriation = exit of problem	Costs of repatriation and/or integration	Protect the culture
		Foreign relations / neutrality		

MPs, as political actors, have to deal with the fact that they are elected by the electorate and are dependent on the electorate for their re-election. Too many foreigners could be considered by the electorate to be a threat to the national identity and culture, the labour market and the economy as a whole, and to the social order. Huntington (1996: 200) summed this up, 'Westerners increasingly fear that they are now being invaded not by armies or tanks but by migrants who speak other languages, worship other gods, belong to other cultures, and, they fear, will take their jobs, occupy their land, live off the welfare system, and threaten their way of life'. Although Huntington wrote this in the last decade of the twentieth century, similar remarks were made in the early twentieth century when Dutch MPs stated that Belgian refugees brought venereal diseases. The 'dilemma' is aptly summarized by van Amersfoort. He says that the following questions have to be answered. Does the country need more people? Is immigration contributing to our economy or at least sectors of it? What are the consequences for social housing programmes and the educational system? What are the consequences for our basic values and norms? In short: what are the numerical, economic, social and cultural consequences (van Amersfoort, 1999: 141). All these can be summarized in the general question about security and the threat immigration poses to a country's security.

Europe

The European Charter for the Protection of Human Rights and Fundamental Freedoms, signed on November 4, 1950, is the most important human rights treaty in Europe. It had taken less than one year from being proposed to its final draft being approved (Boeles, 1995: 174–5). Article 3 of the Treaty states that no-one can be subjected to torture or inhuman or humiliating treatment or punishments. In The Netherlands the judiciary has to take its content into account when human rights cases are put forward. Since the signing of the treaty many protocols have been added, and it has become a prerequisite that states wanting to join the Council of Europe ratify the treaty before they can join. The formalization of these rules seems to be an enormous step forward in the protection of refugees asking for security in Europe. According to Nadig (2002: 3) however, European states have tried to reduce immigration since 1973 in order to avoid more expenditure on an already stressed welfare system. Important here is the signing of the Schengen Treaty in 1985, a form of interstate cooperation which was created outside the European Union (EU) system by countries which were members of the EU.

The Schengen Treaty, originating as a bilateral treaty in 1984 between France and Germany, who were joined in 1985 by The Netherlands, Belgium and Luxemburg, was a result of the fact that interstate trade had to be possible without the hindrance of borders. It was the same year in which political leaders from Europe, North America and Australia held a conference in Geneva at which they discussed how to regain control over immigration, and the formalization of policy measures in this field (Nadig, 2002: 6). The free movement of persons resulting from the Schengen Treaty meant that something had to be devised to regulate the way it applied to non-citizens of the participant countries. Problems included visas, asylum applications and combating illegal immigration. Schengen was not an easy matter, as can be deduced from the fact that it was many years before more countries ratified the treaty. Asylum was a particularly sensitive matter. During the discussion all attention focused on policies such as asylum, illegal immigration and crime fighting (WRR, 2003: 172).[6] The fall of the Berlin Wall meant that mass immigration from Eastern-Europe loomed in the distance. So the Schengen countries reintroduced border checks on non-EC nationals. The Schengen Convention came into operation in 1995, and by that time Italy, Spain, Portugal and Greece had acceded. By then all EU member states had signed two international conventions about asylum and external borders (1990): the Dublin Convention and the External Borders Convention. Dinan states that 'the fact that these conventions were drafted intergovernmentally without reference to the EC's supranational institutions demonstrated the continuing sensitivity of member states of Justice and Home Affairs (JHA) issues' (Dinan, 1999: 443). The 1993 Maastricht Treaty stated that asylum and immigration was part of the EU's third pillar, the policy area in which cooperation in the field of JHA is placed. Decisions in this area have to be taken unanimously. In the 1999 Amsterdam Treaty, asylum and immigration were moved from the third to the first pillar. The Schengen Treaty was, by means of the Schengen Protocol, incorporated into the Amsterdam Treaty, and became part of the EC Treaty. A transition period of five years was agreed upon: until 2004 unanimity was necessary, and the European Commission shared the right of initiative with the member states. In October 1999 the EU Council of Ministers recognized the fact that a right to seek asylum existed and that the common European asylum system should be based upon a complete and an all-embracing interpretation of the 1951 Refugee Convention of Geneva. In June 2002, there was finally a political agreement on minimum standards for asylum in Europe. However, the WRR concluded in 2003 that: 'the biggest problem is that a framework for asylum cases has been developed at European level, but due to national interests and sensitivities this only constitutes a minimal and largely ineffective common policy' (WRR, 2003: 184). Only minimal progress has been made in this respect at the European level. Some progress was made in 2004 when the European Council adopted the Hague Programme, which provided policy aims and strategies. Among these was the field of immigration. The Hague Programme was replaced in 2009 by the Stockholm Programme, which also dealt with immigration policies (European Council, 2009). Yet the fact that it has taken the European states so many years to arrive where they are now illustrates how sensitive this policy area is. One example is that nationals of new eastern European member states do not have the same rights in the labour market in the old member

---

6 WRR: Wetenschappelijke Raad voor het Regeringsbeleid. The Academic Council for Government Policies, an advisory body to the Dutch Government.

states. A common policy equally applicable to all nationals of all member states has not been reached.

The Dutch role in European asylum policy

Andeweg and Irwin (2002: 202–6) see idealism as a constant factor in Dutch foreign policies. Based on this, it could be expected that The Netherlands would play a prominent role within Europe where asylum policies are concerned, and prominent in a way that promotes a generous asylum policy in Europe. The image history gives us is twofold: on the one hand The Netherlands wants, based on its image, a generous asylum policy. On the other hand it has taken initiatives which do not confirm this image, but which clearly show that The Netherlands does not hesitate, where it deems necessary, to let its national interests prevail. The fact that The Netherlands was involved with the establishment of the Schengen system indicates that it at least wants to cooperate. To be able to regulate the traffic of foreigners was important and served Dutch national interests at a time when individual asylum requests were increasing.[7] The Balkan crisis caused The Netherlands, Germany, Sweden and Switzerland to suggest that a form of burden sharing should be introduced in Europe regarding refugees. The other European countries, with fewer refugees, opposed this idea. This is even more apparent when looking at the clause dealing with burden sharing in the Amsterdam Treaty. Burden sharing is exempt from the 2004 deadline by which uniform asylum policies were supposed to have been reached for the EU (WRR, 2003: 179). In December 1998 The Netherlands instigated the founding of a High Level Working Group on Asylum and Immigration (HLWG), which aims at controlling flows of immigration from regions from which high numbers of refugees originate. The HLWG was criticized since it was seen as having a one-sided focus on combating flows of migration and asylum. It was viewed as having a somewhat secretive character (Nadig, 2002: 4–5). Apart from this, the emphasis on control indicated that, since The Netherlands was the initiator, a restrictive refugee policy was desired by The Netherlands.[8]

---

7 Annual asylum requests in The Netherlands rose from 1,330 in 1980 to 43,892 in 2000.

8 In many European countries immigration policies have been a hotly debated issue in the last decades. During the 2009 election campaigns for the European Parliament the political leader of the Dutch conservative liberal party, the Volkspartij voor Vrijheid en Democratie (VVD), announced that the party would be willing 'to make its hands dirty' to solve the problems The Netherlands was facing. Part of this solution was a stricter asylum policy and a tougher policy against crime. The link between asylum and crime was once more seemingly brought forward by a member of the political elite, thus underlining the threat factor (Trouw, May 16 2009: 4).

Europe, or as it sometimes is called 'the European Fortress', is a system of states which are confronted with increasing numbers of migrants and refugees, but which don't consider themselves to be countries of immigration. In the case of The Netherlands this also means that there is no proper integration policy.[9] Invariably, trying to control the flow of immigrants and refugees results in more restrictive legislation. When one country passes restrictive legislation, refugee flows change course in the direction of countries with more lenient legislation. This causes these countries to introduce more restrictive legislation. This phenomenon is called the 'waterbed'.

In September 2002, the Dutch government was worried about electoral developments in the EU, and connected these with illegal immigration (HTK, 2002–2003, 28 604, nr.1). The upcoming entry of Eastern European states into the EU led the government to state that 'it is in the Dutch and European interest to involve these countries in the development of a common asylum and immigration policy in the EU' (HTK, 2002–2003, 28 604, nr.2), the reason being that they are transit countries for illegal immigrants and refugees.

As long as no new system of border control is in place complying with EU rules, existing border controls should be maintained. The combination of a politically sensitive issue and the lack of cooperation on immigration issues within the EU makes it understandable that the Schengen acquis is non-negotiable, and that no exceptions are made for new member states (WRR, 2003: 181). At the end of the century, increases in asylum requests in the Eastern European states were a big problem. The WRR concluded that the Middle and Eastern European states will join the race to the bottom in order to become the least attractive country for asylum seekers, and that a shift in the burden will occur.

There is also an increasing political call for burden sharing at the European level. This may lead to the development of new coalitions of states, for example between Middle and Eastern European countries and countries such as Germany and The Netherlands, which will want to improve cooperation in the field of burden sharing of asylum cases (WRR, 2003: 184). In the mean time, asylum policies are framed within the context of safety and security, a frame that has been applied since the Great War. Emphasis is placed on security and safety for nation states rather than on security and safety for refugees. In this respect not much has changed.

The Netherlands

The first Dutch Immigration Act dates from 1849 and was based on the 1848 Constitution. The subject and title of the Act was 'The regulation of admission and

---

9 The absence of an integration policy meant that integration of many foreigners failed. This played a role in politics and was decisive in the election of 2002. As a consequence a parliamentary committee was formed and published a final report in 2004 (Onderzoek Intergatie Beleid, 2004).

expulsion of aliens (foreigners)'. Until 1914 execution of immigration policies was not according to the law, and was executed in a random manner by local authorities. The large number of Belgian refugees entering The Netherlands in October 1914 caused the Dutch authorities to introduce additional legislation in 1918 to control the influx of refugees. Since then a constant increase of government control on immigration can be seen through new legislation, and policies based on the new laws, the most recent being the Immigration Act of 2000.

Dutch refugee crises

In the twentieth century seven refugee crises can be identified. The first is the influx of one million Belgian refugees into The Netherlands in 1914. Second is the crisis of the 1930s–40s caused by the German Nazi regime. Third is the 1948 Czechoslovakia crisis. Fourth is the 1956 crisis in Hungary. Fifth is the crisis in 1968 when Warsaw Pact troops ended the Prague Spring. The 1980s saw a steep increase in individual asylum applications: the sixth crisis. The increasing number of individual asylum-seekers from non-European countries constituted a new and so far alien cultural and religious element into Dutch asylum policies. The seventh crisis occurred in the 1990s, when the Balkans collapsed. This last crisis had an added element, making it more shocking to the polity and to the general public, as the media showed images reminiscent of Nazi concentration camps.[10] The idea that this could and would not ever occur again in Europe was suddenly shattered and played an important role in the reaction of the general public to the influx of refugees from the Balkans.

All these crises had an effect on the general public and the polity, and consequently on Dutch refugee policies of the time. Yet before 1914 the fact that immigrants entered the country was an accepted fact of life, and no reason to adopt legislation. Instructions given by the central authorities to the lower decentralized authorities, the street level bureaucrats, were minimal, and gave the lower authorities a lot of freedom in executing immigration and refugee policies. One of the effects of the war and the ensuing high number of refugees was that after 1918 organizational measures and structures aimed at dealing with these crises in an expedient way with the aim of preventing or minimizing damage to the perceived interests of the Dutch nation-state and its citizens were increasingly an issue on the political agenda. Refugees were seen as threats to the economy or national security. To illustrate the types of measure introduced by the Dutch authorities, examples of responses to some of the refugee crises follow.

10 Images of emaciated prisoners in prison camps in the former Yugoslavia were similar to the images of prisoners in Nazi concentration camps and were later linked to ethnic cleansing in the Balkan war of the 1990s.

1914: The Great War

After the outbreak of the Great War in August, October 1914 saw the number of Belgian refugees rise to a total of one million. This presented local and national Dutch authorities with an enormous problem. Within three days Dutch representatives contacted the German occupying forces in Belgium to convince them that a joint effort to repatriate the Belgians would benefit both The Netherlands and the occupied Belgian territory, including the German occupying forces. One of the measures taken was the so-called *soft pressure* exercised on the Belgian refugees by housing them separately from the Dutch population e.g. in warehouses in the port of Amsterdam and in camps in the middle of the country away form large towns (Nagtegaal, 2005). They were given (sub-)minimal aid and were thus encouraged to leave, which caused scandal in the media. This policy was effective; within three months, 90 per cent of them had returned to Belgium.[11] While, contrary to Dutch legislation, parliament voted for large sums of money to be used to support the refugees, a large part of the aid given was financed and organized by private organizations. As a reaction to the experience of large numbers of refugees during the Great War, the government introduced additional and stricter immigration legislation in 1918.

1930s: The Nazi crisis

The crisis in the 1930s when large numbers of refugees, mainly of Jewish descent, crossed the German-Dutch borders shows a similar pattern. Unlike the 1914 crisis, Dutch authorities did their utmost not to finance any aid to the refugees. No refugee was to become a financial burden to the Dutch state. Moore (1980: 29) states that 'the government had made it clear from the outset that The Netherlands was only to be a transit point for refugees moving on to other parts of the world, and was not to be considered as a land of settlement'. In 1939 this was voiced by the ministry of Justice, when the relief organization for Jewish refugees, Centrale voor Joodse Vluchtelingen (CJV), requested financial aid. The Ministry of Justice stated: 'in our opinion (such) a subsidizing by the government can only be considered if a refugee committee would be able to prove that all their refugees resident in The Netherlands would be taken up in another country together and under condition that the committee shall not encourage nor ask for new refugees to be admitted to The Netherlands' (Moore, 1980: 94). As a result, private organizations took up the task of helping refugees. The private sector organized and funded the aid. Refugees were housed in refugee camps.[12] Additional legislation was introduced

11 The policy entailed free transport by train to Antwerp.

12 The number of refugees meant that often the aid organizations were unable to house them. A search started to find a place for a refugee camp. When one was found in the middle of the country, again away from concentrations of population, the nearby town opposed this as well as the Dutch queen Wilhelmina who did not want a refugee camp

which made it impossible for German refugees to start their own businesses and be economically independent. This policy was introduced partly because of remarks made by Dutch trade unions which feared that unemployment among Dutch labourers would rise due to foreign workers trying to enter the labour market (Moore, 1986: 56–9). The Ministry of Justice described a refugee as 'an undesirable element for Dutch society and therefore an undesirable alien, who will, therefore, be kept out at the border or if found in the country, escorted to the border' (Government Circular, May 7, 1938). The increasing importance of the refugee issue for the authorities can be illustrated by the fact that no less than four ministries were involved in refugee policies: Justice, Home Affairs, Social Affairs and Economy (Leenders, 1993: 253).

1948: Consequences of World War II and the Czech crisis

The Second World War had a great impact on refugee policies, leading to the introduction of a great many rules regarding human rights, and refugee protection measures. The Dutch government's stance had two sides. On the one hand it took part in the creation of more extensive international rules regarding refugee policies, aimed at better protection for refugees by providing them with more certainty about their legal status and the aid they were entitled to receive. On the other hand, the government was well aware that refugees were a risk to the economy. In 1949, the head of the Immigration Authority (Rijksvreemdelingendienst) wrote to the Minister of Justice about Displaced Persons (DPs), saying that 'the financial risk will have to be taken by the Dutch state. This risk will, in my opinion, not be so great, but on the other hand not much can be said about this with any certainty, because of the very unstable economic situation in our country and abroad' (NA, 5.023.526/194). The Netherlands, in spite of its initiative in the Council of Europe to try to solve the refugee problem in Europe, was not in favour of unconditional help to refugees. Any refugee asking for help was scanned regarding ability to work (Berghuis, 1999: 198/199). There was, moreover, a deeply felt fear of spies entering the country, which led to refugees who were, after an initial interview, considered to be a risk, being held in custody in the 'Court of Roses', a detention centre, where the secret services conducted interviews. The procedure in the 'Court of Roses', was mainly aimed at weeding out malafide, meaning communist, refugees. This procedure was incorporated in a ministerial directive of March 1954 (Berghuis, 1999: 177/178).

close to one of the royal palaces. Ultimately a camp was founded in the north-east, in a remote area of the province of Drenthe. This camp, Westerbork, still inhabited by Jewish refugees at the outbreak of the Second World War, became a concentration camp once Nazi Germany occupied The Netherlands. From here, thousands of Jews were deported to the Nazi extermination camps.

1956: The Hungarian crisis

The 1956 Hungarian crisis meant that while The Netherlands was deeply submerged in the aftermath of decolonization[13] it was confronted with refugees seeking a safe haven. According to Bootsma, the authorities used a system by which the refugee-population 'was creamed off'. This meant that only refugees who could be economically independent, stood a good chance in the labour market, and would not push Dutch workers out of jobs, would be allowed to enter (Bootsma, 1994: 91). In order to co-ordinate all this, an Interdepartmental Committee[14] was established in November 1956. The National Labour Exchange Agency and the Dutch Red Cross Society were also involved. Finances at national level were dealt with by the National Committee for Aid to the Hungarian People (NC). Millions of guilders were collected and managed by local and national private organizations. The Foundation Dutch Federation Aid to Refugees (DFAR) was also involved. These were all private organizations. The Dutch government, however, had no intention of spending any money whatsoever on aid to refugees. Refugees were screened for their political reliability by the Dutch Secret Services and for their potential in the labour market. After having been screened, they were transported to refugee camps which were administered by the Ministry of Social Welfare or by private organizations. Regarding finances, a strict covenant had been agreed upon on November 17, 1956. All costs, including those already accrued by the central government, such as those for the government camps in which refugees were housed, and the costs for welfare paid by local governments, would be claimed at DFAR, which in turn would claim the money from the NC (NA, 2.03.01/3182; Nagtegaal, 2005).

1992: The Balkan crisis

November 1992 saw an estimated 16,000 (ex-)Yugoslav refugees in The Netherlands (HTK, 1992/93, 22704, nr. 9), while at the end of 1993 there were approximately 20,000. The Netherlands had the second highest number of Displaced Persons (DPs) in the European Community, and the fifth highest of all European states (HTK, 1992/93, 22704, nr. 10). Dutch authorities labeled the (ex-)Yugoslavs as DPs, not as refugees. As a consequence their position with regard to rights to be in a state territory, was weaker compared to asylum seekers. In

13 The decolonization meant that well over 200,000 people who were repatriated from the former colonies had to be housed and fed, and to find a place in the labour market (HTK, 1956/1957, 361).

14 The Interdepartmental Commission consisted of a representative of the cabinet of the PM, representatives of the Ministries of Foreign Affairs, Social Affairs, Social Welfare Work, Justice, and the Interior.

1992 emergency legislation, Tijdelijke Regeling Opvang Ontheemden (TROO),[15] created as a temporary exception to regular refugee legislation, was introduced. The government was well aware of the temporary character of this legislation (Muus, 1997: 89). It gave categories of refugees, called displaced persons (DPs) in the Act, temporary permission to be in The Netherlands after thorough scrutiny. The second reason given in parliament for its introduction was that 'normal' procedures, based on existing refugee legislation, would not be delayed, since the TROO meant that if a DP asked for asylum, the request would not be processed, at least for the time being. Thus administrative capacity to deal with applications filed by other nationals would not be diminished. Third, a separate housing system was introduced for those who were allowed to stay under the TROO regime. Facilities for the DPs in TROO camps were poorer than those in regular Dutch reception centres.[16] This measure represented a deviation from regular legal procedures, and standard provision for accommodation, and was intended to encourage departure.

The special measures introduced in 1992 by the coalition were supported by all parties in parliament. They agreed that the situation required these measures, but the question was raised as to whether the introduction of the TROO and the separate legal status connected with it was justified. Most parties preferred reception at regional level, but realized that the situation forced The Netherlands to depart from strict adherence to this (HTK, 1991/91, 22704, nr.3 -6). There were warnings that the temporary status of the TROO and the final status of the DPs would not be debated in due time (HTK, 1991/91, 22704, nr.3, p.7–9). The government was well aware of this (HTK, 1991/91, 22704, nr.3, p.11).

This crisis also had its legislative consequences. Between 1983 and 1992 the annual number of refugees entering The Netherlands had risen from 2,000 to 20,000, so it was a heavy burden on the system to process all the asylum requests and house all the refugees. The New Admission and Reception Policy (NTOM) had been introduced in early 1992, and the TROO meant a deviation from standard procedures. In March 1993 the government announced that processing of Yugoslav DPs would start, and would require an extra effort by the authorities (HTK, 1992/93, 22704, nr.8). A new status was introduced replacing that which obtained within the TROO system, but with the same minimal rights, since it was conditional: Voorwaardelijke Vergunning tot Verblijf (VVTV). It gave increased rights to the holder, over the course of three years, but there was a risk of being sent back if the situation in the home-country improved (Muus, 1997). This status existed until the Immigration Act of 2000.

---

15 Tijdelijke Regeling Opvang Ontheemden (TROO), Temporary Regulation for Reception of Displaced Persons.

16 The reception centres are the housing facilities provided by Dutch authorities for those refugees who, having asked (political) asylum, have been admitted to the procedure and are permitted to wait in The Netherlands for as long as the asylum request is processed by the Dutch authorities.

Transfer

The refugee is part of a community in his own state, but has decided to leave and ask permission to enter the community of another state. Leaving one sovereign territory and entering another means that the authorities of the new nation-state have to use criteria to decide whether to allow entry. Which interests and which criteria play a part in the decision making process? This is determined by social norms and values, which are a frame for language, cultural and political mores, which are translated into legislation and the policies which are executed on the basis of this legislation. New entrants must meet the criteria in order to obtain the same rights as 'old' members. New members of the nation state (citizens)- may influence and change societal norms. This perceived possibility, or danger, of change may be part of the explanation why some 'old' members of the community, be they ordinary citizens, officials or members of government or parliament, are hesitant to admit newcomers.

According to Forsythe, the Dutch see themselves as tolerant and hospitable to foreigners in general, and refugees in particular. In 2000 Forsythe aptly summarized this by saying 'The Netherlands, for example, likes to picture itself as highly international and cosmopolitan. It was the home of Grotius, the father of international law; it was the great trading nation; it was and is a country interested in world peace, for normal trade requires peace; and now it prides itself as a country highly active on human rights' (Forsythe, 2000: 149/150). It is important to emphasize a seemingly insignificant remark Forsythe makes in this quote. He points at the fact that The Netherlands is a great trading nation, and that normal trade requires peace. Peace is threatened by insecurity. This aspect of Dutch policies, the link between peace and trade, and the inclusion of this in a statement about human rights is relevant for Dutch immigration in general and refugee policies in particular. The Dutch image of tolerance and hospitality can be traced through the centuries, yet the main reason for this hospitality seems to be that until 1914 no government support whatsoever was given to immigrants and refugees. They had to fend for themselves and were no burden on society at large, since they did not receive any help from the authorities. This made policies easy, economic, and a matter for the private sector. Since 1914 Dutch authorities have increasingly felt pressure to be actively involved in immigration and refugee policies, since the need for control has become more and more important.

Control measures

Throughout the crises described, consecutive Dutch coalition governments,[17] like any government, took what were meant to be self-protective measures against

17 In Dutch national politics, governments are always coalition governments. Throughout the entire twentieth century, all the major political parties have, in varying constellations, cooperated in these coalition governments. The way these coalition

refugees. Causes differed at different times, but were always connected with a perceived threat to the stability, be it economic or political, of the Dutch nation-state. In 1914 and 1956 factors determining whether refugees would be admitted and taken care of were mainly financial and linked to the economy (Nagtegaal, 2005). The other crises show that Dutch governments attribute high value to financial considerations in deciding whether refugees will be admitted and, if so, in what way they will be supported. The state's protective policies increasingly require the authorities to gather information about applicants in order to be able to screen their applications.

In 1914 central government regained control over refugee policies and framed them within a threat dimension (Nagtegaal, 2004, 2005). Applicants may disrupt the receiving state, its economy, social order and political system. The power to control has since then remained with the central authorities. All legislation, and the rules derived from it, requires that more information about applicants must be gathered in order to enable stricter control by central authorities. The present Immigration Act (2000) is the latest. Together with legislation based upon this act, an extensive system of control has developed.

From paper control to digital control: a risky business

The increasing possibilities for border control arising from technical developments can be compared to the difference between homing pigeons and mobile phones and Twitter. 9/11 did not change this, since technical means of control already existed. These do not dictate the underlying and dominating idea that refugees constitute a threat. Technical means merely facilitate, and the more techniques advance, the more advanced technical means can be used in the control effort.

Already at the end of the nineteenth century and the beginning of the twentieth, new political ideas threatened the establishment as socialists, communists and anarchists from neighbouring countries sought refuge in The Netherlands. This was often not granted, since they were considered to be a threat. At that time, technical means of surveying the entry of foreigners and collecting data were very limited. However, in the twenty first century authorities do have technical means to screen entrants and to build digital data collections (see chapter by Maguire in this book), so that states are better equipped to control their borders. Passports with digital data, visa, fingerprints, iris-scans and other measures enable governments to gather data digitally, and make these available for processing and sharing between national government departments, supranational agencies (such as within the EU) and other organizations involved in the execution of refugee policies. These are primarily intended as self-protective measures by states, but

---

governments framed and executed immigration and refugee policies has always, largely due to the need to agree and consequently form a government, been consistent, with only minor variations. With regard to these policies, differences between political parties have always been negligible.

also, as European states claim, to ensure that so-called 'real' refugees are protected from persecution. There is however a danger in the gathering of these data and the digitalizing of them.

An example of this danger occurred when the Dutch government in 2009, through the Ministries of Justice, Economy and Home Affairs, launched a campaign (www.veiliginternetten.nl) to alert citizens to the fact that they should protect their computers against hackers. In response, Networking4all (www.networking4all.com), a security company, checked the safety of a number of Dutch government websites and found that 570 out of 684 were insufficiently protected. A representative of another company, Certified Secure (www.certifiedsecure.eu), stated that faulty software was an important factor in this. Governments, while digitally gathering data on their own citizens and on asylum applicants, immigrants and foreign workers, need to protect these data. The protective aim of asylum policies (to protect the state and refugees), implies the need for governments to adequately protect data collections. Since The Netherlands has legislation requiring owners of data collections to make sure data are adequately protected, the Dutch government, like any subject of this legislation, is obliged to guarantee that data on citizens and on non-citizens such as asylum seekers and refugees, is adequately protected from outside parties wanting to illegally enter computer systems. So states protecting themselves by the extensive collection of data on asylum applicants, as an instrument of control, need to devise policies protecting the data they collect. This is especially important when collected data concerns asylum seekers giving sensitive information. It is striking that the Dutch authorities, while telling the public to protect data on their computers, is, according to research, not adequately protecting data on government websites and in government data collections. This constitutes a security failure in a field that is extremely sensitive.

Conclusion

Immigration and refugee policies are a politically sensitive and complicated issue. The seemingly new and increased attention to immigration and refugee policies at the start of the twenty-first century in society, politics and the media is, however, framed in a way that is not new at all.

As has been shown, refugee policies involve two conflicting interests. The need for nation states to protect themselves and their citizens on the one hand, and, on the other hand, refugees seeking protection. Nation states have rights based on the fact that they are sovereign nations, and their governments have a duty to protect the state and its citizens. The nation states have transferred part of their sovereignty to another level of authority through international treaties. These treaties give refugees a base upon which they have the right to protection and the right to lodge asylum requests in other states than the state of which they are a citizen. In spite of the fact that states recognize these rights, states see refugees as a threat for a wide variety of reasons, and try to control immigration and asylum requests when and wherever they can. The threat perceived by states includes political, financial-economic and law and order issues, as shown in table 7.1. Since the increase in the threat of terrorism, with 9/11 as a terrible example, this threat has become more dominant in discussions and in the framing of refugee policies, which may have caused the public to think that this problem is new.

Technical means, more than ever, make it possible for governments to control the flow of people across the globe. However, this digitalization of control is merely an extension of what was already present: the threat dimension in refugee policies. The security of refugees seems to have a thin basis in international and national regulations, added to which the quality of digitalization itself poses a threat to the security of refugees. The quality of protection of Dutch government websites has been shown to be questionable.

The basic idea that immigration in general and refugees in particular constitute a threat which has to be controlled has been dominant since 1918. The Netherlands is an example of a nation state where, since the Great War, the framing of immigration and refugee policies has not essentially changed. Whichever coalition is in power, all parties involved look at the interests of the Dutch nation before they consider any asylum request. The arduous way in which European cooperation on immigration and refugee policies has evolved over recent decades indicates that the threat dimension and national interests dominate these policies, and are aimed at the security of nation states, resulting in insecurity for immigrants and refugees.

References

Aleinikoff, T. A. (1995) 'State-centered Refugee Law: From Resettlement to Containment' in Valentine, Daniel E. and Knudsen, John Chr. (eds) (1995) *Mistrusting Refugees*, Berkeley: University of California Press

Amersfoort, H. van (1998) 'Governments and Immigrants: Similarities and Dissimilarities in the Western Experience' in Böcker, A., K., Groenendijk, T. Havinga and Minderhoud, P. (eds) (1998) *Regulation of Migration, International Experiences.* Amsterdam: Het Spinhuis Publishers

Andeweg, R. B. and Irwin, G. A. (2002), *Governance and Politics of The Netherlands*, Basingstoke/New York: Palgrave Macmillan

Arendt, H. (1968) *The Origins of Totalitarianism*, San Diego, New York and London: A Harvest Book – Harcourt Inc

Bendix, R. (1977) *Max Weber: An Intellectual Portrait*, Berkeley, Los Angeles, London: University of California Press

Berghuis, C. K. (1999) *Geheel ontdaan van Onbaatzuchtigheid, het Nederlandse Toelatingsbeleid voor Vluchtelingen en Displaced Persons van 1945 tot 1956*, Amsterdam: Thela Thesis

Böcker, A, K., Groenendijk, T. Havinga and Minderhoud, P. (eds) (1998) *Regulation of Migration: International Experiences*, Amsterdam: Het Spinhuis Publishers

Boeles, P. (1995) *Eerlijke Immigratieprocedures in Europa*, Utrecht: Nederlands Centrum Buitenlanders

Bootsma, M. (1994). 'Ongure elementen? Nederland en de Hongaarse vluchtelingen, 1956' in Graaff, de B. (ed.) *Hun Crisis was de Onze Niet: Internationale Crises en Binnenlandse Veiligheid, 1914–1960*, Den Haag: Sdu

Bos, P. J. L. (2003) *Crisis Management in a Crowded Humanitarian Space: The Politics of Hosting Refugee Influxes*, Stockholm: The Swedish National Defence College

Brochmann, G. (1999) 'The Mechanisms of Control' in: Brochmann, G. and Hammar, T. (eds) (1999) *Mechanisms of Immigration Control, A Comparative Analysis of European Regulation Policies*, Oxford/New York: Berg

Brochmann, G. and Hammar, T. (eds) (1999) *Mechanisms of Immigration Control, A Comparative Analysis of European Regulation Policies*, Oxford/New York: Berg

Broeders, D. (2001) *Immigratie- en Integratieregimes in vier Europese Landen*, Werkdocumenten W

Ceastecker, F. (1998) 'The Changing Modalities of Regulation in International Migration within Continental Europe, 1870–1940' in Böcker, A., Groenendijk K., Havinga, T. and Minderhoud, P. (eds) (1998*) Regulation of Migration: International Experiences*, Amsterdam: Het Spinhuis Publishers

Daniel, E. V. and Knudsen, J. Chr. (1995) *Mistrusting Refugees*, Berkeley: University of California Press

Dinan, D. (1999) *Ever Closer Union: An Introduction to European Integration*, Basingstoke/New York: Palgrave

European Council, (2009) Conclusions, Brussels. December

Forsythe, D. P. (2000) *Human Rights in International Relations*, Cambridge: Cambridge University Press

Hammar, T. (1985) *European Immigration Policies: A Comparative Study*, Cambridge: Cambridge University Press

Huntington, S. P. (1996) *The Clash of Civilizations and the Remaking of World Order*, London: Simon & Schuster

Leenders, Marije, (1993) *Ongenode Gasten, van Traditioneel Asielrecht naar Immigratiebeleid, 1815–1938*, Verloren, Hilversum

Lucassen, L. (1998) 'The Great War and the Origins of Migration Control in Western Europe and the United States (1880–1920)' in Böcker, A., Groenendijk, K., Havinga, T. and Minderhoud, P. (eds) *Regulation of Migration: International Experiences*, Amsterdam: Het Spinhuis Publishers

Moore, B. (1986) *Refugees from Nazi-Germany in The Netherlands, 1933–1940*, Dordrecht/Boston: M. Nijhoff

Muus, Ph. (1997) 'Shifting Borders: The Inclusion and Exclusion of Refugees and Asylum Seekers in The Netherlands' in Muus, Ph. (ed.) (1997) *Exclusion and Inclusion of Refugees in Contemporary Europe*, Utrecht: Ercomer

Muus, Ph. (ed.) (1997). *Exclusion and Inclusion of Refugees in Contemporary Europe*, Utrecht: Ercomer

Nadig, A. (2002) 'Human Smuggling, National Security, and Refugee Protection', in *Journal of Refugee Studies*, Vol. 15, No. 1, pp. 1–25.

Nagtegaal, C. (2005) *Dutch Refugee Policies in 1914 and 1956: Myth of Humanitarian Aid?* Master thesis: Public Administration and Public Affairs, Universiteit Leiden: n.p.

Onderzoek Integratie Beleid, HTK, 2003–2004, 28 689, nrs. 8–9. Den Haag, Sdu, 2004

Russett, B., Starr, H. and Kinsella, D. (2000) *World Politics: The Menu for Choice*, 6th edn, Boston/ New York: Bedford/St. Martin's

Weiss, T. G., Forsythe, D. P. and Coate, R. A. (2001) *The United Nations and Changing World Politics*, 3rd edn, Boulder/Oxford: Westview Press

Wetenschappelijke Raad voor het Regeringsbeleid (2003) *Slagvaardigheid in de Europabrede Unie. Rapporten aan de Regering, nr. 65*, Den Haag: WRR/Sdu Uitgevers

Other Sources

NA – (Dutch) National Archives
Proceedings of Dutch Houses of Parliament
HEK – Senate
HTK – Second Chamber
Newspaper: TROUW, May 16 2009: 4

Chapter 8

Unaccompanied Minors in Europe: Between Immigration Control and the Need for Protection

Bernd Parusel[1]

Introduction

Throughout recent years, several Member States of the European Union (EU) have experienced the arrival of significant numbers of unaccompanied minors from 'third countries', that is children and young persons below the age of 18 from non-EU countries seeking refuge in Europe without any accompanying parents or legal guardians. Unaccompanied minors are a multifaceted and diverse group. Some minors arrive in Europe to be reunited with their family but do not fall under official family reunification schemes, others are children who travelled with one or more relatives irregularly and lost them along the way, some are born in Europe and have parents who are undocumented. Unaccompanied minors also include children who are sent by their families to Europe in search of better conditions or who have fled from wars, crises or persecution and seek asylum (PICUM, 2009: 5).

Especially from 2007 and 2008 onwards, numbers have increased substantially in some EU Member States. For example, unaccompanied minors applying for asylum made the headlines in Finland in 2008 as the number of applicants rose from less than 100 applicants in 2007 to 706. In Sweden, numbers climbed from 1,264 asylum applications in 2007 to 1,510 in 2008. Germany saw the arrival of 763 unaccompanied underage asylum seekers in 2008, compared to only 180 in 2007. One of the main receiving countries in the EU is the United Kingdom, with 4,285 applicants in 2008 (3,645 in 2007) (European Migration Network, 2010: 124).

The arrival of unaccompanied minors (and the recent increase) is a challenge for both the EU as a whole and the governments of the individual Member States. On the one hand, the EU has built its common asylum and migration policy around the principle that in-flows of irregular migrants must be restricted and controlled. On the other hand, they have implemented acts of law and policies according to which children enjoy special protection. Unaccompanied foreign minors cannot, for instance, be detained or removed in the same way as adults.

1 All views expressed in this chapter are those of the author and do not necessarily represent the views of the institutions he works for.

Quite the contrary, children and young persons seeking refuge without any parents or legal guardians need, and are entitled to, more help and support than adult migrants. The authorities concerned must, among other things, ensure that these minors are treated in a manner suitable to children or young persons. In the case of minors who are applying for asylum, the bodies responsible for interviewing them and assessing their claims have to proceed in a particularly careful and sensitive manner.

Until now, there has been rather little empirical research and relatively poor data on unaccompanied underage migrants in the EU. As far as published books and papers are available, they most often concern themselves with legal issues relevant for asylum practitioners, social workers and care-givers as to what has to be taken into account when dealing with minors, or advocacy papers published by NGOs working on refugees and child protection matters, such as Save the Children, the UNHCR, Amnesty International or national refugee councils. Nonetheless, even if reliable data and facts are scarce, unaccompanied minors have been high on the agenda of national and EU policy makers. From 2009 onwards, for instance, EU institutions have been in the process of reviewing legal instruments that were established within the framework of the common policy on immigration and asylum. The 'Hague Programme', which was adopted by the European Council in 2004 and which set out the most important policy aims and strategies for the development of the EU's internal affairs policies including migration policy, has been replaced by the 'Stockholm Programme', adopted by the Heads of State or Government of the Member States in December 2009 (European Council, 2009: 9). Among many other issues within the framework of common policies relating to 'freedom, security and justice' in the EU, the 'Stockholm Programme' concerns itself with unaccompanied minors, who are referred to as 'a particularly vulnerable group which requires special attention and dedicated responses'. More specifically, the Stockholm Programme calls for the 'exchange of best practices and common action in such areas as smuggling of minors, cooperation with countries of origin, the question of age assessment, identification and family tracing, and the need to pay particular attention to unaccompanied minors in the context of the fight against human trafficking'. Furthermore, a 'comprehensive response at EU level' should be developed, combining 'prevention, protection and assisted return measures while taking into account the best interests of the child' (Council of the European Union, 2009: 68). Further to this, some of the already existing EU directives in the area of asylum, such as the 'Asylum Procedures Directive'[2] and the 'Reception Conditions

2 Council Directive No. 2005/85/EC of 1 December 2005 on minimum standards on procedures in Member States for granting and withdrawing refugee status. This directive introduced a minimum framework for the functioning of asylum procedures in the EU, providing among other things binding rules on the rights of asylum seekers during those procedures.

Directive',[3] which contain special provisions concerning minors, such as their representation by a legal guardian or their placement in suitable accommodation, are in the process of being amended. In this context, the European Commission and the Parliament intend to strengthen some protection standards, which have been regarded as insufficient or too vague.

With these discussions as a starting point, this chapter aims at examining some of the challenges that arise from the entry into, and reception of, unaccompanied minors in the EU – challenges requiring policy makers to strike an often rather difficult balance between the interest of the EU member states in controlling and restricting immigration on the one hand, and on the other the particular need of minors for protection and the 'best interest of the child', which has to be used as the main criterion for decision-making in respect of all actions taken by the state in relation to children and young persons. Put in other words, the chapter seeks to describe the concrete effects on unaccompanied minors of the tensions between international law and EU legislation dealing with the well-being of children and the migration control and security approach, manifesting itself in tightened borders, compartmentalization and limited access of foreign nationals to protection. On the basis of such an analysis, the chapter will then outline some perspectives for policy change at the EU level. The central argument is that, while the migration control framework represents a distinctive obstacle to the effective protection of unaccompanied migrant children, the presence of international human rights law, advocacy work and the involvement of non-governmental actors in EU policy development might offer some opportunities, although limited, for improving the well-being of migrant children.

To a large extent, this chapter is based on the comparative study 'Policies on Reception, Return and Integration Arrangements for, and Numbers of, Unaccompanied Minors' (European Migration Network, 2010), which has recently been carried out in the framework of the European Migration Network (EMN).[4] Twenty-two of the total of 27 EU Member States participated in this project, which is part of the goal of the EMN to provide up-to-date, objective, reliable and comparable information on migration and asylum, with a view to supporting policy-making in the EU. The study was initiated especially in order to assist the European Commission in drafting an Action Plan on unaccompanied minors (European Commission, 2010) and in developing ideas for supporting safe reception arrangements in either host (EU Member State) countries or their countries of return. As far as no particular references are mentioned in this chapter,

3 Council Directive No. 2003/9/EC of 27 January 2003 laying down minimum standards for the reception of asylum seekers. This directive contains, among other aspects, rules on the accomodation of asylum seekers, access to education and the labour market, health care etc.

4 The author of this chapter has been responsible for the German contribution to this comparative project (Parusel, 2009). Some national reports of other Member States as well as general information on the EMN can be accessed at: http://emn.sarenet.es

the findings are based on this study and the individual national reports associated with it, which have been drafted by the 'National EMN Contact Points' of the EU Member States.[5]

Unaccompanied minors in Europe – an overview

There are few reliable statistics on how many minors from third countries are present in the EU. This is, in part, due to the fact that many minors enter the EU on an irregular basis without being detected by government authorities. Moreover, even if most minors sooner or later come into contact with some official authority or institution, there is sometimes no systematic gathering of statistical data. The most reliable data sources, though deficits exist even in this regard, are the national asylum statistics of the EU member states. As the processing of asylum claims by foreign nationals usually falls within the exclusive competency of a single national authority, data can more easily be gathered on a national level.

Table 8.1 shows the numbers of unaccompanied minors (UAM) who have filed an application for asylum in selected EU countries during the years 2006, 2007 and 2008 respectively. It should be kept in mind, however, that these statistics do not represent the overall number of unaccompanied minors entering the EU during these years, because only a certain proportion of these minors actually claims asylum. Some enter the EU on an irregular basis and then turn to relatives or friends who are already present, instead of claiming asylum. Furthermore, in some EU Member States, foreign minors can be granted a legal status even if they do not apply for asylum.

Even though the numbers in Table 8.1 should be interpreted with caution, they indicate that most EU Member States shown have recently witnessed an increase in the number of underage foreign migrants asking for asylum. Only Belgium, France, Ireland, the Czech Republic and the Slovak Republic have recorded slight declines in 2008 compared to 2006.

Looking at the most frequent countries of origin, the pattern is in many cases similar to that of adult asylum seekers. For the EU as a whole, most unaccompanied minors come from Afghanistan, Iraq, and Somalia. In Germany, the most frequent countries of origin of unaccompanied minors in 2008 were Iraq, Vietnam, Afghanistan, Guinea, and Ethiopia. In Sweden, a majority came from Iraq, Afghanistan, Somalia, the Palestinian territories and Eritrea (Migrationsverket, 2009: 13). The most frequent countries of origin of minors travelling to the UK in 2008 were Afghanistan, Iraq, Iran and Eritrea.

---

5 These National Contact Points (NCPs) are often located in national Interior Ministries, national agencies responsible for asylum and/or immigration, external scientific bodies or international organizations such as the International Organisation for Migration (IOM).

Table 8.1 Unaccompanied minor asylum applicants in selected EU Member States

	2006	2007	2008
Austria	414	516	697
Belgium	491	555	470
Bulgaria	*n.a.*	23	13
Czech Republic	92	56	36
Denmark	110	93	297
Finland	112	90	706
France	571	459	410
Germany	186	180	763
Greece	*n.a.*	44	296
Hungary	61	66	176
Ireland	131	94	98
Italy	*n.a.*	*n.a.*	573
Malta	180	330	300
Portugal	3	7	6
Spain	*n.a.*	12	13
Slovak Republic	138	157	72
Sweden	820	1 264	1 510
United Kingdom	3 450	3 645	4 285

Notes:

1. n.a. means 'not available'.
2. Figures from Eurostat are rounded to the nearest 5.
3. Figures for Germany for 2006 and 2007 only refer to unaccompanied minor asylum applicants below the age of 16. Sixteen and 17 year old unaccompanied minors have only been statistically acknowledged from 2008 onwards.
4. In some countries, such as Belgium, Greece, Italy, Portugal and Spain, only a minority of unaccompanied minors apply for asylum.

Sources: Eurostat, EMN national reports, UNHCR

Many unaccompanied minors arriving in the EU and claiming asylum state that they are fleeing and seeking protection, such as from wars, civil wars, crises, riots and torture in the country of origin, disturbances and conflicts, or equally from poverty and natural catastrophes, plus (impending) political persecution, or dangers by reason of their ethnic or religious affiliations, or the association of a family member with political or other activity, as well as the threat of honour-related violence.

For example, many children and youths who lodge an application for asylum in Germany are fleeing because their parents too are fleeing or have fled. Another reason is the fear of impending harmful practices (e.g. genital mutilation of girls and young women), as well as the fear of forced marriage of girls and young women, and of sexual abuse or forced prostitution. Other reasons mentioned in asylum hearings are to avoid slavery or child labour (Afghanistan, Angola, Kenya), persecution as a result of refusal to enter national service, or fear of conscription (Russia), collective punishment for the crimes of a family member (Ethiopia) or forced recruitment as child soldiers in African countries such as Angola, Sierra Leone, Somalia, or Guinea (Parusel, 2009: 19–20).

Further to this, the experiences of asylum officials in EU Member States indicate that there are several, varied and interconnected reasons for unaccompanied minors to enter the Member States. Apart from fleeing from harmful practices, wars or poverty, there can be 'pull-factors' such as a wish to join family members or relatives already living in the EU, the presence of a diaspora or foreign community in a specific EU country, or the aspiration to get a good education or to work. Furthermore, officials in Austria, Finland, Ireland and Poland have observed that unaccompanied minors can be victims of trafficking for the purpose of sexual exploitation or forced labour (European Migration Network, 2010: 39).

In the framework of the EMN study on unaccompanied minors, the participating EU Member States identified certain patterns with regard to the migration of unaccompanied minors to Europe. They also found that there are links between different motivations of unaccompanied minors to travel to the EU, and that unaccompanied minors may enter the EU for more than one reason or switch from one motivation to another.

According to Maltese officials, most immigrants arriving in boats from North Africa are young men in their twenties, but often there are also a few women, children and unaccompanied minors. Many do not have the intention to land or stay in Malta. In other words, they arrive by accident or by mistake. Upon entry, almost all apply for asylum, including the unaccompanied minors, who give as their main reasons for fleeing, religious persecution, family reunification, poverty, and human rights abuses such as child prostitution, child marriage, slavery and recruitment as child soldiers. The intention, however, seems more to be to seek asylum on the European mainland. Even after being granted asylum, some unaccompanied minors leave the island believing a better life awaits them on the continent (EMN – National Contact Point for Malta, 2009: 12).

In the case of Belgium, the comparatively stable number of unaccompanied minors having arrived in the country in recent years is in contrast to an overall decrease in the number of illegal entries into the country. With the accession of the 'EU-10' Member States[6] in 2004, it was anticipated that the number of unaccompanied minors coming to Belgium would decrease, since minors from these states were no longer recorded as third country nationals. This, however, did not happen. Instead, entering Belgium as unaccompanied minors seems to allow

6 The so called 'EU-10 Member States' that joined the EU in 2004 are Estonia, Latvia, Lithuania, Poland, the Czech Republic, the Slovak Republic, Hungary, Slovenia, Malta and Cyprus.

young immigrants from third countries to bypass other, more complicated and potentially lengthy entry procedures, in order to enter for the purpose of studies, family reunification, adoption or medical treatment (Belgian Contact Point of the European Migration Network, 2009: 15).

In Spain, the increase in the number of unaccompanied minors is considered to be linked to the intensified surveillance of the EU's external borders and to reinforced efforts by the government to refuse entry and to remove illegally staying migrants.[7] So the restrictive immigration regime in Europe may have boosted the entry of minors, whose return, given their age and child protection laws, is more difficult to implement (Vidal, 2007: 5). According to the Spanish EMN report, mafia gangs have intensified the smuggling of unaccompanied minors to Spain to counteract their decrease in income caused by the effectiveness of the fight against illegal migration (Spanish EMN National Contact Point, 2009: 27). While Spain states that most unaccompanied minors arriving in the country migrate for economic reasons (Spanish EMN National Contact Point, 2009: 23), Portuguese officials note that 'it is above all the desire to escape from situations of serious violations of human rights – particularly persecutions motivated by race, ethnic origin or gender – armed conflicts, forced military recruitment and trafficking for sexual exploitation that induce them to enter Portugal' (Serviço de Estrangeiros e Fronteiras, 2009: 9).

Ireland and Malta have noted that the foreign minors themselves often do not know why they have been sent to the EU. The motivation behind their migration is likely to be that of the parent or guardian. Meanwhile, Portugal found that the choice of the destination country is not always intentional and depends on random factors. In fact, it is not uncommon to find situations where an unaccompanied minor is unaware that he or she is in Portugal, since the destination was determined by the available means of transport and by support provided in the country of origin by adults, who are often unknown to them (Serviço de Estrangeiros e Fronteiras, 2009: 9).

Family reunification can represent another reason for unaccompanied minors to come to the EU. There is evidence that some unaccompanied minors arrive on an irregular basis in order to join their parents or other family members. As the normal legal procedures for family reunification are sometimes complicated and can take a long time, entering as an unaccompanied minor is attempted in order to try another – unbureaucratic – way of joining relatives or family. Another route identified is for the minor to come ahead of the family in order to be the first one to try to obtain asylum and, once recognized as a refugee, apply for family reunification. In the context of the EMN study, Finland, for instance, has found that when so-called 'anchor children' are granted a residence permit, applications by relatives for family reunification are likely to follow.

A further relevant motivation may also be a simple aspiration for a better life in Europe. This can be either an aspiration of parents or of the minor himself or

7 On surveillance and borders, see the chapter by Maguire in this volume.

herself. Unaccompanied minors may have been 'chosen' by their parents to travel to Europe and to access an education not available in their home countries. Often parents have high expectations for the unaccompanied minor, particularly with regard to remittances. A lack of perspectives in the country of origin, calamitous living conditions, scarce resources and poverty combined with a search to financially support the family via remittances can also be considered as 'push factors'.

Furthermore, some unaccompanied minors enter a Member State but do not intend to stay there, instead their intention is to continue to another Member State.[8] In Belgium, for example, most of the unaccompanied minors apprehended are actually on their way to the United Kingdom because they perceive that it offers more favourable employment opportunities, along with better benefit payments, access to health care and better social conditions compared to other EU Member States. The Czech Republic has found that it is primarily considered to be a transit station for unaccompanied minors who have found themselves there rather by accident on their way to Germany, the United Kingdom or other Member States (Ministry of the Interior of the Czech Republic, 2009: 11).

Last but not least, unaccompanied minors can also be victims of trafficking, which means that they are smuggled to an EU Member State for the purpose of sexual exploitation or illegal or domestic labour (European Union Agency for Fundamental Rights, 2009). Authorities in the Czech Republic, for instance, have become aware of criminal organizations which specialize in smuggling unaccompanied minors into the country. A specific example is a seasonal increase in the number of children coming from Romania, Bulgaria and the Slovak Republic for the purpose of prostitution, especially in summer (Ministry of the Interior of the Czech Republic, 2009: 10). Other Member States have reported that unaccompanied minors often disappear from care facilities. The disappearances suggest that those children might be victims of trafficking networks.

The need for protection

While all irregular migrants face the risk of poverty and exclusion, unaccompanied children are in an especially vulnerable position in terms of their ability to access rights and protection. This has been widely recognized at the European and international levels. From a number of international agreements and conventions, specific requirements to protect unaccompanied minors who seek refuge in the EU arise. Furthermore, several directives in the framework of the common EU asylum policy as well as the so-called 'Dublin-II regulation'[9] specifically refer to unaccompanied minors, including provisions regarding the protection of the rights of children.

---

8 On secondary movements of asylum seekers and refugees within the EU, see Schuster's chapter in this volume.

9 Council Regulation 343/2003 of 18 February 2003 establishing the criteria and mechanisms for determining the Member State responsible for examining an asylum

International law relating to the protection of minors

International law relating to refugees is based on the Geneva Convention on Refugees of 1951, which has been signed by more than 130 countries. The Convention does not contain any special regulations relating to underage refugees; yet, since it does not recognize any age limits, the general legal protection proceeding from it also extends to children and youth. It should be inferred from the purpose of this Convention that there is a requirement to interpret it in a manner suitable for children. Thus, unaccompanied minors must have access to asylum procedures in all countries which have ratified the Convention. In the EU, almost all Member States have received asylum applications by unaccompanied minors. Whereas in some countries, such as Sweden and Germany for example, nearly all unaccompanied foreign minors apply for asylum, in some others only a minority lodge an application for asylum. This is particularly true for Mediterranean countries such as Italy, Spain and Portugal. Whether or not an asylum application is lodged certainly depends on the individual situation of the minor, that is whether he or she has suffered from conditions that can lead to recognition as a refugee in line with the Geneva Convention. However, there are also differences between EU Member States regarding the accessibility of asylum procedures.

The Hague Convention on the Protection of Children of 1961, another important element of international law pertaining to minors, puts on record that the country in which a minor is habitually resident is, as a matter of principle, responsible for the protection of that minor and of his or her property. According to authorities in Germany, habitual residence exists if a minor has been living in the country for six months, or if it can clearly be recognized from the start of his or her residence that his or her life is in future to be based in this country. In accordance with Article 2, the responsible authorities and courts of law are under an obligation to take such measures for the child's protection as may be provided for under the domestic laws of that country. This includes, for example, finding accommodation for the child in a youth welfare institution, the appointment of a legal guardian and the regulation of the child's residence. Most EU Member States have laws in this regard, requesting authorities to arrange accommodation and guardianship. According to the EMN study on unaccompanied minors, guardians are often appointed from government agencies (such as the Youth Welfare Service in Austria and Germany, the Child Protection Services in Spain) and/or from an NGO (the Red Cross in Spain or the NIDOS foundation[10] in the Netherlands). For the United Kingdom, which does not currently provide an unaccompanied child with a legal guardian, the welfare and safety of an unaccompanied minor is the responsibility of the local authority, under the same arrangements as any other child who is in the care of the local authority (European Migration Network, 2010: 54). The national report from Greece to the European Migration Network states that the competent authorities shall take

application lodged in one of the Member States by a third-country national.

10 The Foundation NIDOS is the Dutch guardianship institution.

the appropriate measures to ensure the minor's necessary representation through the appointment of a legal guardian. This provision, however, is not properly implemented due to problems such as underfunded authorities, especially in the border areas, no follow up of the cases of unaccompanied minors represented by guardians, a lack of proper information for unaccompanied children concerning their rights vis-à-vis the legal guardian (Centre for European Constitutional Law / European Migration Network, 2009: 14).

According to Article 9 of the Hague Convention, in cases of urgency the simple fact of the minor's residence within that territory is sufficient for the country to be under an obligation to take measures for the minor's protection. The Convention has been ratified by 13 states, predominantly European ones.[11]

Probably the most important instrument of international law relating to minors, the UN Convention on the Rights of the Child, was approved on 20 November 1989 by the General Assembly of the United Nations. It has been ratified by 140 states. According to Article 2, the contracting states shall respect and ensure the rights set forth in the Convention to each child within their jurisdiction without discrimination of any kind, irrespective of the child's or his or her parent's or legal guardian's race, colour, sex, language, religion, political or other opinion, national, ethnic or social origin, property, disability, birth or other status. Article 3 determines that in all actions concerning children, whether undertaken by public or private social welfare institutions, courts of law, administrative authorities or legislative bodies, the best interests of the child shall be a primary consideration. According to Article 9, a child shall not be separated from his or her parents except when such separation would be in the best interests of the child. According to Article 10, applications for the purpose of family reunification shall be dealt with in a positive, humane and expeditious manner.

According to Article 12, in the context of the free expression of opinions, the child shall in particular be provided with the opportunity to be heard in any judicial and administrative proceedings affecting the child, either directly, or through a representative or an appropriate body.

As one of the most crucial provisions affecting government policies towards unaccompanied minors, Article 20 establishes that children who are temporarily or permanently deprived of their family environment, or in whose own best interests cannot be allowed to remain in that environment, shall be entitled to special protection and assistance provided by the State.

The EU 'acquis'

The institutions of the EU have been dealing with unaccompanied minors since 1997, when the Council passed a resolution concerning 'unaccompanied minors

11 Austria, France, Germany, Italy, Latvia, Lithuania, Luxemburg, Netherlands, Poland, Portugal, Spain, Switzerland and Turkey.

who are nationals of third countries' (Council of the European Union, 1997). The purpose of this resolution was to establish guidelines for the treatment of unaccompanied minors, with regard to matters such as the conditions for their reception, stay and return and, in the case of asylum applicants, the handling of applicable procedures. Articles covered admission, asylum and return procedures and, in the Annex, measures to combat trafficking and illegal entry. The resolution was not legally binding on the Member States.

In the framework of the more recent EU-harmonized policy on immigration and asylum, provisions for unaccompanied minors have been laid down in several directives. The above-mentioned 'Reception Conditions Directive', laying down minimum standards[12] for the reception of asylum seekers, includes the requirement for Member States to:

- ensure representation of the unaccompanied minor by appointing a legal guardian, a person acting on behalf of an organization responsible for the care and well-being of minors, or any other appropriate representation;
- ensure that unaccompanied minors are placed with adult relatives, with a foster-family, in reception centres with special provisions for minors or in other suitable accommodation. Unaccompanied minors aged 16 or over may be placed in accommodation centres for adult asylum seekers, whilst, as far as possible, siblings are kept together and changes of residence of unaccompanied minors shall be limited to a minimum;
- endeavour to trace the family members of unaccompanied minors as soon as possible with due regard for their safety;
- ensure that those working with unaccompanied minors shall receive appropriate training.

Comparable provisions for unaccompanied minors are also given in the 'Qualification Directive',[13] which defines minimum standards for the qualification of third country nationals or stateless persons as refugees or as persons who otherwise need international protection, and in the 'Asylum Procedures Directive', serving to establish minimum standards on asylum procedures.

The so-called 'Dublin-II Regulation' lays down the criteria and mechanisms for determining the Member State responsible for examining an asylum application lodged in one of the Member States by a third-country national. There are a number of rules specific to unaccompanied minors. For example, Article 6 states that the Member State responsible for examining his/her application is the Member State

12 Minimum standards define a range of basic provisions, which are binding for all Member States while allowing them to introduce or maintain more favourable provisions for the persons concerned.

13 Council Directive 2004/83/EC of 29 April 2004 on minimum standards for the qualification and status of third country nationals or stateless persons as refugees or as persons who otherwise need international protection and the content of the protection granted.

where a member of his/her family is legally present. If no family member is legally present, the Member State where the asylum claim is lodged is responsible. In addition, if the unaccompanied minor has a relative or relatives in another Member State who can take care of him/her, Member States shall if possible unite the minor with his or her relatives.

The 'Family Reunification Directive'[14] determines the conditions for the exercise of the right to family reunification by third country nationals residing in the Member States. In the case of an unaccompanied minor who is a recognized refugee, Member States are required to authorize the entry and residence of his/her first-degree relatives in the direct ascending line. Some Member States also authorize the entry and residence of a legal guardian or any other member of the family, if the refugee has no relatives in the direct ascending line or if these cannot be traced.

The 'Return Directive',[15] which was adopted by the European Parliament and the Council in December 2008, obliges a Member State, before deciding to issue a return decision to a third country national without legal residence, to grant unaccompanied minors assistance by appropriate bodies other than the authorities enforcing return. Furthermore, before removing an unaccompanied minor from the territory of a Member State, the authorities have to be certain that he or she will be returned to a member of his or her family, a guardian or an adequate reception facility in the country of return. Where necessary, Member States are required to extend the period for voluntary departure by an appropriate period, taking into account the specific circumstances of the individual case, such as the length of stay, school attendance and the existence of other family and social links. With regard to detention, the Return Directive provides, among other safeguards, that unaccompanied minors and families with minors shall only be detained as a measure of last resort and for the shortest appropriate period of time.

In sum, those instruments are intended to guarantee certain minimum standards, and they demonstrate that there is an understanding of unaccompanied minors as a vulnerable group in need of special protection. Other aspects of the reception of unaccompanied minors however, such as certain details in the context of the recognition of child-specific forms of persecution in asylum procedures, age assessment techniques, or the duties and responsibilities of legal guardians, are not determined by EU law but remain within the national competence of each Member State. Besides, existing directives only set a range of minimum standards, the scope of which sometimes remains rather vague and ambivalent. The provisions of the Reception Conditions Directive concerning the accommodation of minors include a telling example: While Member States must ensure that unaccompanied

---

14 Council Directive 2003/86/EC of 22 September 2003 on the right to family reunification.

15 Directive 2008/115/EC of the European Parliament and of the Council of 16 December 2008 on common standards and procedures in Member States for returning illegally staying third-country nationals.

minors are placed with adult relatives, with a foster-family, in reception centres with special provisions for minors or in other suitable accommodation, exceptions are permissible for unaccompanied minors who are 16 or 17 years old. They can, if national law foresees this, be placed in accommodation centres for adult asylum seekers where, in principle, the level of services and protection provided is lower.[16] In a similar vein, some Member States refrain from refusing entry at the border and from forced removals of unaccompanied minors, whereas others, such as Austria, France, Germany, Portugal and the Netherlands, depending on the circumstances, do refuse entry and/or enforce returns (European Migration Network, 2010: 45–7).

Migration control and the security approach

Since the beginning of the 'Europeanization' of asylum and refugee policies in the EU in the 1980s, the EU policy framework in this area has been described by researchers as focusing on restrictive measures and migration control. It has been argued that immigration is perceived by policy makers as representing a risk, and that the legal instruments in place emphasize the potentially destabilizing effects of migratory movements to societies in the EU, instead of attempting to maximize the benefits that migration can trigger and to effectively protect migrants in need of a safe haven. Consequently, the constructive principle of the common asylum policy has been labelled 'Fortress Europe', with several authors drawing attention to the tendency of policy makers and national interior ministry officials to 'securitize' migration (Huysmans, 1995; Geddes, 2000; Lavenex, 2001; see also the chapter by Karyotis in this book). Obvious indications are, among others, reinforced external border surveillance, the interception of boats and other means of transport, visa restrictions, limited access of third country nationals to protection, efforts to intensify the forced return of irregular immigrants and rejected asylum seekers, along with burden sharing mechanisms such as the Dublin-II system, which in practice rewards Member States for restrictive migration policies (Parusel, 2010: 116).[17]

Whilst several provisions in EU legislation on migration emphasize the need to protect vulnerable groups of migrants such as unaccompanied minors, these are also strongly affected by deterrent migration control policies. This section gives some concrete examples of situations in which clashes between migration control and the need for protection approaches become evident.[18]

To begin with, it is very hard for minors who are fleeing from their home countries to travel to the EU and to enter EU territory, as it is a fundamental requirement for underage nationals of third countries, just like adults, to be in

16 One example for this is Germany, see Parusel 2009: 32.

17 Concerning the Dublin System, see also the chapter by Schuster in this volume. On the securitization of migration, see chapter by Karyotis in this volume.

18 These examples are by no means exhaustive.

possession of a passport and, in most cases, a visa. The latter must be applied for at a diplomatic mission in the subject's country of origin. Unaccompanied minors, however, often do not have an opportunity to apply for a visa. In many countries there is, by reason of crises or acts of war, no functioning administrative structure capable of issuing a valid passport, and the embassies of possible countries for flight are not always accessible or within reach. Also minors, by reason of their age, may not fulfil the preconditions for the issuing of a visa (preconditions such as joining family members living abroad, or working or studying in the EU). As a result, their entry will generally take place on an irregular basis, either by air, by land or by sea. It is common for escape agents to be involved.[19]

If a minor who is travelling alone is not able to produce the requisite visa at the time of his or her attempt to enter the EU, then the border authorities are entitled to refuse entry. This happens to a varying degree. While some EU Member States usually grant unaccompanied minors access to their territory, some do refuse entry at border crossing points. In Germany, for instance, 174 unaccompanied minors were detected at the external land, air or sea borders in 2008. While a large majority (125) were handed over to a local youth welfare office inside Germany, 8 were refused entry and 12 were returned immediately after their irregular entry (Parusel, 2009: 27). In Portugal, 64 unaccompanied minors were refused entry at the external borders at Lisbon airport in 2008. In France, unaccompanied minors arriving at a maritime, air or rail border are, seen from a legal perspective, in an almost identical situation to adults. While unaccompanied minors present inside France cannot be deported, no special procedure on the basis of the age of a person is taking place at the external borders. Thus, unaccompanied minors who are not admitted at the border on the basis of valid travel documents can be refused entry or indeed sent back.

As with the procedure applicable to irregular entry by land, so-called 'airport procedures' are also applied to minors in some countries, such as Germany, Belgium and Switzerland. In Germany, this means that third-country nationals who are attempting to enter the territory via an international airport, and who are applying for asylum, have to pass through a fast-tracked asylum procedure in the transit area, and are accommodated there for the duration of the procedure. If no application for asylum is lodged, then as a rule entry into Germany will be refused. Only children under the age of 14 who are seeking asylum are usually spared the 'airport procedure'. Usually in such cases, entry into Germany (that is, permission to leave the transit area) will be granted.

In order for administrations and local authorities to be able to decide how to proceed in each individual instance, it is necessary to ascertain the age of young persons attempting to enter the territory of a Member State. This often turns out to be problematic. People who have entered illegally frequently carry

19 The involvement of 'escape agents' or 'people smugglers' does not always mean that criminal organizations carry out the illegal entry into EU territory. On occasion, distant relatives or family acquaintances of the unaccompanied minor in question also act as escape agents.

no identity papers, and border guards often mistrust statements made by these people concerning their actual age, that is claiming to be underage. In order to determine whether a young foreign migrant really is an unaccompanied minor, many Member States have implemented procedures to assess the age of a person. In this regard, there are big differences across the EU (Masson, 2008: 7; European Migration Network, 2010: 75–83). Whereas some Member States rely on medical or radiological examinations (e.g. x-ray of the carpus, inspection of the teeth), others carry out a simple visual inspection or let social workers interview the minors in order to assess their personal maturity. There is, however, widespread dissatisfaction with many of these proceedings, as both medical tests, inspections and interviewing techniques are imprecise. Sometimes, the persons concerned do not agree with the results of such examinations. It can also happen that different public institutions dealing with minors (such as youth welfare offices, foreigners' authorities, medical services, asylum authorities among others) disagree over the alleged age of a foreign juvenile. Moreover, medical examinations such as x-ray can be considered to interfere with a persons' physical integrity as radiation is a health hazard.

It has also been argued that the asylum procedures which Member States have put into practice in order to determine whether a third country national seeking protection really is a refugee, are not accessible enough and do not always sufficiently take into account child-specific forms of persecution and violence such as (impending) female genital mutilation or forced (underage) marriage. A general criticism is that asylum procedures have been designed for adults. For children and young persons, it can be significantly harder to make clear, or present in a structured manner, the precise reasons for their flight in conversations with people who work for authorities. The ability to describe the circumstances of one's flight in a credible manner presupposes a certain level of maturity (Cremer, 2006: 51). Consequently, NGOs have demanded that the rooms where interviews take place be equipped in a child-friendly way, and that asylum officials receive better training in order to be able to conduct the interviews in a less formal and more child-friendly manner (Ingham *et al.*, 2008: 3–4). Another criticism is that some national authorities responsible for the assessment of asylum applications tend to grant subsidiary protection to unaccompanied minors more frequently than refugee status, with subsidiary protection sometimes awarding weaker rights in terms of residence and access to social benefits than refugee status.[20] Although there

20 Across Europe, the recognition rates of unaccompanied minors differ significantly from the recognition rates of adults. Often much higher proportions of subsidiary protection are granted to unaccompanied minors than to adults, while the share of asylum applicants being granted refugee status is sometimes lower. Subsidiary protection is a form of protection that may be awarded to persons who do not fulfil the requirements for recognition as refugees but who have put forward substantive reasons that they would, upon returning to their respective countries of origin, be at risk of suffering serious harm such as the imposition of the death penalty, torture or a serious individual threat to their life or

are deficits, one also should take into consideration that unaccompanied minors in some EU Member States are more likely to be granted some form of protection (refugee status, subsidiary protection or national forms of a humanitarian right of residence) than adult asylum-seekers. In Germany, for example, the protection rate for unaccompanied minors in 2008 was 43.3 per cent, and 37.7 for all asylum seekers. In Finland, approximately 90 per cent of all unaccompanied minors receive favourable decisions.

Different approaches and deficient policies at the national level also exist in terms of providing legal guardianship and suitable accommodation. EU law provides that a guardian must be appointed after, rather than before, unaccompanied children have filed an asylum application (Save the Children, 2009). This can be seen as a 'protection gap'. Moreover, it has been observed in some Member States that legal guardians are often overloaded and/or lack knowledge of the asylum procedure and of legal aspects such as the right of residence. As far as accommodation is concerned, some Member States place unaccompanied minors in specialized child care facilities where guardianship and a high level of social assistance can be provided and access to the education system and health care be organized, whereas others offer accommodation only in, sometimes poorly equipped and run-down, reception centres for (adult) asylum seekers (Ingham *et al.*, 2008: 15). Sometimes, accommodation actually implies detention. In Malta, for example, after being interviewed by officials, all irregular migrants including unaccompanied minors are taken to a detention centre (European Migration Network, 2010: 110). Furthermore, while forced returns are carried out by some Member States (e.g. Spain, Greece, Netherlands), others, such as the Czech Republic or Portugal, generally refrain from such measures as long as the affected persons are underage (Save the Children, 2009; European Migration Network, 2010: 114–22).

Policy development

In light of the tensions between immigration control and the need to protect minors, or state-security versus the personal security of migrants, there is an ongoing debate in the EU on possible action to increase protection efforts and to address existing gaps. Yet there is a lack of reliable data and comparable evidence, and the margins for legal action are rather narrow, as European policy makers do not want to create new 'pull-factors', that is to possibly encourage young irregular migrants to try to get access to the EU, by establishing more favourable conditions. Thus, political reform concerning minors will require policy makers to find a new balance between the 'best interest of the child' and the security and migration

physical integrity due to armed conflict. For recognition rates of adult asylum seekers in the EU Member States, cf. Albertinelli/Juchno, 2009.

control approach. Several initiatives have been brought forward in this regard in 2008 and 2009.

Following an evaluation of the implementation of the Reception Conditions Directive for asylum seekers, for example, the European Commission tabled in December 2008 a proposal for its re-casting (European Commission, 2008a), including among other aspects a provision to ensure an adequate standard of living for 'persons with special needs' by, for example, requiring Member States to set up mechanisms for identifying persons with such needs as soon as an application for asylum has been lodged and by ameliorating the conditions for accessing health care.[21] Concerning the level of protection to be afforded to minors, the proposal specifies certain factors to be taken into consideration when assessing the 'best interests of the child'. For example, it forbids the detention of minors, unless it is in their best interests, and it obliges Member States to trace the family members of unaccompanied minors as soon as an asylum application is lodged. While the European Parliament has supported the idea of introducing a mechanism or procedure in the Member States to identify persons with special needs, some governments have voiced opposition against such a new procedure, arguing that it could be costly and lengthy.

Furthermore, the Commission published a proposal to amend the Dublin Regulation in December 2008 (European Commission, 2008b). Amongst the proposed changes is the addition of a new article requiring Member States to co-operate closely with each other in assessing the best interests of the child, and in particular to take account of certain factors such as family reunification possibilities, the minor's well-being and social development, safety and security considerations, as well as the views of the minor, in accordance with his or her age and maturity. It was also proposed to make the reunification of an unaccompanied minor with a relative who can take care of him or her a binding criterion for determining which Member State is responsible for the person in question, and clarify that, in the absence of a family member or another relative, the Member State responsible is the one where the applicant lodged his or her most recent asylum application, if this is in his or her best interest.

Meanwhile, several working groups and preparatory bodies of the Council of Ministers of Interior and Justice of the EU Member States held discussions on the issue of unaccompanied minors during 2009 and 2010. At Ministerial level, it was declared in September 2009 that all Member States would benefit from the development of common approaches and increased cooperation with the countries of origin of unaccompanied minors, including cooperation to facilitate their return. Aspects identified as requiring particular attention included the exchange

21 Persons with such needs are, according to the Commission proposal, minors, unaccompanied minors, disabled people, elderly people, pregnant women, single parents with minor children, victims of trafficking, persons with mental health problems and persons who have been subjected to torture, rape or other serious forms of psychological, physical or sexual violence.

of information and best practice, the question of age assessment methods, family tracing, and the need to pay particular attention to minors in the context of the fight against human trafficking. To this end, the European Commission published an 'Action Plan' on unaccompanied minors in May 2010 (European Commission, 2010). While not imposing any binding actions or measures on the Member States, it draws attention to some unresolved problems, calling for improved data collection on unaccompanied minors, new measures for the prevention of unsafe migration and trafficking of children, the development of child protection systems in third countries, enhanced reception systems and access to procedural guarantees within the EU, and monitoring of the quality of guardianship and legal representation of minors, among other issues. The Action Plan also suggests that decisions by Member State authorities on the future of each unaccompanied foreign minor entering their territory should be taken within the shortest possible period of time, and that minors who cannot be returned should be granted some form of legal status.

At this stage, it is difficult to predict the long-term outcome of the legislative and political processes to which the Action Plan of the Commission might give fresh impetus. The European Commission, the Parliament and the Member States' governments have all in many instances indicated that the overall aim of strengthening the EU's capacities to control immigration prevails. This includes the point of view that migration is primarily a security concern, and that 'pull-factors', which could attract irregular immigrants, must be further reduced. In practice, this means that reception conditions for asylum seekers, access to protection and social benefits in the EU, as well as the redistribution of asylum seekers between EU Member States should deter potential immigrants rather than attract them. When it comes to unaccompanied minors, however, the control and deterrence approach is to a certain extent counterbalanced by international human rights standards, such as the UN Convention on the Rights of the Child. Moreover, human rights NGOs and other bodies such as the EU Agency for Fundamental Rights are promoting better protection of minors and other vulnerable groups. By participating in the opinion making processes of the relevant political actors at the EU level, these organizations and agencies try to influence the outcomes.

The EU Commission, which frequently relies on academic, NGO and interest group expertise in order to increase its legitimacy and authority in the policy making process, and to strengthen its credibility in negotiations with the Parliament and the Member States (Boswell, 2008: 477–9), has given NGOs preferential access to certain discussions and policy structures dealing with unaccompanied minors. One example is the 'Eurasil' Network of asylum practitioners, a forum where asylum officials from the Member States meet several times a year in order to exchange experiences related to asylum seekers from certain countries of origin or other asylum issues. These meetings are often co-organized or indeed chaired by NGOs or IGOs such as the UNHCR. In May 2009, such a meeting was held on the topic of unaccompanied minors. Participants discussed child-specific grounds for asylum, child-adequate asylum procedures and methods of age determination. The meeting was prepared and carried out by representatives from the EU Commission in close collaboration with UNHCR and Save the Children.

In addition to NGO involvement in certain structures dealing with the exchange and gathering of experiences and knowledge, NGOs also receive Community funding from EU financial instruments within the framework of the 'Solidarity and Management of Migration Flows' programme, the European Refugee Fund, the European Return Fund or the 'Action programme for administrative cooperation in the fields of external borders, visas, asylum and immigration' (ARGO) to carry out or to act as partners in projects related to asylum seekers and, more specifically, the reception of unaccompanied minors in EU Member States. This not only enables them to do active fieldwork, but also to assemble valuable in-depth knowledge on matters pertaining to reception conditions and asylum procedures. On this basis, NGOs can act as qualified and valued experts vis-à-vis European and national officials and policy makers. By participating in hearings and arranging meetings, workshops or conferences in the context of the alternating Presidencies of the European Council, they can influence the policy agenda, too. In September 2009, for example, an event organized under the auspices of the Swedish Presidency of the European Union by Save the Children on unaccompanied and separated children addressed protection issues for these children. Intended as a discussion on policy and practical concerns, it gathered participants from the Swedish government, the European Commission, MEPs, the UN Committee on the Rights of the Child, UNHCR, Unicef, Human Rights Watch and the Separated Children in Europe Programme alongside national officials. Whether such opportunities for NGOs to voice their concerns and influence discussions at official meetings can increase the chances of resolving some of the problems unaccompanied minors are confronted with and of strengthening protection measures, however, remains to be analysed.

Concluding remarks

Looking at the issue of unaccompanied minors from a wider perspective, the increase in the numbers of such minors arriving in the EU in recent years can be seen by policy makers as an embarrassment. The presence of these minors reminds them that migration control strategies which are based on 'compartmentalization', measures of deterrence and walling-off have their limitations. Admittedly, it has been possible to drastically restrict the access of adult irregular migrants to protection, to establish detention systems, and to foster both voluntary and forced return. However, as this chapter has demonstrated, it is significantly more complicated to apply restrictive measures to minors, because they enjoy special protection due to international, European and national child protection laws. Thus, even if policy makers do not welcome the presence of unaccompanied minors in the EU, these children and juveniles need to be dealt with in a child-adequate way.

Moreover, public opinion is also more favourable towards vulnerable persons such as minors than towards irregular adult immigrants. Whereas the public has repeatedly turned against both 'legal' and 'illegal' immigrants in many EU countries, with mass media sometimes portraying them as dangers to public security and social or cultural cohesion, the approach towards unaccompanied minors tends to be more sympathetic. This may enable human rights organizations and refugee advocacy groups to act with greater confidence and, as a consequence, to achieve greater influence on the agenda of government officials and politicians.

Recent proposals for new legislation in the EU aimed at strengthening the protection of vulnerable groups may therefore have a positive impact on certain aspects of the reception and treatment of unaccompanied minors in Europe. On the other hand, there is little reason to expect that the governments of EU Member States will abandon or even relax the migration control and security framework, which has been institutionalized and gradually reinforced over many years. As a result, it is uncertain whether new initiatives to address protection gaps and to enhance the protection of minors really will lead to significant changes and more favourable conditions. It may be more realistic to expect that future EU policies will focus on the prevention of migration to Europe and, if considerable numbers of unaccompanied minors nonetheless continue to arrive, to facilitate their return to their countries of origin.

References

Albertinelli, A. and Juchno, P. (2009) 'Asylum Applicants and Decisions on Asylum Applications in Q2 2009', in *Eurostat Data in Focus*, 39/2009, Brussels: Eurostat

Belgian Contact Point of the European Migration Network (2009) 'Unaccompanied Minors in Belgium – Reception, Return and Integration Arrangements', Brussels

Boswell, C. (2008) 'The Political Functions of Expert Knowledge: Knowledge and Legitimation in European Union Immigration Policy', in *Journal of European Public Policy*, Vol. 15, No. 4, pp. 471–88

Centre for European Constitutional Law / European Migration Network (2009) 'Policies on Reception, Return, and Integration Arrangements for, and Numbers of, Unaccompanied Minors – an EU Comparative Study – Report on Greece', Athens, August 2009

Council of the European Union (1997) 'Council Resolution of 26 June 1997 concerning Unaccompanied Minors who are Nationals of Third Countries', in *Official Journal*, C 221 of 19 July 1997, pp. 23–7

Council of the European Union (2009) 'The Stockholm Programme: An Open and Secure Europe Serving and Protecting the Citizens', Doc. No. 17024/09, Brussels, 2 December 2009

Cremer, H. (2006) *Der Anspruch des unbegleiteten Kindes auf Betreuung und Unterbringung nach Art. 20 des Übereinkommens über die Rechte des Kindes*, Baden-Baden: Nomos

EMN – National Contact Point for Malta (2009) 'Unaccompanied Minors in Malta – Their Numbers and the Policies and Arrangements for their Reception, Return and Integration', Valletta

European Commission (2008a) 'Proposal for a Directive of the European Parliament and of the Council Laying down Minimum Standards for the Reception of Asylum Seekers', COM(2008) 815 final, Brussels

European Commission (2008b) 'Proposal for a Regulation of the European Parliament and of the Council Establishing the Criteria and Mechanisms for Determining the Member State Responsible for Examining an Application for International Protection Lodged in one of the Member States by a Third-country National or a Stateless Person', COM(2008) 820 final, Brussels

European Commission (2010) 'Communication from the Commission to the European Parliament and the Council, Action Plan on Unaccompanied Minors (2010–2014)', COM(2010) 213/3, Brussels

European Council (2009) 'European Council 10/11 December 2009 – Conclusions', Brussels

European Migration Network (2010) 'Policies on Reception, Return and Integration Arrangements for, and Numbers of, Unaccompanied Minors – an EU Comparative Study', Brussels

European Union Agency for Fundamental Rights (2009) 'Child Trafficking in the European Union: Challenges, Perspectives and Good Practices', Vienna: FRA

Geddes, A. (2000) *Immigration and European Integration: Towards Fortress Europe?*, Manchester: Manchester University Press

Huysmans, J. (1995) 'Migrants as a Security Problem: Dangers of 'Securitizing' Societal Issues', in Miles, R., Thränhardt, D. (eds) *Migration and European Integration*, London: Pinter

Ingham, D. *et al.* (2008) 'The Protection of the Rights and Special Needs of Irregular Immigrant Minors and Asylum Seeking Children', Vienna: European Union Agency for Fundamental Rights / Eurasylum Ltd

Lavenex, S. (2001) *The Europeanisation of Refugee Policies: Between Human Rights and Internal Security*, Aldershot: Ashgate

Masson, B. (2008) 'La situation et le traitement des mineurs isolés étrangers (MIE) en France', in *E-migrinter*, 2/2008, pp. 6–12

Migrationsverket (2009) *Kort om Migration 2008*, Norrköping: Migrationsverket

Ministry of the Interior of the Czech Republic (2009) 'Policy on Reception, Return, Integration Arrangements for, and Numbers of, Unaccompanied Minors in the Czech Republic', European Migration Network

Parusel, B. (2009) 'Policies on Reception, Return and Integration Arrangements for, and Numbers of, Unaccompanied Minors in Germany', *Working Paper 26*, Nuremberg: Federal Office for Migration and Refugees

Parusel, B. (2010) *Abschottungs- und Anwerbungsstrategien: EU-Institutionen und Arbeitsmigration*, Wiesbaden: VS Verlag für Sozialwissenschaften

PICUM (2009) 'Undocumented Children in Europe: Invisible Victims of Immigration Restrictions', Brussels: Platform for International Cooperation on Undocumented Migrants

Serviço de Estrangeiros e Fronteiras (2009) 'Reception, Return and Integration of Unaccompanied Minors in Portugal', Portugal National Contact Point for the European Migration Network

Save the Children (2009) 'Background Paper to Conference "Addressing the Protection Gap for Unaccompanied and Separated Children in the EU: Role of the Stockholm Programme"', Brussels: Save the Children

Spanish EMN National Contact Point (2009) 'Policies on Reception, Return and Integration Arrangements for Unaccompanied Foreign Minors – Spain', Madrid: Permanent Observatory for Immigration

Vidal, N. E. (2007) 'Social Construction of Neglect: The Case of Unaccompanied Minors from Morocco to Spain', *Working Paper 2007-007*, Rostock: Max Planck Institute for Demographic Research

Chapter 9
The Gendered Dimension of Security and Migration

Giovanna Campani

Introduction

In today's Europe, the concept of 'security', already unclear from the academic point of view because of its multiple definitions (i.e. 'national security', 'human security' and 'public security'), is blurred by ideological discourses which mix border control issues with those of migratory pressure, terrorism and perceived threats to cultural, economic and physical safety (see chapter by Karyotis in this book). The rhetoric produced by the growing 'populist' forces increases the perception of 'insecurity' connected to migration in many European societies, often leading to manipulation of statistical data on crime. Moreover, the notion of 'human security', which comprehensively encompasses the various social and personal dimensions linked with international and national security, has hardly been incorporated in political discourse and academic research on migration.

Given these premises, this chapter considers how a gender-approach can contribute to deconstructing the 'security rhetoric' in political discourse, focusing on the migration-security nexus, and analysing the multiple impacts of the policies labelled 'security' on migrants. 'Security-rhetoric' analysed through a gender approach reflects images and symbols belonging to past authoritarian ideologies (including fascism) that are revitalized by new populist forces. The chosen examples concern the Italian case.

Viewing the migration-security nexus through a gender perspective is like using a mirror, revealing how this link is manipulated and how it is embedded in a populist view of 'people', society and citizenship. The case of Italy, where various populist centre-rights parties are in power (including the *Lega Nord* (Northern League) and the *Popolo delle Libertà* (Peoples of the Liberties)) is a useful example. The rhetoric of these parties in relationship to 'immigrants who rape women', the 'protection of our women' and the paternalism towards Muslim women reveal a continuity with the legacy of fascism in attitudes to foreigners and colonized people. Study of the statistical data and analysis of the rhetoric in the Italian political arena clearly show that the security-migration nexus has nothing to do with 'facts', but is part of the re-construction of an authoritarian and anti-democratic ideology that is embedded in the national-populist and fascist traditions.

While irregular migration of men is considered a threat to national security (rising crime, terrorism, black work), that of women is strongly associated with questions of morality and sexuality (prostitution). Women can also be seen as victims, especially victims of trafficking. This narrative of victimhood reserved to women produces a rhetoric that considers irregular migrant women as forced to migrate (mainly to work in prostitution) and irregular migrant men as a threat to society (potential criminals). The fact that irregular migration, often provoked by extremely restrictive migratory policies, exposes all migrants to insecurity and vulnerability is hidden.

This chapter aims to show how the study of gendered trajectories questions the generally accepted statement that irregular migrants represent a special threat to security. It contrasts and confronts the gendered anti-immigrant rhetoric on security with the concrete experiences of migrant women, who perceive their position as more and more 'insecure' precisely because of the 'security policies'. The impact of 'security policies' is analysed through the biographical trajectories of a selected sample of migrant women who have been contacted in the frame of a EU-funded research project, FEMILPOL[1]. The biographies reveal the existing gap between people's lives, gendered security rhetoric and 'security measures', even those measures which claim to protect 'vulnerable' immigrant women.

It is 'human insecurity', precarious social environment, poverty and unsustainable social reproduction, that pushes women to migrate. Through migration, they are searching for the core elements of 'human security', namely a better life for them and their families; but the new environment can present new forms of human insecurity. Irregular migrant women continue to experience various dimensions of human insecurity in their migratory trajectories (similarly and differently from men) (see papers by Pajnic, Togral and Lazaridis and Konsta in this book). These narratives provide evidence that 'security' rhetoric in political discourse and 'security' measures, harsher border control and pressure on irregular migrants, rarely play a positive role in assuring better human security for immigrants or natives.

Security: a 'nebula concept'

The word 'security' is derived from the Ancient Greek 'Se-Cura', which literally translates as 'without fear', therefore the state of being secure, or the actions employed to achieve that state, i.e. to be secure is to be without fear of harm. When we pass from the literal meaning to the socio-political concept of security, it seems problematic to give a clear definition, independent of political discourse and the interplay between the various political and social actors. 'Security' seems

1 The FeMiPol project, financed by the VIth framework program (2006–2008), explores and analyses the impact of integration policies on the position of migrant women within EU countries. Accessed at: http://www.femipol.uni-frankfurt.de/.

to be a 'nebula concept' with mobile borders, a crucible of meanings, expression of priorities, interests and expectations in a continuous interaction, which is, at the end, oriented by the dominating political discourse (Sigona, 2007).

Stivachtis writes: 'In both academic and policy-making communities, prevailing views of security have been shaped by distinct conceptions of what it means to be secure. In fact, what constitutes a security problem has been a subject of academic discussion for many decades (Nye and Lynn-Jones, 1988). All these perspectives proceed from different premises, establish different priorities, and generate different types of predictions. Moreover, what scholars and policy-makers consider being national security issues is not fixed but varies over time.' (Stivachtis, 2008: 1). For instance, during the first half of the twentieth century and particularly during the Cold War, security acquired a narrow military definition.

According to Stivachtis (2008) and Buzan (Buzan, 1991a; Buzan *et al.*, 1998) the concept of security applies to five sectors: military, political, economic, societal and environmental. They correspond to the political definitions of security: international security, national security, human security and public security. The sectors are, to an extent, interdependent (i.e. changes in one sector, whether positive or negative, affect the other sectors), but the importance given to one or another depends on policies, which are more often expressions of 'visions' of security than responses to measurable threats. The nexus between migration and security is obviously a problematic issue, considering the different forms of migration (i.e. economic migration, temporary work, asylum, illegal migration); but the national identity ideology and the model of integration shape it more often than the objective facts do. For instance, a nation-state whose integration policies are based on multiculturalism deals with the migration security-nexus in a different way from a nation-state whose policy is hostile to multi-ethnicity, and considers immigration as a threat to national identity or religion.

In the post-Cold War world, the idea that the international community should aim at reconciling state security and human security in line with the democratic values of the post World War Two world and the Declaration of Human Rights (developed as concepts of human security by international organizations, especially by the United Nations Development Programs (UNDP, 1994)) represents a new definition of security, to be implemented in a global world, in an era in which the Cold War is over (and the focus on military security),[2] nation-states are losing much of their control of the economy, and circulation of goods, people and ideas is much more frequent. National security alone cannot guarantee national, regional

2 Human Development Report 1994: New Dimensions of Human Security: http://hdr.undp.org/en/reports/global/hdr1994/. The 1994 Report introduces a new concept of human security, which equates security with people rather than territories, with development rather than arms. It examines both the national and the global concerns of human security, seeking to deal with them through a new paradigm of sustainable human development, capturing the potential peace dividend, a new form of development co-operation and a restructured system of global institutions.

and global stability. This can result only from the spreading of human security, which means freedom from want and freedom from fear for all persons. Human security is a people centred-view of security, in which individuals' rights have to be taken account of. International organizations acknowledge this, but it is rarely accepted by nation-states.

Reconciling state security and human security

The notion of human security is especially important for the migration-security nexus. Given the fact that deprivation of basic human rights and of essential life conditions are likely to result in an increase in the number of migrants, especially refugees and asylum seekers, migrants are certainly not to blame when they escape a situation of human insecurity in order to reach a certain level of security. In this context security migration policies based on a strict notion of national security (protecting 'our citizens' against the 'others') provoke insecurity both for migrants and natives.

> Migrants' insecurity becomes more severe with the increasing rigidity of border control, while residents' insecurity grows as the state fails to effectively implement border control at feasible costs. Moreover, under certain conditions, migrant control by the state's domestic security apparatus increases migrants' insecurity, while the inability of the police forces to provide such control increases the insecurity of the citizens of the receiving state (Miller, 1998: 15–27).

This doesn't mean the abandonment of national security policies, in the sense of the right of individual states to protect their territory. The State must monitor flows of persons across the borders for state security (to protect the State from potential health risks or from terrorist threats), but national security policies should be combined with the principles of human security. International organizations such as the Council of Europe and the United Nations, and the transnational civil society, are trying to impose this perspective on refugees and asylum-seekers. In her report on the human rights of migrants, the Special Rapporteur of the Commission on Human Rights of the United Nations notes that:

> In exercising their sovereign right to regulate the entry, stay and movement of migrants and their policy on immigration, asylum and refuge, States should bear in mind the international obligations they have assumed in the area of human rights. In other words, States party to the International Covenant on Civil and Political Rights, the International Convention on the Elimination of All Forms of Racial Discrimination, the Convention on the Elimination of All Forms of Discrimination against Women and the Convention on the Rights of the Child must guarantee to anyone who is in their territory and subject to their jurisdiction the rights recognized in those legal instruments (UNGA, 2004: 11).

This is not always the case, as the Rapporteur observes in the same document:

> Despite the fact that these basic rules of human rights are applicable to non-citizens, the actual circumstances of migrants, particularly those with irregular administrative status, do not match the ideas reflected in the aforementioned instruments (UNGA, 2004: 11).

The populist threat

Reconciling state security and human security in relationship to migration is not an easy task. As Koser (2005: 2) states:

> Irregular migration poses very real dilemmas for states, as well as exposing migrants themselves to insecurity and vulnerability. Most states have, nevertheless, failed to manage or control irregular migration effectively or efficiently. What is therefore required are new, more effective and coherent approaches to address the issue of irregular migration, that recognize both the concerns of states in this respect and the need to protect the rights of irregular migrants.

One of the reasons why this aim has become such a challenge in Europe is the growth of 'populism', imposing its own agenda to many European countries' national politics. The security rhetoric evoking a migration threat is at the core of the 'populist' forces' message, but even non-populist parties are trying their best to recapture the outright populists' themes. At the same time, the media, in their search for an audience, are often playing the sensationalist card (see papers by Krasteva and Tsoukala in this volume). Consequently, migratory policies are determined more by ideological emotions than by the goal of managing and controlling irregular migration effectively or efficiently.

For twenty years, the European 'populist trend' has been investigated by scholars, while the term 'populism' itself has crossed the boundaries of academic debates into the political arena, and into common parlance. In-depth analyses of the theorizations of right wing populism (Betz, 1994; Betz, and Immerfal, 1998) feature scholarship focusing on the radical right parties (Kitschelt, 2007), the new populist right (Laycock, 2005; Mudde, 2004), the far right parties (Taggart, 2004), right-wing populism (Helms, 1997), radical right populism (Rydgren, 2003), or simply the populist parties (Abts and Rummens, 2007; Fella and Ruzza, 2009). Recent cross-national studies went further and distinguished between neo-fascist parties, which are anti-democratic and anti-capitalist, and radical right-wing parties, which combine opposition to immigration with pro-capitalist, neo-liberal economic positions, social conservatism, and basic acceptance of democracy. Another suggested distinction extends the taxonomy to include neo-liberal populism and national-populism (Betz, 1994).

If most scholars consider that the term populism is appropriate to describe present European political trends, others are more sceptical about its scientific value. Sartori (1970, 1984) speaks of conceptual stretching, a notion used also by Collier and Mahon (1993): the allocation and name-giving of the term 'populism' would be stretched and 'adjusted' to denote too large an array of political tendencies, from the Austrian Freedom Party with former leader Jörg Haider, to the Front National, led by Jean-Marie Le Pen; the Bulgarian Ataka of Volen Siderov to Northern League fronted by Umberto Bossi; from the Belgian extreme right-wing Vlaams Belang to the Polish League of the Families; from the Italian leader Silvio Berlusconi, with his party Forza Italia, to Vadim Tudor of the Great Romania Party; from the Dutch Lijst Pim Fortuyn to the True Finns[3]. These forces, parties, groups occupy various positions on the national political stage in EU Member States; some are extra-parliamentary groupings with no institutional legitimacy; yet others are present in national parliaments, but they are avoided as coalition partners in forming governments; others manage to be both present in parliaments and to play a role in the building of governments. Even if some of these parties express nostalgia for authoritarian regimes, not all populist parties and movements present at EU level demand radical constitutional changes. Consequently, some scholars prefer to talk about 'populisms' in the plural.

In spite of these reservations, populism can be a useful term to describe the growth of political parties whose discourse, going beyond the traditional right-left opposition, is centred on the division between the people as such and the corrupt elite, combined with nationalism, xenophobia and racism, in order to divide 'us' from the 'others'. With regard to right wing populist parties, which constitute the great majority of populist parties, most of the literature (Betz, 1994, Helms, 1997, Betz and Immerfal, 1998, Albertazzi and McDonnell, 2008) identifies their main arguments as ranging from Euroscepticism, opposition to immigration and immigrants, and rejection of multiculturalism, to the defence of what they call 'traditional values', which are most commonly identified with a specific understanding of Christianity and the 'patriarchal' family. These positions are presented as the will of the 'people', whom the populists pretend to represent, positioned against the establishment, which generically represents the technocrats, the bureaucrats (especially the EU bureaucracy) and the corrupt national elites. Populists may have different interpretations on matters of economy, such as neo-liberal populists and national-populist (Betz, 1994), but they have in common their anti-immigration stance and preference for a national workforce.

In the populist discourse, the security argument is combined with the nationalist, xenophobic, racist discourse that wants to exclude or even persecute the 'others', and emphasizes the threats. Consequently, a distinction must be made between a democratic approach that attempts to combine national security and human security, in line with international organizations and trans-national civil

3 Albertazzi, D. and McDonnell, D. (eds) (2008) analyse populism by looking at the conditions facilitating its emergence and success in specific national contexts.

society, and the national-populist discourse that combines security with identity discourses (in an escalation from xenophobic to openly racist, as it is the case of Italy). The populist discourse on security makes simple what is complex: most studies show that, while irregular migration represents a problem for the state that is supposed to control its borders, there is no clear link between irregular migrants and criminality. The link between migration (settled communities, diasporas etc.) and criminality is also complex: it changes according to periods, phases, countries, ethnic groups, criminal networks and so on. Only an approach taking into account this complexity can offer social and political answers, but this is not the main interest of populist forces, who use security arguments to gain electoral consensus, posing emotional and aggressive arguments against the 'others'.

The Italian case: a nightmare for Europe[4]

The combination between populism and ineffective migratory policies (both in term of border control and integration) is typical in Italy. Between 2001 and 2006 and since 2008, populist forces (the PDL, Party of Freedom of Mr. Berlusconi, and the Northern League) have been in power; the last law on immigration was prepared by the Northern League and the post-fascist party, while irregular migrants continue to represent a regular stock of over 500,000 persons!

The forms of Italian populism match the legacy of fascism with localism. Since 1948, political parties inspired by fascism have always been tolerated in Italy, in violation of the Constitution that forbids the reconstruction of a fascist party. Nevertheless, they were systematically excluded from governmental coalitions. Things changed in 1994, when Silvio Berlusconi incorporated the fascist party MSI (Italian Social Movement) into his right-wing alliance. In order to become fully legitimized, the MSI became National Alliance, formally breaking with fascist tradition and embracing democracy. Still, fascism-nostalgia is a common feature among members of the National Alliance Party. Together with the post-fascists, Silvio Berlusconi incorporated in his government a clearly populist movement, the Lega Nord (Northern League), which built its electoral support playing the card of the ethnic identity of a presumed 'people of the North', an historical aberration that could find appeal only in a country where, according to the linguist Tullio De Mauro, only 20% of the population has the linguistic, cultural and technical instruments to understand contemporary society.[5] The Northern League ethnic

4 The expression 'Italy has become a nightmare for Europe' was used by the Spanish movie director Pedro Almodovar in 2008, after the elections that brought the Berlusconi coalition into power.

5 'Five Italians in 100 between 14 and 65 cannot distinguish one letter from another, one number from another. Thirty-eight can do it, but with difficulty. Thirty-three have no problem reading, writing, understanding numbers, but cannot understand a written text concerning collective facts important for their everyday life (e.g. newspaper articles),

identity discourse has found favourable ground in a period during which the nation state has been weakened by globalization, and in a country where regional differences remain extremely important 150 years after unification.

Together with hostility to Southern Italians, the Northern League has developed an anti-immigrant (as well as anti-Muslim) security rhetoric. The anti-immigrant security rhetoric has progressively become the main feature of the movement. In the previous Berlusconi government (2001–2006) the League had succeeded in imposing a Law that made it extremely difficult for immigrants to get a regular permit. In the present Italian government, the Northern League has control of the Ministry of the Interior; the security rhetoric and the focus on the threat of immigrants represent a pretext for imposing an ideology (racism, but also sexism) and for prosecuting an authoritarian change in the organization of society and state. This was done through the introduction of new legal provisions, introduced by Law 94/2009 of 15 July 2009: *'Measures in the field of public security'*. The 'crime of illegal immigration' was introduced, together with detention in an identification and expulsion centre for up to 180 days (previously 60 days); the arrest and the imprisonment for up to three years for owners of properties rented to foreigners who are not officially registered at the time of stipulation of the contract, or renewal of the rental contract; the obligation for non-Community citizens who intend to marry an Italian citizen (or citizen of the Community in Italy) to have a regular permit; application for citizenship following matrimony with an Italian citizen may be made only after two years (previously six months). Moreover, the Security Decree has authorized 'patrols' (*ronde*), namely citizens' associations, to patrol local areas and notify security forces of social unrest. Participants will be registered and must first be trained by ex-officers of the law.

Merlino writes:

> On the basis of a 'persistent and extraordinary influx of non-EU citizens' and 'with the aim of strengthening the activities of fighting and management of the phenomena', the government prorogued and extended to the entire Italian territory the state of emergency. The alarming declaration of a national state of emergency together with the adoption of measures such as the introduction of the crime of illegal immigration have engendered sentiments of insecurity (in relation to the constructed image of the 'illegal' migrant) and xenophobia across the country. It seems that this 'situation of emergency' has been produced (or at least heightened) by the government itself (Merlino, 2009: 3).

---

graphics or icons. Only 20% of the adult Italian population possess the minimum instruments of reading, writing and calculating needed to be at ease in a contemporary technological society ...', quote from 'Analfabeti d'Italia' Tullio De Mauro in *Internazionale*, No. 734, 7 March 2008.

The introduction of the crime of 'illegal immigration' goes against democratic principles: the immigrant becomes a criminal because s/he is an immigrant. It is particularly interesting to notice that the same populist forces that have now introduced these laws have always opposed the introduction of a law such as the sponsorship laws which exist in various democracies such as Canada, and which would permit the legal entry of immigrants having friends and relatives in Italy. Instead they are forced to enter irregularly (family reunification is limited to very close relatives). At the same time, black work and the informal economy are tolerated and, as a result, Italy continues to have a constant stock of around 500,000 irregular migrants, over-exploited by employers.

Obviously the notion of 'human security' is absent from the Italian debate. As a consequence of its 'security policies', the Italian government sends many people back to Libya who should be entitled to humanitarian protection or asylum, and the UNHCR has denounced this.

The gendered rhetoric of populist discourse on security

In the context of the populist threat, the debate among Italian academics studying migration is polarized between those who attempt to construct an 'objective' notion of security through data and indicators essentially taken from criminological statistics, and the ones who regard all 'security' issues as ideological. The latter theorize that insecurity is purely perceived: people are afraid of immigrants because of the changing socio-political and economic landscape, and because of the government's propaganda. They tend to refer to Bauman's (2000, 2005) and Beck's (1998, 2008) analyses of people's loss of identity in the face of globalization, in order to explain the causes of 'perceived insecurity' in our societies, precisely when statistics show that crime is diminishing. It must also be stressed that their approach to Bauman and Beck doesn't take into account the complexity expressed by the two scholars. According to scholars such as Dal Lago (1999), for some reason, Italian society, more than other European societies, had been characterized by the spread of 'fear', fear for the future, of a precarious life, of change in life-styles, a feeling of existential and personal uncertainty. We may wonder, however, why in Italy 'fear' would have been more widespread than in other European societies such as Spain or Portugal, which have also known great social changes.

The debate on the criminological statistics is characterized by a division into two opposing 'factions': the first one insists that foreigners commit more crimes than Italians, and the second attempts to show that foreigners do not commit more crimes than Italians. This debate seems rather absurd in the Italian context, where three regions are controlled by Italian organized crime (Sicily, Calabria and Campania). Nobody can deny that foreigners constitute part of the prison population:[6] As of 31 July 2008 there are 20,439 foreign prisoners in the 205 Italian prisons (19,375 men and 1,074 women), representing almost 37 per cent of the entire prison population. According to the association Antigone, around 29.5 per cent of these cases arise from crimes against property (the same percentage as for Italians), 24.8 per cent are drug-related and 19.2 per cent are crimes against the person. However, this data gives little information without carrying out complex analysis through the 'intersectionality' approach (gender, class, ethnicity). To show the irrelevance of a mechanistic interpretation, we could look at the data about Italians in German jails, where they are overrepresented: over 1,100 Italians are detained in Germany, many for drug-related crimes. Does this mean that Italians are by definition criminals? All these phenomena need a global, complex approach, which takes into account the specificities of the crimes, the categories of the persons jailed, and the existence of criminal networks which can exploit vulnerable groups. In other words, there needs to be a global approach, inspired by an idea of democracy and human rights (including the rights of the victims of crime), in order to construct a cohesive society. The populist approach had been interested in 'problem-solving', but in obtaining consensus through emotional arguments based on the opposition between 'us' and the 'others' around an ethno-national idea of identity. The criminalization of immigrants is part of this agenda. Consequently, crimes committed by immigrants are put at the centre of populist rhetoric. In cases of particularly horrible crimes such as rape, populist rhetoric can evoke all the arguments traditionally used in nationalist and fascist ideologies.

The question of rape has been crucial in the implementation of 'security policies' in Italy. A few horrible crimes committed by immigrants (mainly by Rumanians) have raised public emotions, but right-wing politicians have exploited this in order to construct anti-immigrant feelings, while media, both right-wing and moderate, have played the emotional card, contributing to the creation of a climate hostile to all immigrants. A strongly gendered rhetoric has been used in politicians' discourses and in media representations. On October 30 2007, Giovanna Reggiani was brutally attacked by a Rumanian immigrant in a peripheral area of Rome and died a few days later from her injuries.

> The following day Italian government (at the time still centre-left) issued a decree granting prefects, local representative of the interior ministry, the power to expel any European citizen deemed to be a threat to public security (Norocel, 2009, quoting Fois, 2009: 2).

6 The sociologist Barbagli (1998, 2003) has analysed all the possible data, trying to disprove both the idea that immigrants are over-represented in crime and the idea that there is no problem at all. His aim is to try to show that there is a real problem of criminality for some groups of immigrants, but he insists on the complexity of the issue and tries to avoid simple equations.

A few months later, in spring 2008, a 14 year old girl was raped by two Rumanians near Rome. This second event happened in the middle of an electoral campaign that saw the victory of the right-wing coalition of Silvio Berlusconi. In this context, the reactions of politicians and media have been marked by '*the resurrection of stereotypical Balkan, which is oftentimes assimilated to the lack of reason and rationality*', as Christian Norocel (2009) states, quoting Kambourov (2003). Moreover, the populist exploitation of these events posits Italian women as passive carriers of the Italian inheritance –a common interpretation in nationalist rhetoric (Norocel, 2009).

These horrible crimes are, in fact, part of a larger picture characterized by a very high level of violence against women that has also to do with the position of Italian women. In 2007, the National Institute of Statistics published the first research on violence against women in the family and outside. A sample of 25,000 women aged between 16 and 70 were interviewed all over the national territory in the second half of 2006. The results indicate that 6,743,000 women between 16 and 70 years are victims of physical or sexual violence during their lifetime (31.9% of the age group). Five million women have been victims of sexual violence (23.7%), 3,961,000 of physical violence (18.8%). Around 1 million women have been victims of rape or attempted rape (4.8%). 14.3% of women have been victims of physical or sexual violence by a present or previous partner. 17.3% just from the previous partner. 24.7% of women have been victims of violence by another man. According to this research, foreigners committed only 10% of the rapes. Moreover, people who are not acquainted with the women commit only 6%. In 2006, rapes or attempted rapes have been around 200 a day, 74 thousand in one year.

The migration-security link is just one, even a relatively marginal issue, in a terrible context of violence against women. Still, in the rhetoric of the populist forces in power in Italy, foreigners are the only ones to blame for the violence. In fact, in February 2009, the Minister of the Interior, Maroni, quoting data from Barbagli (2003), declared in a press conference that 40% of rapes are committed by foreigners and 60% by Italians. This contradicts the quoted research, but impacts on public opinion, bearing in mind that foreigners are only 6% of the resident population. Maroni adds that the majority of rapists are Rumanians, followed by Moroccans, and that 68.9% of victims are Italian. In fact, Maroni manipulates the data. Barbagli has produced data indicating that 40% of rapes had been committed by foreigners, but he has also declared that most rapes happen inside the community (often inside the family), and has also insisted that 30% of rape victims are foreign women. In fact, according to Barbagli, foreigners rape mainly foreign women and Italians rape Italian women. The populist discourse is not interested in showing the complex gender, class and ethnicity situation in the rape issue. The populist discourse wants to reinforce an old idea: foreigners rape our women. This is the constant discourse of the parliamentarians of the Northern League. The examples offered are endless: the network is filled with videos of Northern League parliamentarians and senators campaigning against rape, expressing racist opinions against immigrants: 'we have to stop the rape provoked by the immigrants, we have to cut them, to smash them (the immigrants, the "clandestini" …)'.

> Rumanians are genetically predisposed to rape (Pergiorgio Stiffoni, Senator of the Northern League).

> Immigration has provoked the growth of crimes against women, rapes, violence of all kinds, FGM and segregation … Italy has been brought back to the Middle Ages (Claudio d'Amico, Northern League parliamentarian).

In order to protect the good Italian women, another depute of the Northern League, Matteo Salvini, proposes metro coaches in Milan are reserved only for women, then specifies that they should be reserved to the Milanesi, isolating the immigrants (men and women).

The gender argument is also used to justify the exclusion of some immigrant groups such as Muslims. Because they mistreat women, they should not be allowed to come to Italy.

It must be noted as well that many Italian women follow the populist discourse. Marshall Macluhan once noted, 'If a fish could speak, water is the last thing it would identify as part of its environment'. All women live in a world where male violence against women is around them, as water for fish, without identifying it. For this reason Italian women accept the rhetoric that treats violence as coming from outside the national or ethnic group (from the foreigners, the immigrants).

The subjective perception of security by immigrant women

Women represent about half of the immigrant population in Italy, but they represent a tiny percentage of the ones who commit crimes. On the contrary, they are victims of violence and rape. The populist law introducing the 'crime of illegal immigration' will criminalize immigrant women, who are certainly not responsible for the crimes some immigrant men commit. The fifteen interviews we are considering below, collected in the already quoted European Framework 6th project, FEMILPOL, on the integration processes of immigrant women, tell the stories of women who entered Italy illegally. What would have happened to them if the recently introduced law on security had existed at the time?

The fifteen women interviewed were not trafficked. They came to Italy of their own will, generally with the agreement of the family. Given the present migratory laws, it is rare to arrive in Italy with a regular stay permit. In our sample, the women entered Italy with counterfeit documents, or with a tourist visa, or even without any kind of permit. It must be remembered, however, that even those who come with a contract risk experiencing periods of non-regular status from time to time, because of Italian immigration law that binds stay permits to work permits.

The narratives show that women migrated for reasons of 'human insecurity', that is an insecure social environment (poverty), and 'unsustainable social reproduction', which means that they couldn't have a family and raise children. Through migration, they are looking for an aspect of human security: a better life for them and their families. This applies to women from Latin America, Eastern Europe and the Balkans, regardless of their origin. In our sample there are women from very poor backgrounds, while others come from what may be considered lower middle class or even middle class.

There are four Brazilian women in the sample. Two of them, Adriana and Paula, speak of their difficulties in their Brazilian favelas, with no money, no education, no work and broken families. Paula mentions, at the beginning of the interview, that she has 'no studies', and, because of this, could get no promotion in Brazil. Still, the other two Brazilians, who come from a better milieu, also felt that there was no future for them. Peruvian and Ecuadorian migrants share the same feeling, even if they had studied as far as the first years of University. A situation of insecurity was also common for the women of Eastern Europe, coming from the Ukraine of the economic transition, post-Ceausescu Romania, or Albania.

Irregular migration can be very expensive. When Judith, a Peruvian lady, came to Italy in 2002, she travelled initially with an illegal Peruvian passport, for which she had to pay '*between 7,000 and 7,500*' dollars. After that she got a passport of another country which cost her the same amount of money. She travelled across Ecuador and arrived in Italy as a tourist. She started working irregularly, but after about a year and a half she obtained the stay permit, thanks to the help of her first employer. Two years ago her daughter came to Italy without a visa, which made it less expensive – about 10,000 euros. She doesn't specify how they managed to get the girl into the country. She was irregular for one year, but now she's also 'legal', having a work contract.

The condition of irregularity affects their work position. Paula from Brazil travelled to Italy with a tourist visa, telling the Border Guard that she would be staying at her cousin's house for a week. At the time of the interview she had a residence permit, but she doesn't explain when or how she obtained this. She states that at the beginning, when she didn't have her documents, she was treated very differently by the house owners. They made her pay more, used every excuse to throw her out of the apartment, even if she had already paid the rent, and forbade her basic things such as listening to music or using the washing machine. Paula believes that they did this so that they could pretend that nobody, or at least no foreigner, was living in the apartment.

From the interview we understand that Adriana (also from Brazil) didn't have any documents at the beginning of her stay in Italy. After a while, when her youngest son had a serious road accident, she got a 'residence permit for health reasons' for 10 months. Adriana states that after this event her life got even worse. She had to take out a policy of 300 euros to bring an Italian to trial, and she had (and still has) to pay for the medical care for her son. She also had to pay another 300 euros for a tourist visa for her son, when the police caught him without a valid document. Unfortunately her permit didn't allow her to work, so she didn't know how to earn the money for the insurance and the medical costs, and she became a prostitute.

A Ukrainian woman, Mascia, entered Italy with a tourist visa. During the first years she was working without a permit to stay. About four years and three jobs later, when she had to go back to Ukraine for the wedding of her daughter and one of her ex-pupils, she had to stay there because she couldn't get a tourist visa to return to Italy. She stayed one year in Ukraine before finally obtaining her visa, thanks to the help of friends. She went back to Florence and started working for and living with a family in San Miniato, which helped her to get a permit of stay. She states that it wasn't difficult to get this first permit and to keep it, but it is a great problem to renew it. They always renewed it 'for free' before, but now one has to pay 75 euros and send the necessary documents to Rome, which are later sent back to Florence. After that one has to wait for one year, and as soon as the permit arrives one has to start renewing it again. Mascia now has the expired permit of stay and the post receipt that proves that she's renewing it. She couldn't find another job, because nobody gives a contract to someone without a permit. She complains that immigrants without documents can stay 'calm' in Italy, while those with documents have to deal with paying contributions and renewing their permit of stay, and have to wait one year for each renewal. She believes that it should take less time to get a new permit, and that one should have to renew it just once in a few years instead of every year.

An Albanian woman, Zamfira, had been without documents for 4 years. It was a difficult period as it was not possible to do anything without documents, 'only cleaning and domestic works, illegally ...'.

Costanza and her children arrived in Italy from Albania in 1999, all irregularly. The lack of a permit of stay caused them problems. When Costanza had to go to Albania for a funeral, it was a dangerous situation because if she had been found with her false document she couldn't have re-entered Italy, and her husband and children would have been constrained to come back to Albania. These worries caused a lot of suffering. Another difficulty arising from the lack of valid documents was that people took advantage of their problematic situation. For example, they had to pay a lot of money for a room. Since they urgently needed a place for their family, and this was very difficult to find, they had to accept the offer. There were also problems at work. When Costanza's husband was working as a builder and broke his arm by falling down during work, he didn't get any money for medical care.

It is clear that for irregular migrants, the new environment can present new forms of human insecurity, while offering, in some cases, better life conditions. Irregular migrant women continue to experience various dimensions of human insecurity (similarly and differently from men) in their migratory trajectories. From these biographical narratives it appears that 'security' rhetoric in political discourse and the ensuing 'security' measures, harsher border controls and pressure on irregular migrants, rarely play a positive role in assuring better human security for either immigrants or natives.

Lorena from Ecuador describes this situation very well, giving a very detailed, concrete picture of the immigrants' situation in Italy:

> The situation for immigrants, here in Italy, is very difficult, as a law by which we could regularize ourselves, all of us foreigners, doesn't exist … so we keep being clandestine immigrants. Many times, we do not have the right to health service or many times we don't have the right to find good work. Because in all the jobs, well, the 'stay permit' is requested, the permit of being allowed to work in this town.

At the same time, she understands the necessity of migration controls, to avoid the 'wrong people' coming to Italy. She thinks that Italians should be attentive about who could enter the country:

> To say the truth, I think that something's right, that also these people, these Italians who open the doors of their country to us, well … They also must take care of their backs, not to recruit wrong people, or people with bad habits or bad ways of living. I think this is right, but at the same time … we find ourselves in an uncomfortable position … because there's a great need to come to this country, to look for a new form of life. Because, when we come here, many times we don't find a job, some people have slept on the street, go to eat in those 'refectories for poor', in the public dining-rooms where they get food for free, they just have to register …

The security finally comes with the stay permit: this represents a moment of final stability and happiness.

This is the case now for Zanfrina. The regular stay permit gives her more chance of finding a job and of obtaining all the normal worker's rights:

> Having Italian documents, such as permit for work, I have better opportunities to find a job, all of my 'contributions' to be paid, to have the right to compensation in case I get 'licence' from a work.

Since 2002 the family of Costanza has had a permit of stay, which has a lot of positive consequences for them. First of all, they have regular job contracts with care assistance. They work regularly, they can go on holidays, in other words they 'are like' Italians, and they are very happy because of that.

However, this permit has to be renewed every two years, and there is always a risk of losing it, if there is no work. Lorena, in fact, explains that she now has to renew her stay permit every two years, showing that she has a regular job contract:

> This July, it was the second time that I have renewed my stay permit, and I have always renewed it with my work. This permit has to be renewed every 2 years, it has to be always renewed, even if the person has work supporting her/him.

Conclusions

Taking gender as an important variable in the security issue, it is possible to deconstruct and criticize the ideological nexus between migration and security. The gender approach reveals the difficulty of constructing an 'objective' notion of security through data and indicators taken from criminological statistics in the presence of populist rhetoric which plays on traditional nationalistic gendered imaginary.

The chapter also shows that the issue of security is strongly gendered, and that immigrant women cannot be blamed for any threat or crime. Instead they become victims of 'security policies'.

From the life histories of migrant women in Italy, it emerges that their subjective idea of security and insecurity during their migratory trajectories and settlement is close to the notion of 'human security': a mix of dreams, wishes, provisory plans for an uncertain future, dominated by their affection for their families and children, who should be safe and comfortable. The 'subjective' idea of security shared by these women (a million light years away from any will to commit a crime) is confronted by their categorization as irregular migrants (clandestines), as people who may become criminal, according to the new Italian legislation, just because they entered with a tourist visa that had expired or with a fake passport.

References

Abts, K. and Rummens, S. (2007) 'Populism versus Democracy', in *Political Studies*, Vol. 55, pp. 405–24

Albertazzi, D. and McDonnell, D. (eds.) (2008) *Twenty-First Century Populism: The Spectre of Western European Democracy*, Basingstoke: Macmillan

Barbagli M. (1998) *Immigrazione e criminalità in Italia*, Bologna: Il Mulino

Barbagli, M. (2003) *Rapporto sulla criminalità in Italia*, Bologna: Il Mulino

Baumann, Z. (2000) *Liquid Modernity*, Cambridge: Polity Press

Baumann, Z. (2005) *Liquid Life*, Cambridge: Polity Press

Beck, U. (1998) *Risk Society*, Cambridge: Polity Press

Beck, U. (2008) *World at Risk*, Cambridge: Polity Press

Betz, H.-G. (1994) *Radical Right-wing Populism in Western Europe*, New York: St. Martins Press

Betz, H.-G., and Immerfal, S. (eds.) (1998) *The New Politics of the Right: Neo-Populist Parties and Movements in Established Democracies*, New York: St. Martins Press

Buzan, Barry (1991a) *People, States and Fear*, 2nd edn, London: Harvester Wheatsheaf

Buzan, Barry, Waever, Ole, and De Wilde, Jaap (1998) *Security: A Framework for Analysis*, Boulder: Lynne Rienner

Collier, D. and Mahon, J. (1993) 'Conceptul Sketching Revisited', in *American Political Science Review*, Vol. 87, No. 2, pp. 845–95

Dahrendorf, R., (2003) 'Acht Anmerkung zum Populismus', in *Transit*, Vol. 25, pp. 156–63

Dal Lago, A. (1999) *Non-persone*, Milan: Feltrinelli

Fella, S. and Ruzza, C. (2009) *Reinventing the Italian Right: Territorial Politics, Populism and 'Post-Fascism'*, London: Routledge

Fois, G. A. (2009) 'The EU and International Context' in Newell, J. L. (ed.) *The Italian General Elections of 2008*, Basingstoke, Palgrave Macmillan

Helms, L. (1997) 'Right-wing Populist Parties in Austria and Switzerland: A Comparative Analysis of Electoral Support and Conditions of Success', in *West European Politics*, Vol. 20, No. 2: pp. 37–52

Kambourov, D. (2003) 'The Balkans and the Blackmale of Europe' in Novikova, I. and Kambourov, D. (eds) *Men in the Global World: Integrating Postsocialist Perspectives*, Helsinki: Kikimora Publications

Kitschelt, H. (2007) 'Growth and Persistence of the Radical Right in Postindustrial Democracies: Advances and Challenges in Comparative Research', in *West European Politics*, Vol. 20, No. 5, pp. 1176–206

Koser, K. (2005) Irregular Migration, State Security and Human Security A Paper prepared for the Policy Analysis and Research Programme of the Global Commission on International Migration, September

Laycock, D. (2005) 'Visions of Popular Sovereignty: Mapping the Contested Terrain of Contemporary Western Populisms', in *Critical Review of International Social and Political Philosophy*, Vol. 8, No. 2: pp. 125–44

Merlino, M. (2009) The Italian (In)security Package. Security versus Rule of Law and EU Fundamental Rights, Challenge, Liberty and Security, Working Paper 14

Miller, Mark J. (1998) 'International Migration and Global Security' in Poku, Nana K. and Graham, David T. (eds) *Redefining Security*, New York: Praeger

Mudde, C. (2004) 'The Populist Zeitgeist', in *Governance and Opposition*, Vol. 39, No. 3, pp. 541–63

Norocel, C. (2009) 'How About Taking Gender in the Theorising of Populism? Berlusconi's Right Wing Populism: The Patriarch and His Family', Paper presented at the 9th Conference of the European Sociological Association, Research Network 32, Political Sociology

Nye, Joseph S. Jr. and Lynn-Jones, Sean M. (1988) 'International Security Studies: A Report of a Conference on the State of the Field', in *International Security*, Vol. 12, pp. 5–27

Rydgren, J. (2003) 'Meso-level Reasons for Racism and Xenophobia: Some Converging and Diverging Effects of Radical Right Populism in France and Sweden', in *European Journal of Social Theory*, Vol. 6, No. 1, pp. 45–68

Sartori, G. (1970) 'Concept Misformation in Comparative Politics', in *American Political Science Review*, Vol. 64, No. 4, pp. 1033–53

Sartori, G. (ed.) (1984) *Social Science Concepts*, London: Sage

Sigona, N. (2007) 'The 'Latest' Public Enemy: Romanian Roma in Italy', Paper presented at the Council of Europe Expert Committee on Roma, Sinti and Travellers, Strasbourg, 8 November

Stivachtis, A. Y. (2008) 'International Migration and the Politics of Identity and Security', in *Journal of Humanities and Social Sciences*, Vol. 2, No. 1, 2008, accessed at: http://www.scientificjournals.org/journals2008/articles/1387.pdf.

Taggart, P. (2004) 'Populism and Representative Politics in Contemporary Europe', in *Journal of Political Ideologies*, Vol. 9, No. 3, pp. 269–88

Taguieff, P. A. (2002) *L'Illusion Populiste*, Paris: Berg International Editeurs

UNDP (1994) UNDP's Human Development Report

United Nations General Assembly (2004) Report on the Human Rights of Migrants submitted by the Special Rapporteur of the Commission on Human Rights, Geneva: UN

Chapter 10
Turning Immigrants into Security Threats: A Multi-Faceted Process

Anastassia Tsoukala

Creating social enemies is a longstanding social, transcultural phenomenon that has been of increasing appeal to social sciences scholars. In criminological circles, social problem construction became of interest in the late 1960s, that is, when certain tenants of social interactionism, with their ensuing Marxist revisionism, drew upon the labelling theory and put forward the concept of moral panic to analyse media representation of the deviant 'Other' (Cohen, 1972; Hall *et al.*, 1978; Cohen and Young, 1981). In spite of the frequent criticisms it raised in academia (Young, 1987, 2007b; Goode and Ben-Yehuda, 1994; Schlesinger and Tumber, 1994; Stabile, 2001), the moral panics model turned out to be highly influential among scholars wishing to highlight the transformation of specific subgroups or categories of people into threatening figures for the maintaining of the well-being of the community.

Therefore, while analysis of social problem construction rapidly spread across both sides of the Atlantic to encompass many different cases, ranging from youth delinquency and urban unrest to illegal immigration and terrorism (Huysmans, 1995; Palidda, 1997; Marshall, 1997; Dal Lago, 1999; Wacquant, 1999; Ceyhan and Tsoukala, 2002; Tsoukala, 2007, 2008a; Bonelli, 2008; Huysmans and Buonfino, 2008; Le Goaziou and Mucchielli, 2009), more often than not criminological and sociological analysis of the role played by media in shaping the dangerous 'Other' was fully or to a certain extent drawing upon the moral panics model (Hall, 1978; Whannel, 1979; Champagne, 1991; Jenkins, 1992; Welsh *et al.*, 2002; Tsoukala, 2002, 2008b, 2009a; Poulton, 2005). Many works have thus sought to address criminalization of immigrants and asylum seekers in Europe by highlighting the binary, stereotyped media representation of these people and the way they were being correlated with various law and order issues (Butterwegge, 1996; Maneri, 1997; Cotesta, 1999; Triandafyllidou, 1999; Mai, 2002; Erjavec, 2003). Yet, notwithstanding their qualities, these analyses of press discourses are limited in time and space to the extent that, due to obvious methodological reasons, they only cover specific territories during specific periods of time.[1] Recently, however,

---

1 In-depth analyses of relevant press discourses usually focus on short periods, ranging from a few weeks to some months. When they encompass a few years, they rest upon short sample periods (Maneri, 1997; Triandafyllidou, 1999). When they encompass longer

certain scholars have started questioning the time dimension of the moral panics construct. In his work on antisocial behaviour, Waiton (2008) is drawing attention to the emergence of permanent states of social anxiety, while Feeley and Simon (2007) point out that in some cases moral panics tend to be institutionalized, thus becoming deeply entrenched in social and political life. Concerning the framing of 'social enemies' in the long-term, research on media discourses on football-related violence over the last four decades (Tsoukala, 2009a) uncovers that, once framed as threatening, an issue may be subject to essentially static media coverage regardless of its dynamics and other domestic specificities. Comparative research on media representation of terrorists and football hooligans in the UK (Tsoukala, 2008b) shows, however, that framing of the threatening 'Other' relies heavily on the position held by the targeted 'Other' in domestic social and political fields.

When it comes to immigration, it is broadly acknowledged that since the late 1980s, that is, when international migratory movements became an issue for all post-bipolar European countries, immigration and asylum-related matters have been rapidly politicized and turned into major security issues. Accordingly, immigrants and asylum seekers became the target of many different threat-fuelling public discourses that reached public opinion both through spectacular ephemeral moral panics and the ordinary low-key reporting of fear- and anxiety-producing issues. At a political level, defining foreigners as a source of multiple social problems that eventually justified strict immigration policies has been often analysed as a predictable reaction of political elites who, fearing loss of sovereignty due to Europeanization and globalization, sought to reassure their electorate by ranking high in their political agendas both protection of borders and law and order issues (Huysmans, 1995; Palidda, 1997; Dal Lago, 1999; Bigo, 2002; Tsoukala, 2005). At the social level, this widespread social problem construction is, however, also to be understood in relation with a rise in social and existential insecurity due to globalization, growing socio-economic exclusion among working class and low/middle-class strata, erosion of the welfare state, and rapid technological progress (Castel, 2003; Bauman, 2007; Young, 2007a). In this worrisome context, anxiety grows not only because all prior landmarks seem to be drifting away but also, and above all, because liability is usually being deterritorialized, thus making impossible claim-making against those deemed responsible for insecurity, be it domestic or EU officials, transnational companies, banks or stock markets. Blame is thus displaced, to be put on those who visibly contribute to downgrading quality of life in disadvantaged or at-risk-of-decline urban areas. Scapegoating may then target immigrants as well as other folk devils, ranging from unruly youth to drug users and petty-criminals.

Similarly to those folk devils, immigrants may be easily targeted because they can be related to law and order issues and acted against (Critcher, 2009). Yet, unlike those folk devils, usually rooted in the margins of social life, immigrants

periods of time, such analyses tend to go beyond the usual methodological constraints to paint social reality with broad strokes (Butterwegge, 1996).

do not remain static in social and political terms. Their position in domestic social and political fields may evolve significantly due to integration in mainstream society, changes in domestic labour markets, variations in political decisions, and so on. Consequently, though they represent one of the clearer cases of moral-panic targets and long-term social problem construction, their evolving position in domestic communities and polities could arguably give rise to differentiated media representation. Could we assume then that media coverage of immigration and asylum-related issues may vary in the long-term according to an array of vested interests? In such a case, what articulations, if any, are to be found between former negative stereotypes and new images of the 'Other'?

To answer these questions, this chapter will rest upon past and recent research on coverage of immigrants and asylum seekers in the French and Greek press. Situated respectively in Northern and Southern Europe, these countries are also thought to be representative of old and recent immigration countries. These two national case studies are further believed to be interesting due to domestic specificities. Greece, one of the more ageing EU countries, is characterized by a large shadow economy, is fourth in asylum requests in the EU, and hosts around one million immigrants amongst roughly eleven million nationals. In 2008, poor treatment of asylum seekers and failure to provide effective asylum procedures were severely criticized by domestic, European and international bodies. Last, from the mid-1990s onwards (that is, when immigration came to be seen as a key social and political issue), most Greek media adopted an overtly xenophobic stance, which actually refutes the idea that anti-immigrant discourses are associated with the duration of the presence of foreigners on national territory and/or reflect colonialist stereotypes (Tsoukala, 2002). France, one of the less ageing EU countries, recorded in 2008 the highest number of requests for asylum in the EU, has the largest Muslim community in Europe, and many of its cities experience serious and longstanding urban unrest in their peripheries.

In the French case, research relies on analysis of 495 articles published in *Libération* and *Le Figaro* from 1 January 2007 to 31 October 2009. The sample was collected via the electronic database Factiva. In the Greek case, research rests upon analysis of 865 articles published in *Eleftherotypia* and *I Kathimerini* during the same period. The sample was collected via these newspapers' online archives. In both cases, the research is based on a thematic content analysis of all relevant articles. The period of time chosen for this purpose encompasses both passing moral panics and ordinary coverage of the issue. However, differences between these types of coverage are not highlighted. They are not thought to be relevant in an analysis that seeks above all to seize (dis)continuities in the long-term coverage of the issue. This blurring of differences is not believed to induce key methodological biases because, as Woodiwiss and Hobbs have recently reminded us (2009), the moral panic concept is also useful in analysing long-term social problem construction. All newspapers are upmarket nationally distributed dailies that are also believed to be illustrative of domestic political scenes. *Libération* and *Eleftherotypia* are left-wing, while *Le Figaro* and *I Kathimerini* are respectively right-wing and centre-right dailies. Due to the limits of this chapter, the presentation of each theme relies on a selection of quotations (in single quotation marks) and references that are seen as representative of its whole coverage. For the same practical reasons, only media representation of the issue in the late 2000s will be analysed in detail. Its context, media discourses in the late 20th century, will be presented in a summary way.

Media representation in the 1980s and 1990s: a prevailing multiform threat

Previous research on French and Greek media representation of immigrants and asylum seekers in the 1980s and 1990s has repeatedly highlighted the prevalence of negative stereotypes (Bonnafous, 1991; Battegay and Boubeker, 1993; Perotti, 1993; Karydis, 1996; Vamvakas, 1997; Tsagarousianou, 1997; Charon, 1998; Gastaud, 1998, 2000; Tévanian and Tissot, 1998; Koukoutsaki, 1999; Tsoukala, 1999a, 1999b, 2002, 2005; Moschopoulou, 2005; Daghmi, 2007). While defining the grounds upon which would eventually rest the alleged 'otherness' of the target groups, these stereotypes were also fuelling fear and unease with regard to what was seen as a source of multiform threat posed on host societies. Schematically speaking, narratives on 'threatening' immigrants and asylum seekers were being structured around five, frequently overlapping points.

Firstly, immigrants were being presented as dangerous due to their high involvement in crime. In the Greek case, fear-fuelling discourses, adopted by practically all newspapers, were systematically turning nationality into an absolute constant of the identity of immigrants arrested for serious and even minor offences. Reporting of crimes rightly or wrongly attributed to foreigners rose exponentially, while denunciation of rapidly emerging organized crime networks was usually combined with assumptions that such networks were run by foreigners who were operating both inside and outside the country. Though academics constantly denounced inherent faults in the very construction of police and prison statistics, as well as deliberate misuses by journalists, official data were regularly being presented as key evidence of (ir)regular migrants' dangerousness. In targeting above all Albanians, from the mid-1990s onwards this media coverage of foreign criminality fuelled several waves of Albanophobia and ensuing claims for strict coercive policies across the country (Karydis, 1996; Koukoutsaki, 1999; Lazaridis, 1999; Tsoukala, 1999b, 2002; Moschopoulou, 2005). In France, the upmarket press has usually avoided such stigmatizing discourses. Nonetheless, in echoing an increasingly security-oriented political agenda, dangerousness of immigrants was being sustained indirectly, that is, by demonizing the rise in crime and urban unrest in many disadvantaged urban areas that were mainly inhabited by first-, second- and third-generation migrants (Champagne, 1991; Rinaudo, 1995; Derville, 1997; Collovald, 2001; Peralva and Macé, 2002; Bonelli, 2008). Frequent recourse to deviance amplification mechanisms, ranging from rhetorical figures and sensational headlines to misuse of police statistics, rapidly associated large groups of immigrants and descendants of immigrants with many law and order issues. Urban unrest, petty crime, small-scale trafficking and antisocial behaviour in these areas were thus being turned into key security issues, and seen as revelatory of weak, if not impossible, integration into mainstream society.

The argument of cultural inassimilability, which was prevalent in French media discourses, went well beyond unruly youth in urban peripheries to cover Muslims, who in their vast majority were also immigrants or descendants of immigrants. In the 1980s, the seriousness of the threat posed to French secular democracy by Muslim cultural 'otherness' has been mainly illustrated by the broad dispute over what was thought to be a key symbol of backwardness and totalitarianism, wearing the head scarf in public places (Boyer, 1993).

The early 1990s have been characterized by a growing confusion between immigrants from Northern Africa, nationals originating from Northern Africa, and fundamentalist Muslim terrorists. Following terror attacks perpetrated in 1995 by an Algeria-born young man, who had been raised in France, people originating from Northern Africa came to be generally seen as potential terrorists, members of a genuine fifth column, regardless of their nationality (Rabah, 1998; Cesari, 1999). Spreading of broad suspicion rested then upon a series of overlapping categories, in which people targeted could be indistinguishably asylum seekers, regular or irregular migrants, foreigners or nationals belonging to a given ethnic community, foreigners or nationals belonging to a given religious community. Henceforth, the combined effect of the alleged security and cultural threats posed by 'foreigners' was so powerful that triumphant discourses on the positive impact of multiculturalism in the aftermath of the Football World Cup 1998 victory that was won by a manifestly multiethnic national team, eventually remained marginal (Gastaut, 2008) and could by no means counter the already well established image of dangerousness.

While in both countries immigrants were further presented as harmful to domestic economies because they were reducing employment opportunities for nationals, and fuelling the informal economy (Marie, 1999; Lyberaki and Pelagidis, 2000), discourses on demographic threat diverged significantly. In France, immigrants and asylum seekers were not thought to threaten the 'racial purity' of the community. In contrast, Greek journalists appeared so alarmed by the massive arrival of (ir)regular migrants in Greece that they frequently labelled it an 'invasion'. Warlike images of a country under siege thus alternated with aquatic metaphors of floods and dams under pressure that were further heightening the impression of a country under imminent risk, and illustrating collective fears over loss of 'demographic purity' since, as was regularly repeated, the birth rate remained very low[2] (Tsoukala, 1999b, 2002).

Security, cultural, economic and demographic threats did not always go together with distrust. Quite unsurprisingly, due to the combined effect of recent migratory

2 Population annual growth rate (1990–2000): 0.7% (accessed at: http://www.unicef.org/infobycountry/greece_statistics.html).

movements and a traditionally weak welfare state, Greek journalists paid little, if any, attention to welfare issues. Their French counterparts, however, regularly depicted immigrants and asylum seekers as morally dubious people. Articles about persons who had cheated or were attempting to cheat government officials in order to get into the country (bogus asylum seekers, bogus marriages with nationals, bogus family reunifications) and/or were cheating welfare services, once settled in the country, were thus at the heart of the ordinary representation of the issue.

Media representation in the late 2000s: usefulness versus threat

When compared with the aforementioned pattern, media representation of the issue in the late 2000s differs significantly in that it is solidly structured around a twofold axis, consisting of desecuritizing and securitizing discourses. While the former revise some of the previous negative stereotypes and denounce present immigration and asylum policies, the latter keep on fuelling the threat in many (in)direct ways. Within this pattern, French press discourses often differ depending on the political orientation of each medium. In contrast, Greek press discourses usually tend to transcend such political cleavages; when this does not occur, differences are highlighted.

Revision of former negative stereotypes

First of all, the immigration-economy nexus is henceforth framed in terms that seek to shake away all prior threatening correlations. Far from being seen as liable to reduce employment opportunities for nationals, immigrants are thought to be vital for preserving welfare systems and whole economic sectors. In Greece, their role in upholding agriculture, construction and domestic work is broadly acknowledged, while their impact on the real estate market, as renters or buyers of many shops in the centre of Athens and buyers of many old apartments and houses that nationals would have never bought, is thought to be pivotal for the whole sector (*Kathimerini*, 2007a, 2007e, 2008a). Seen as the only solution for preserving the welfare state in a rapidly ageing country, they are further depicted as significant 'contributors to GNP rise' (*Kathimerini*, 2008e) and job creators for thousands of nationals (*Kathimerini*, 2007b, 2009b; *Eleftherotypia*, 2008a, 2008b).

At first encounter, France seems well ahead of Greece in recognizing the positive impact of immigrants on the domestic economy and welfare state. In fact, the turn was institutionalized in 2006 following the introduction of the 24 July 2006 Immigration and Integration Law that established 'chosen' as opposed to 'imposed' immigration[3]. On a closer look, however, it becomes clear that public

3 'Chosen' immigration welcomes migrants whose working skills correspond to specific needs of the French labour market. Along with them, highly qualified or talented migrants may also benefit from advantageous conditions for entry and short residence.

debate about the foreign labour force is more restrictive than in Greece. In a euphoric demographic context, it is frequently stated that the foreign workforce is only welcome in a selective way. It is therefore expected that immigrants, who may be subject to a quota policy, should only be recruited in specific economic sectors that lack unskilled or highly skilled staff (*Figaro*, 2007e, 2007j, 2008a).

In both countries, then, revision of the former negative stereotype is not unproblematic. In accepting immigrants as strict economic units, it is assumed that their presence on the territory is tolerated as long as, and to the extent that, they are deemed necessary to cover domestic needs. Media discourses therefore draw a line between 'welcome' immigrants, chosen due to their skills or even lack of skills, and the 'others', who seek to 'impose' themselves on domestic labour markets, and thereby continue to be seen as unwanted. Quite predictably then, the selective opening of French borders is meant to be accompanied by a stricter clampdown on fraud, broader recourse to biometrics, and closer cooperation with countries of origin (*Figaro*, 2007e). Besides, in both countries it is clear that being 'welcome' under such conditions is not understood in the full social and human sense of the term. While in Greece demographic decline entails pragmatic acceptance of immigrants on a permanent basis but has little effect on public opinion,[4] in France 'welcome' migrants are only entitled to temporary work permits that, by definition, hinder integration into mainstream society and reduce life projects to short-term scheduling.

Revision of prior negative stereotypes goes beyond economic-threat-focused discourses to also encompass the immigration-demography nexus. Always absent in France, representation of immigrants as a demographic threat is rapidly waning in Greece where, as mentioned before, acknowledgement of demographic decline turns immigrants into 'anti-ageing elixir' (*Kathimerini*, 2009b). Yet, while threat-fuelling terms such as 'invasion', 'flood' or 'tide' of migrants and asylum seekers are no longer prevailing, demographic acceptance of the 'Other' still seems problematic. In a society that keeps on considering the number of foreigners on its territory worrisome[5], fear-appeasing discourses occasionally rest upon the implicit racist contention that 'Albanians form the only ethnic group akin to Greeks' and that 'osmosis between them and Greek nation is well-evidenced over the centuries' (*Kathimerini*, 2008e).

The impact of desecuritizing discourses is further weakened by scarce positive representation of immigrants and asylum seekers. Articles relating success stories of first-generation migrants, or presenting immigrants as ordinary honest working people, are extremely rare in Greece (*Kathimerini*, 2008b; *Eleftherotypia*, 2008d) and virtually absent in France. In fact, in both countries the prevailing image is that of people who suffer from poverty, exclusion and discrimination in ghettoized

4 According to a 2009 poll, 52 per cent of respondents believed immigrants to be harmful to the domestic economy (*Kathimerini*, 2009g).

5 According to a 2009 poll, 69 per cent of respondents believed there were too many immigrants in the country (*Kathimerini*, 2009g).

neighbourhoods. Of course, it would be unfair to blame journalists for accurately reporting problems encountered by immigrants in everyday life. Yet, it should be borne in mind that occasional compassion expressed over those who die at sea trying to reach European coasts (*Libération*, 2007e; *Eleftherotypia*, 2009g) cannot counter negative feelings raised by regularly drawing attention to those who managed to reach these coasts only to live 'amongst rubbish' (*Eleftherotypia*, 2008d). Nor can it counter worry raised by the apparently ceaseless arrival of immigrants and asylum seekers in Greek islands that makes 'even good-willed local societies desperate' (*Kathimerini*, 2009c). Such a focus on migration-induced problems ends up depicting immigrants and asylum seekers as the (in)direct source of multiple and frequently intermingled problems which host societies have to face in the name of multiculturalism, demographic rise and economic development.

Denouncing immigration and asylum policies

Growing deterioration in the treatment of asylum seekers in Greece and a sharp rise in racist aggression across the country, along with racism awareness seminars amongst media professionals produced significant shifts in the reporting of immigration and asylum policies. Denouncement of degrading living conditions and of abusive treatment of irregular migrants and asylum seekers in detention centres, frequently qualified as 'places of torment', or in Patras harbour where numerous irregular migrants wishing to cross the Adriatic Sea are gathered (*Eleftherotypia*, 2007c, 2008e, 2009d, 2009f), thus goes together with broad coverage of relevant hard criticisms made by EU bodies as well as domestic, European and international human rights organizations. In dismissing the taken-for-granted idea that Greeks are not racists, and the ensuing qualification of anti-migrant attacks as ordinary brawls or even acts of self-defence, journalists now define such attacks as the xenophobic and racist reactions of a conservative and indrawn community. Racism and physical violence on the part of police officers in particular are being vehemently criticized, especially by *Eleftherotypia*, and seen as a symptom of longstanding authoritarian trends in policing deviant people, whether foreigners or political activists (*Eleftherotypia*, 2007a, 2008h, 2009a).

Denouncement of physical violence is coupled with disapproval of the exploitation of the foreign workforce. In contrast to earlier coverage of the issue, immigrants are no longer thought to contribute to the shadow economy. They are depicted as victims of Greek employers, who only offer them black market jobs, refuse to pay them and threaten instead to hand them over to the police, and eventually remain unpunished due to the complacent attitude of labour inspectorate agents (*Eleftherotypia*, 2008f, 2009b, 2009c; *Kathimerini*, 2008i). In-depth analyses of the position of immigrants in the local labour markets reveal multiple discrimination in recruitment, wages and working conditions, leading to the conclusion that 'we want them [here] so that they do hard work [...] but we want them to remain in a status of illegality or semi-illegality [in order to keep their wages low]' (*Kathimerini*, 2009a).

Claims for fair treatment in labour markets are however outnumbered by calls for less strict asylum, immigration and naturalization policies (*Eleftherotypia*, 2007b, 2008c, 2008g, 2009e; *Kathimerini*, 2008h, 2009f). In sharp contrast to public opinion that remains massively hostile to foreigners,[6] claims for rational government policies transcend political divisions, to rest upon both humanitarian and pragmatic grounds. Asylum seekers and second-generation migrants who, in spite of them being born and raised in Greece, cannot be granted citizenship and may even be deported when they reach adulthood, are therefore a subject of rising concern. But, at the same time, it is often repeated that only full integration of immigrants could preserve the domestic welfare state (*Eleftherotypia*, 2008a, 2008b).

Concerns related to French asylum and immigration policies remain, of course, similar to the ones observed above but differ from them due to domestic specificities. Quite unsurprisingly for an old immigration country, claims for less strict government policies have been an integral part of media discourses in the 1980s and 1990s. Increasing politicization of immigration and asylum related questions has, however, deeply affected media coverage of these issues. In fact, representation of government policies depends heavily on the political orientation of each medium. Therefore, whilst *Libération* firmly denounced the introduction of DNA tests for candidates seeking residency under family reunification statuses (2007d), *Le Figaro* frequently pointed out that both countries and individuals concerned are respectively willing to introduce and be subject to such tests (2008b, 2008c). Massive deporting of irregular migrants according to deportation quotas is hardly criticized by *Le Figaro* whereas *Libération* sees in it an ordinary expression of symbolic politics that, moreover, is likely to produce abuse (2009a). Similarly, while *Libération* supports claims for amending the law so that people who, in the name of solidarity, offer shelter and food to irregular migrants cannot be prosecuted as traffickers (2009b, 2009c),[7] *Le Figaro* avoids analysis of the issue (2009d). Attitudes also diverge with regard to the way authorities should handle the 'Calais-issue', that is, the gathering in Calais of hundreds of irregular migrants who live rough while seeking a way to cross the Channel. The government decision to dismantle makeshift camps was thus welcomed by *Le Figaro*, as likely to put an end to these 'shameful gatherings' (2009b), but considered by *Libération* as symbolic politics that would only displace the problem (2009d).

In some cases, diverging coverage may imply marginalization or even obliteration of an issue. *Libération* thus is virtually alone in addressing the tragedy of irregular migrants who die at sea while seeking to reach European coasts (2007e, 2008d). It is alone too when denouncing certain controversial government policies, such as summoning irregular migrants to present themselves at police headquarters in order to be regularized and, once in the premises, having them arrested and deported (2008b, 2008e). It is also the only medium to denounce the

6 According to a 2009 poll, 72 percent of respondents were favourable to deporting irregular migrants (*Kathimerini*, 2009g).

7 For a EU-wide overview of crimes of solidarity, see Fekete (2009).

establishment of a genuine presumption of guilt over allegedly bogus marriages, to criticize relevant discretionary administrative and police practices, and to call for less strict family reunification policies (2007c, 2008g, 2009e). When it comes, however, to addressing the exploitation of irregular migrants in labour markets, both media denounce broad hypocrisy over those 'illegal taxpayers' (*Figaro*, 2009a) and call for a rational and fair policy (*Figaro*, 2007l; *Libération*, 2008c).

Fuelling the threat

While revisions of formerly prevailing negative stereotypes remain ambivalent, and claims for less strict immigration and asylum policies may be put forward in a highly selective way, immigrants and asylum seekers continue to be associated with an array of security-related problems. In both countries, this association is mainly sustained by the conservative press. Being structured around worries of loss of control over public space, insecurity- and fear-fuelling discourses refer to worrisome situations and specific involvement in crime[8].

In the former case, a fact that is often drawn attention to is that indigent, homeless irregular migrants, along with drug users, are seriously downgrading several areas in the centre of Athens, thus chasing away local people and harming tourism (*Kathimerini*, 2008g, 2009e). Fears about neighbourhoods being turned into ghettos become a key issue, while scholars see in minor clashes among migrants and police the beginning of a process likely to lead to urban unrest similar to that observed in French disadvantaged suburbs[9] (*Kathimerini*, 2008d).

In this worrying context, the image of poverty-stricken newcomers cooperating with other (ir)regular migrants in criminal activities becomes overwhelming, leading to widespread belief that foreigners are responsible for a rise in crime[10]. Concern over foreign criminality reached an apex in 2008. Denouncement of the heavy involvement of foreigners in drug trafficking, sex trafficking and other forms of organized crime thus went together with 'ethnic mafia'-based mapping of the centre of Athens. Readers learnt that small-scale drug trafficking, sex trafficking and counterfeit-related businesses were run by Africans, and burglaries were committed by Bulgarians and Romanians while large-scale drug trafficking and other organized crime activities were reserved to Albanians, who may also

8 For an analysis of the way concern about foreign criminality leads to new restrictive policies, see Campani's contribution to this volume.

9 From the 1970s onwards, many different suburban communities, inhabited by a high proportion of first-, second- and third-generation migrants, are witnessing declining urban life conditions, discriminatory public policies, high unemployment, and growing crime rates. Since the late 1970s, discontent in these disadvantaged areas has regularly fuelled urban unrest.

10 According to a 2007 poll, 65 percent of respondents held migrants responsible for the rise in crime (*Kathimerini*, 2007d). In a 2009 poll, migrants were thought to be responsible for the rise in crime by 76 percent of respondents (*Kathimerini*, 2009g).

cooperate with Greek, Romanian, Bulgarian, Russian and Turkish criminals (*Kathimerini*, 2008c, 2008f).

Unlike other European countries, in Greece politicians and security officials have never considered the country to be a prime target for Islamic terror attacks (Tsoukala, 2009b). Accordingly, media discourses do not seek to associate Muslim migrants with terrorism. Yet, though there is no imminent threat, terrorism still serves as a fear-fuelling perspective in that it is seen as a highly probable outcome of the unsuccessful or hindered integration of Muslim migrants into mainstream society. From this standpoint, reform of immigration and naturalization law should not be introduced in the name of the future enrichment of the host society, but in order to preserve society from future terror attacks (*Kathimerini*, 2009d).

In France, a country that has been targeted by Muslim terrorists in the past and continues to be included by Al-Qaida among key Western targets, concern over terrorism is always vivid. It is reminded that Europe remains a prime target for terror attacks (*Figaro*, 2007i), and intelligence officers recall foiled terror attacks in France, to conclude that 'it is not a question of if, but where, when and how' (*Libération*, 2007b). Threat assessments never exclude the possibility that attacks may be perpetrated by foreigners. But, more often than not it is assumed that attacks could equally be perpetrated by Muslims living in France (*Figaro*, 2007h). The spectre of the 'enemy within' is thus hovering over coverage of an array of issues related to nationals, immigrants and asylum seekers alike. Nationals going abroad to study in fundamentalist universities thus become a source of concern because, though 'it is not certain that these 'students' wish to perpetrate attacks', intelligence officials fear that some of them may provide logistical support to terrorists planning to attack France' (*Figaro*, 2007f). Under the all-encompassing term 'third-generation migrants', immigrants and 'not-so-perfectly-integrated nationals' are thought to be particularly vulnerable to radical Islamist preachers because their recurrent 'identity crisis is likely to be appeased through the acceptance of a trans-cultural and transnational 'Muslim' identity' (*Figaro*, 2007i). This threatening identity may be heightened by fundamentalists recruiting in prisons, where practically 50 percent of inmates are Muslims (*Figaro*, 2008e), but also, and above all, by radical Islamist preachers reaching youth in disadvantaged suburbs via internet or satellite television (*Figaro*, 2007k).

Youth in disadvantaged urban areas continue to rest at the heart of security-related media discourses. In the aftermath of the 2005 urban riots[11], attention is not focusing so much on urban unrest but on ordinary crime. Alarmist discourses about the emergence of a 'brutal generation' that takes to 'gratuitous violence' are accredited by senior police officers who further denounce lax judges, and the ensuing undermining of police work and reduction in the number of sentences imposed on juvenile delinquents (*Figaro*, 2007b). Longstanding worries about

---

11 In October and November 2005, France witnessed a nation-wide series of urban riots. These were thought to be so serious that a state of emergency was declared on November 8.

urban areas being turned into lawless zones are now fuelled by assumptions that 'juvenile delinquency is rapidly expanding, as if it were under capillary action', to reach other urban areas and 'even far away villages' (*Figaro*, 2007c, 2007g). Voices arguing that the 2005 riots should be seen as 'a claim for fair treatment' rather than evidence of failed integration (*Libération*, 2007a),[12] and attributing the rise in insecurity to economic inequalities, exclusion, and discriminatory treatment on the part of public authorities and employers (*Libération*, 2008a), remain so marginal that they cannot counter conservative and socialist calls for strict clampdowns on juvenile delinquency in these areas (*Figaro*, 2007a, 2009c).

Widespread concern over security threats posed by allegedly crime-driven migrants or imperfectly-integrated descendants of migrants intermingles, especially in France, with a broadly shared belief that 'foreigners' represent a key cultural threat to society. In Greece, in spite of a rapid rise in racism and xenophobia, the issue of whether, and to what extent, foreigners are likely to threaten domestic civilization is not raised in public. EU and domestic polls do uncover general hostility to multiculturalism but also significant tolerance to religion-based cultural differences. What arguably reflects the importance allotted to worship-related matters in a conservative, religion-based society is well illustrated by the findings of a 2009 poll revealing that 56 and 71 percent of respondents were respectively favourable to the construction of a mosque and Muslim cemetery in Athens (*Kathimerini*, 2009g). These findings are consistent with those of a 2007 poll showing that 68 percent of those interviewed did not consider migrants to be threatening to orthodox faith (*Kathimerini*, 2007d). On the other hand, a 2007 Eurobarometer survey reveals that 47 percent of respondents believed people of different ethnic origin did not enrich national culture (Eurobarometer, 2007). Present analysis of media discourses shows, however, that journalists have not sought to fuel hostility to cultural 'otherness'. Nor have they adopted culturalist views. Nonetheless, apart from occasional reporting about ethnic cultural festivals, press coverage of multiculturalism as a process likely to enrich everyday life remains marginal.

In France, rising multiculturalism and prevailing political correctness have somewhat weakened extremist political discourses on foreigners as a source of cultural threat. Racially-driven arguments do not guarantee electoral gains any longer, even among far-right voters. Such arguments are all the more unwelcome in that they are likely to chase away ethnic voters living in disadvantaged urban areas where far-right and populist parties obtain high scores in general and municipal elections. In this respect, it is telling that in the second half of the 2000s, the far-right Front National started communicating on their multicultural image, while the opening of Muslim private schools was not a source of concern to them (*Figaro*, 2007d). The waning of race-based discriminations and culturalist stances in the media landscape was further heightened in the aftermath of the 2005 urban

12 For recent analysis of the way violence and a high rate of criminality could be seen as a sign of integration in anomic conditions, see Young, 2007a; Hall *et al.*, 2008.

riots which, among other things, normalized people of colour as journalists on private and public television channels, and as actors in advertisements. These shifts in public representation of cultural 'otherness' did not, however, wipe away media discourses on the threat that cultural 'Others', whether ethnic nationals or immigrants, were allegedly posing to society.[13]

Once again, the threat is thought to stem from Muslims. In a context of growing concern over the preservation of the secular state, fears of losing what is seen as the cornerstone of the French Republic focused in 2003 on the right to wear the head scarf in public schools. Public debate ended with the introduction in 2004 of a law that prohibited the wearing of conspicuous religious signs in public schools. Ever since, the idea that Muslims threaten French cultural identity appears recurrently in the media in relation to many different issues.

In 2008, for example, following a court ruling that absence of virginity could be a ground for annulment of marriage, the question of whether such a ruling jeopardized the secular state drew media attention for a long time (*Figaro*, 2008d; *Libération*, 2008f). Debate over Muslim cultural 'otherness' reached a new apex in 2009, when public authorities considered prohibiting burqa in public places.[14] Hostility to burqa rests upon a threefold argument. While the garment is presented as a 'sign of extremist beliefs rather than religious ones', its prohibition is thought to be consistent both with protection of equal rights for men and women (*Figaro*, 2009e) and compliance with ordinary administrative procedures. In the latter case, its incompatibility with administrative rules is illustrated by the problems it causes when taking photos for passports or identity cards, or when celebrating marriages (*Figaro*, 2009f). Counter-discourses pointing out that most of the burqa-wearing women are young French citizens who have converted to Islam, seeking in this way to affirm their identity, and who should be free to wear this garment precisely in the name of their civil liberties (*Libération*, 2009f; *Figaro*, 2009e), struggle to appease fears of a rapid expansion of the phenomenon. Burqa-wearing women are estimated at several thousands (*Figaro*, 2009e), and burqa-wearing, henceforth a symbol of 'growing obscurantism', is associated with creeping 'religious pressure put on young children' at school, and the calling into question of biology courses (*Figaro*, 2009f). Fear-fuelling discourses drop sharply when intelligence officers affirm that burqa is only worn by 367 women, most of whom wear it willingly (*Libération*, 2009g), to be raised again when socialist mayors and MPs express the view that the garment jeopardizes 'fundamental values of the French Republic', and call for republican values to be restored (*Libération*, 2009h).

13 On cultural difference-as-a-threat narrative, see also Togral's contribution to this volume.

14 For an analysis of the French debate over the burqa in 2010, see Krasteva's contribution to this volume.

Concluding remarks

When seen in light of prior media coverage of immigration and asylum issues, comparative analysis of recent French and Greek press discourses reveals significant discontinuities and continuities. Of the former, the most important one is related to the emergence of desecuritizing discourses that seek to wipe away previously prevailing stereotyping of immigrants as economic and demographic threats. Notwithstanding domestic specificities, this shift is to be attributed both to demographic needs and structural changes in labour markets, which turn immigrants into desirable economic and demographic units. Desecuritizing thus is the outcome of a move that seeks to protect certain domestic interests that are consensually accepted as vital for the well-being of host societies. Quite predictably then, its broadcasting transcends political divisions. This necessity-oriented scope limits, however, its positive effects on the public representation of target groups. The latter may indeed appear as less threatening, but are not believed to be welcome in the full sense of the term either. Their ambivalent framing sets them in a vague space of temporary tolerance where, if needed, acceptance may be easily replaced, once again, by rejection.

The uncomfortable, in-between status of immigrants renders them all the more vulnerable to ongoing negative stereotyping that correlates them with an array of security and cultural threats. In line with former prevailing media coverage of immigrants and asylum seekers, securitizing discourses keep on reproducing the image of 'foreigner' as a source of key social threats, thereby fuelling widespread fear and anxiety over security and identity related issues.

This shifting (de)securitizing media representation of immigrants and asylum seekers suggests that social construction of threat obeys strict rational criteria that incorporate vested interests and needs of the host society.[15] Far from always being structured in binary terms, threat-fuelling discourses are flexible so that they can adjust themselves to the specificities of evolving social and political fields. In this complex process, threat is not dismissed; it is merely redefined, to be focused on security and identity. As mentioned above, from the late 1980s onwards these elements also lie at the heart of political and social concerns in rapidly changing European societies. Threat-fuelling discourses are therefore highly politicized and closely related to political prioritization of interests and values. Consequently, while in both countries broadcasting of security-threat-focused discourses depends heavily on the political orientation of each medium, broadcasting of cultural-threat-focused discourses only exists in France, that is, a country where such claims are also being put forward by mainstream political elites.

Ongoing interplay between desecuritizing and securitizing discourses should not, however, shift our attention away from the fact that the decreasing visibility of immigration-related economic and demographic threats has not implied any significant changes in the policy-making realm. Asylum policies are becoming

15 On this point, see also Karyotis's contribution to this volume.

ever more restrictive. Increasing restrictions on visa issuing, family reunification rights, mixed marriages, and naturalization law confirm that in both countries immigrants remain to some extent 'unwelcome'. Regular fuelling of discourses about failed integration and ensuing involvement in crime and terror attacks further shows that descendants of immigrants remain to some extent 'not-so-full members' of the community. This in turn suggests that, in spite of its effects in real life, desecuritizing is mainly to be understood as a rhetorical move. Demographic threat in particular is denied in media discourses, while it actually keeps on fuelling expansion of suspicion to all foreigners and large groups of nationals.

Assuming that scapegoating occurs at times of social disorganization (Erikson, 1966), ongoing demonizing of 'foreigners' uncovers a deep social and political crisis of European societies that, among other things, encourages them to enhance the symbolic imperviousness of their borders, and to redefine the boundaries of their collective identity by (re)designing who is entitled to belong to the community. In this respect, security-threat- and identity-threat focused media discourses should be seen as embedded in social and political processes of crisis management. Their persistence and occasional coexistence with desecuritizing discourses indicate then that the latter are adjustments to temporary requirements of long-term crisis management plans that, at best, struggle to set themselves free from the mythical image of 'foreigner-free' societies.

References

Battegay, A. and Boubeker, A. (1993) *Les images publiques de l'immigration*, Paris: L'Harmattan

Bauman, Z. (2007) *Liquid Times*, Cambridge: Polity

Bigo, D. (2002) 'Security and Immigration: Toward a Critique of the Governmentality of Unease', in *Alternatives. Global, Local, Political*, Vol. 27 (suppl.), pp. 63–92

Bonelli, L. (2008) *La France a peur*, Paris: La Découverte

Bonnafous, S. (1991) *L'immigration prise aux mots*, Paris: Kimé

Boyer, H. (1993) 'Tchador: Les mots de « l'Affaire »', in *Mediascope*, 4, pp. 67–72

Butterwegge, C. (1996) 'Mass Media, Immigrants and Racism in Germany: A Contribution to an Ongoing Debate', in *Communications*, No. 2, pp. 203–20

Castel, R. (2003) *L'Insécurité Sociale*, Paris: Seuil

Cesari, J. (1999) 'L'Islam: figure de l'ennemi intérieur et extérieur', 6th Congress of the French Political Science Association, Rennes

Ceyhan, A. and Tsoukala, A. (2002) 'The Securitization of Immigration in Western Countries: Ambivalent Discourses and Practices', in *Alternatives. Global, Local, Political*, Vol. 27 (suppl.), pp. 21–39

Champagne, P. (1991) 'La construction médiatique des « malaises sociaux »', in *Actes de la Recherche en Sciences Sociales*, No. 90, pp. 64–75

Charon, J.-M. (1998) 'La presse quotidienne régionale et l'immigration', in *Les Cahiers du Journalisme*, No. 4, pp. 40–44

Cohen, S. (1972) *Folk Devils and Moral Panics*, London: Routledge

Cohen, S. and Young, J. (eds) (1981) *The Manufacture of News*, London: Constable

Collovald, A. (2001) 'Des désordres sociaux à la violence urbaine', in *Actes de la Recherche en Sciences Sociales*, No. 136/137, pp. 104–13

Cotesta, V. (1999) 'Mass Media, conflitti etnici e identità degli Italiani', in *Studi Emigrazione*, No. 135, pp. 443–70

Critcher, C. (2009) 'Widening the Focus: Moral Panics as Moral Regulation', in *British Journal of Criminology*, Vol. 49, No. 1, pp. 17–34

Daghmi, F. (2007) 'L'image de 'l'immigré' dans la presse française' in Boyer, H. (ed.) *Stéréotypage, stéréotypes*, Paris: L'Harmattan, pp. 109–19

Dal Lago, A. (1999) *Non-persone*, Milan: Feltrinelli

Derville, G. (1997) 'La stigmatisation des jeunes de banlieue', in *Communication et langages*, No. 113, pp. 104–17

Eleftherotypia (2007a) 'Oi vrikolakes kai o esoterikos echthros', 23 June

Eleftherotypia (2007b) 'Ratsismos defteris genias', 7 July

Eleftherotypia (2007c) 'Friki sto Elladanamo', 30 October

Eleftherotypia (2008a) 'I moni lisi', 15 February

Eleftherotypia (2008b) 'Emeis kai o Ali', 6 May

Eleftherotypia (2008c) 'Edo gennithikame, edo zoume alla paramenoume xenoi', 3 June

Eleftherotypia (2008d) 'Ton metanaston oi koinotites', 14 June

Eleftherotypia (2008e) 'Stoivagmenoi se kolastiria', 21 June

Eleftherotypia (2008f) 'Krisi sto merokamato tis oikodomis', 8 November

Eleftherotypia (2008g) 'Apo tin mavri ipeiro sti mavroEllada', 2 December

Eleftherotypia (2008h) 'Topos vias kai vasanismou i diefthinsi allodapon', 4 December

Eleftherotypia (2009a) 'Oxi alloi nekroi sto allodappon', 12 January

Eleftherotypia (2009b) ''Mavroi' sti zoi kai sti douleia', 17 February

Eleftherotypia (2009c) 'Volikes exisoseis', 16 March

Eleftherotypia (2009d) 'I Kaboul tis Patras', 28 March

Eleftherotypia, (2009e) 'Pogrom en opsei me prota thimata tous metanastes', 30 March

Eleftherotypia (2009f) 'Kleinei to kolastirio', 30 October

Eleftherotypia (2009g) 'Mavros epilogos tis tragodias', 30 October

Erikson, K. T. (1966) *Wayward Puritans*, New York: Wiley

Erjavec, K. (2003) 'Media Construction of Identity through Moral Panics: Discourses of Immigration in Slovenia', in *Journal of Ethnic and Migration Studies*, Vol. 29, No. 1, pp. 83–101

Eurobarometer (2007) *Discrimination in the European Union*, Brussels: European Commission

Feeley, M. and Simon, J. (2007) 'Folk Devils and Moral Panics: An Appreciation from North America' in Downes, D., Rock, P., Chinkin, C. and Gearty, C. (eds) *Crime, Social Control and Human Rights*, Cullompton, Willan Publishing

Fekete, L. (2009) 'Europe: Crimes of Solidarity', in *Race & Class*, Vol. 50, No. 4: 83–97

Figaro (2007a) 'La sécurité, un défi pour la République, un chantier pour les territoires', 2 January

Figaro (2007b) 'Dans les banlieues, le sentiment d'insécurité reste élevé', 10 January

Figaro (2007c) 'Ces maux qui empoisonnent la vie de Vitry-le-François', 6 February

Figaro (2007d) 'Êtes-vous favorable à la création de collèges musulmans?', 7 March

Figaro (2007e) 'D'abord intégrer les migrants par le travail', 20 March

Figaro (2007f) 'Les 'missionnaires' français de l'islam radical', 28 March

Figaro (2007g) 'Insécurité dans le '9–3': le rapport qui dérange', 4 June

Figaro (2007h) 'La menace circonscrite en France', 2 July

Figaro (2007i) 'L'ensemble de l'Europe est devenu la cible d'apprentis terroristes', 3 July

Figaro (2007j) 'L'objectif fixé à Hortefeux: 50% d'immigration économique', 10 July

Figaro (2007k) 'Mettre l'islam au centre de l'organisation du culte musulman', 30 August

Figaro (2007l) 'Employeur recherche clandestins', 24 November

Figaro (2008a) 'La commission Attali veut relancer l'immigration', 10 January

Figaro (2008b) 'Des étrangers demandent l'application du test ADN', 7 February

Figaro (2008c) 'Tests ADN: le Cameroun intéressé', 19 May

Figaro (2008d) 'Mariage annulé de Lille, la polémique enfle', 31 May

Figaro (2008e) 'Les islamistes recrutent dans nos prisons', 11 September

Figaro (2009a) 'Au coeur de la France qui triche', 28 April

Figaro (2009b) 'De l'autre côté de la "jungle"', 2 May

Figaro (2009c) 'Pas de quartier pour la délinquance', 21 May

Figaro (2009d) 'La gauche fait fausse route', 10 June

Figaro (2009e) 'Polémique sur l'interdiction de la burqa', 19 June

Figaro (2009f) 'A Vénissieux, terre d'expansion de la burqa', 1 July

Gastaut, Y. (1998) 'Le multiculturel au miroir des médias', in *Projet*, No. 255, pp. 35–44

Gastaut, Y. (2000) *L'immigration et l'opinion en France sous la Ve République*, Paris: Seuil

Gastaut, Y. (2008) *Le métissage par le foot*, Paris: Autrement

Goode, E. and Ben-Yehuda, N. (1994) *Moral Panics*, Oxford: Blackwell

Hall, S. (1978) 'The Treatment of Football Hooliganism in the Press' in Ingham, R. (ed.) *Football Hooliganism*, London: Interaction

Hall, S., Critcher, C., Jefferson, T. and Clarke, J. (1978) 'Policing the Crisis', Basingstoke: Palgrave Macmillan

Hall, S., Winlow, S. and Ancrum, G. (2008) 'Criminal Identities and Consumer Culture', Cullompton: Willan Publishing

Huysmans, J. (1995) 'Migrants as a Security Problem: Dangers of 'Securitizing' Societal Issues' in Mils, R. and Thrännhardt, D. (eds) *Migration and European Integration*, London: Pinter
Huysmans, J. and Buonfino, A. (2008) 'Politics of Exception and Unease: Immigration, Asylum and Terrorism in Parliamentary Debates in the UK', in *Political Studies*, Vol. 56, No. 4, pp. 766–88
Jenkins, P. (1992) *Intimate Enemies*, New York: Aldine de Gruyter
Karydis, V. (1996) *The Criminality of Migrants in Greece*, (in Greek), Athens: Papazissis
Kathimerini (2007a) 'Apo metanastes dimiourgeitai ena sta tesera kainourgia noikokiria', 5 January
Kathimerini (2007b) 'Oi metanastes tononoun tin oikonomia', 26 January
Kathimerini (2007c) 'Gerasmenes oi chores tis EE', 26 January
Kathimerini (2007d) 'Tholi eikona Ellinon gia tous metanastes', 15 February
Kathimerini (2007e) 'Simantiki i simvoli ton metanaston sti diatirisi tis ipsilis zitisis', 6 October
Kathimerini (2008a) 'Otan oi metanastes sozoun kai tin agora emporikon akiniton', 13 January
Kathimerini (2008b) 'Me ton idrota tous, afentika se xeni gi', 27 January
Kathimerini (2008c) 'Allodapoi, o efialtis ton allodapon', 10 May
Kathimerini (2008d) 'To proto ghetto stin protevousa', 22 May
Kathimerini (2008e) 'Oi Alvanoi metanastes prosferoun stin Ellada', 5 August
Kathimerini (2008f) 'Oi 'ethnikes' omades mafiozon tis Athinas', 7 September
Kathimerini (2008g) 'Oi plateies to vradi ginontai apeilitikes', 14 September
Kathimerini (2008h) 'Prosfiges: o thriamvos tis egataleipsis', 7 December
Kathimerini (2008i) 'Oi dio taxeis metanaston tis Athinas', 21 December
Kathimerini (2009a) 'Ta dikaia tou metanasti', 5 February
Kathimerini (2009b) 'Elixirio antigiransis tis Elladas i athroa eisroi metanaston', 12 March
Kathimerini (2009c) 'To provlima asfaleias me lathrometanastes', 1 April
Kathimerini (2009d) 'Proklisi gia tin elliniki asfaleia', 26 April
Kathimerini (2009e) 'I Athina allazei alla sto cheirotero', 8 May
Kathimerini (2009f) 'Emeis kai oi metanastes', 17 June
Kathimerini (2009g) 'Megaliteri anisichia gia metanastes', 5 July
Koukoutsaki, A. (ed.) (1999) *Images of Crime* (in Greek), Athens: Plethron
Lazaridis, G., Poyago-Theotoky, J. and King, R. (1999) 'Retirement Migration to Corfu, Greece' in King, R. and Connell, J. (eds) *Small Worlds, Global Lives*, Cassell Academic Press
Le Goaziou, V. and Mucchielli, L. (2009) *La violence des jeunes en question*, Nîmes: Champ social
Libération (2007a) 'Nous sommes passés à une racialisation de l'identité française', 3 April
Libération (2007b) 'Notre pays est explicitement visé, la question est de savoir quand', 13 April

Libération (2007c) 'Hérault: haro sur les unions mixtes', 16 April
Libération (2007d) 'ADN, le "détail" de Fillon suscite la polémique', 7 October
Libération (2007e) 'Le désespoir: les migrants', 22 December
Libération (2008a) 'Les banlieues ont-elles besoin d'un énième plan?', 22 January
Libération (2008b) 'Quand les préfectures piègent les sans-papiers', 14 April
Libération (2008c) 'Action "coup de poing" des sans-papiers en Ile-de-France', 15 April
Libération (2008d) 'Les brûleurs de frontières', 28 May
Libération (2008e) 'La préfecture, un guet-apens de plus en plus fréquent', 10 June
Libération (2008f) 'L'islam est-il incompatible avec la laïcité?', 12 July
Libération (2008g) 'Couples mixtes, le règne de l'arbitraire', 13 December
Libération (2009a) 'Expulsions: Besson se trompe d'objectif', 6 April
Libération (2009b) 'Besson contredit les associations', 8 April
Libération (2009c) 'Délits de solidarité: le Gisti répond à Eric Besson', 21 April
Libération (2009d) 'Chasse aux migrants dans la "jungle" avant l'arrivée de Besson', 22 April
Libération (2009e) 'Sale printemps pour les couples mixtes', 23 May
Libération (2009f) 'Une manière d'adopter le stigmate par lequel on est discriminé', 19 June
Libération (2009g) 'Moins de 400 femmes porteraient le voile intégral en France', 30 July
Libération (2009h) 'Burqa, crise de maires', 1 October
Lyberaki, A. and Pelagidis, T. (2000) *The Fear of the Foreigner in the Labour Market* (in Greek), Athens: Polis
Mai, N. (2002) 'Myths and Moral Panics: Italian Identity and the Media Representation of Albanian Immigration' in Grillo, R. D. and Pratt, J. C. (eds) *The Politics of Recognising Difference: Multiculturalism Italian Style*, Aldershot: Ashgate
Maneri, M. (1997) 'Les médias dans le processus de construction sociale de la criminalité des immigrés. Le cas italien' in Palidda, S. (ed.) *Immigrant Delinquency*, Brussels: European Commission
Marie, C. -V. 'Emploi des étrangers sans titre, travail illégal, régularisations: des débats en trompe-l'œil' in Dewitte, P. (ed.) *Immigration et intégration*, Paris: La Découverte
Marshall, I. H. (ed.) (1997) *Minorities, Migrants and Crime*, London: Sage
Moschopoulou, A. (2005) *Criminality of Migrants* (in Greek), Athens: Sakkoulas
Palidda, S. (ed.) (1997) *Immigrant Delinquency*, Brussels: European Commission
Peralva, A. and Macé, E. (2002) *Médias et Violences Urbaines*, Paris: La Documentation française
Perotti, A. (1993) 'L'immigration en France – revue de presse', in *Migrations société*, No. 26, pp. 65–73

Poulton, E. (2005) 'English Media Representation of Football-related Disorder: "Brutal, Short-hand and Simplifying"?', in *Sport in Society*, Vol. 8, No. 1, pp. 25–47
Rabah, S. (1998) *L'islam dans le discours médiatique*, Beirut: Al-Bouraq
Rinaudo, C. (1995) 'Le procès de publicisation d'un « quartier à problèmes »', in *Migrations Société*, No. 42, 75–90
Schlesinger, P. and Tumber, H. (1994) *Reporting Crime*, Oxford: Clarendon Press
Stabile, C. (2001) 'Conspiracy or Consensus? Reconsidering the Moral Panic', in *Journal of Communication Inquiry*, Vol. 25, No. 3, pp. 258–78
Tévanian, P. and Tissot, S. (1998) *Mots à maux*, Paris: Dagorno
Triandafyllidou, A. (1999) 'Nation and Immigration: A Study of the Italian Press Discourse', in *Social Identities*, Vol. 1, pp. 65–88
Tsagarousianou, R. (1997) 'Mass Communications and Nationalism: The Politics of Belonging and Exclusion in Contemporary Greece', in *Res Publica*, Vol. 2, pp. 271–80
Tsoukala, A. (1999a) 'The Perception of the 'Other' and the Integration of Immigrants in Greece', in Geddes, A. and Favell, A. (eds) *The Politics of Belonging*, Aldershot: Ashgate
Tsoukala, A. (1999b) 'Le discours grec sur la criminalité des immigrés', in *Hommes & Migrations*, No. 1218, pp. 77–89
Tsoukala, A. (2002) 'Le traitement médiatique de la criminalité étrangère en Europe', in *Déviance et société*, Vol. 26, No. 1, pp. 61–82
Tsoukala, A. (2005) 'Looking at Immigrants as Enemies' in Bigo, D. and Guild, E. (eds) *Controlling Frontiers*, Aldershot: Ashgate
Tsoukala, A. (2007) 'The Terrorism-immigration Nexus in the Post-September 11th Era: An Analysis of the Political Discourses in the British Press' in Berggren, E., Likic-Brboric, B., Tokzös, G. and Trimikliniotis, N. (eds) *Irregular Migration, Informal Labour and Community in Europe*, Maastricht: Shaker
Tsoukala, A. (2008a) 'Defining Terrorism in the Post-September 11th Era', in Bigo, D. and Tsoukala, A. (eds) *Terror, Insecurity and Liberty*, London: Routledge
Tsoukala, A. (2008b) 'Boundary-creating Processes and the Social Construction of Threat', in *Alternatives. Global, Local, Political*, Vol. 33, No. 2, pp. 137–52
Tsoukala, A. (2009a) *Football Hooliganism in Europe. Security and Civil Liberties in the Balance*, Basingstoke: Palgrave Macmillan
Tsoukala, A. (2009b) 'Terrorist Threat, Freedom, and Politics in Europe' in Noxolo, P. and Huysmans, J. (eds) *Security and Insecurity: Community, Citizenship and the 'War on Terror'*, Basingstoke: Palgrave Macmillan
Vamvakas, V. (1997) 'Albanian Migrants in the Daily Press: Representing Clandestine Lives' (in Greek), in *Dokimes*, Vol. 5, pp. 7–40
Wacquant, L. (1999) 'Des 'ennemis commodes'', *Actes de la recherche en sciences sociales*, No. 129, pp. 63–7
Waiton, S. (2008) *The Politics of Antisocial Behaviour*, New York: Routledge

Welsh, M., Price, E. and Yankey, N. (2002) 'Moral Panic over Youth Violence: Wilding and the Manufacture of Menace in the Media', in *Youth and Society*, Vol. 34, No. 1, pp. 3–30

Whannel, G. (1979) 'Football, Crowd Behaviour and the Press', in *Media, Culture, and Society*, Vol. 1, pp. 327–42

Woodiwiss, M. and Hobbs, D. (2009) 'Organized Evil and the Atlantic Alliance', in *British Journal of Criminology*, Vol. 49, No. 1, pp. 106–28

Young, J. (1987) 'The Tasks Facing a Realist Criminology', in *Contemporary Crises*, Vol. 11, pp. 159–83

Young, J. (2007a) *The Vertigo of Late Modernity*, London: Sage

Young, J. (2007b) 'Slipping Away – Moral Panics Each Side of the "Golden Age"' in Downes, D., Rock, P., Chinkin, C. and Gearty, C. (eds) *Crime, Social Control and Human Rights*, Cullompton: Willan Publishing

Chapter 11
Visibilization of the Invisible: Gender, Islam, Insecurity[1]

Anna Krasteva

When combined, gender, Islam and radicalism produce an explosive mixture. The objective of this chapter is to analyse the political logic of the production of gendered insecurity by the French public discourse on the basis of this mixture, of an amalgam of Islam and radicalism on the one hand, and national identity and *laïcité*[2] on the other.

This chapter examines security through gender perspectives for two reasons. The first reason is the need to reinforce the gender aspect in security studies, this being often underestimated. Grant (1991) and Smyth (1995) have demonstrated that women have been left out of the notion of security because such an approach relies on the distinction between the public, a male domain, and the private, to which women are relegated. It does not deal with 'objective insecurity', with the economic and social problems of the worldwide trend of spiralling insecurity for the most vulnerable groups in society, namely the poor, women and children (Smyth, 1995). It deals with the (over)production of gendered figures of otherness by the political discourse which forges social perceptions and feelings of insecurity.

The second reason is the predilection of the political discourse for the female to be viewed as a symbolic resource for conceiving national identity.

> Women are not only the biological reproducers of an ethnic group, but also the 'cultural carriers' who have the key role in passing on the language and cultural symbols to the young. In nationalist discourses women serve as the symbolic embodiment of national unity and distinctiveness. They nurture and support the (male) warrior-citizen. Such symbolism legitimates the political inferiority of women: they embody the nation, while the man represents it politically and militarily (Lutz *et al.*, 1995, quoted by Castles and Miller, 2003: 38).

Logically therefore, radical otherness, contrariety, denial of the own ought also to belong to the feminine gender. The amalgam of gender, security, Islam, xenophobic and populist discourses is characteristic of a number of European countries. Just a

1 This chapter was written during the author's stay at the Institute of Advanced Studies in Nantes, France, January–June 2010.

2 The separation of the church and the state.

few cases of the rape of Italian women by immigrants trigger a wave of extremist comments and restrictive policies (a case considered by Campani in her chapter in this book). A different aspect of gender is shown by the case of 'super mother' of ten, Barbara Rosenkranz, an extreme right wing candidate in the Austrian presidential elections (April 26 2010). She uses this essentialist capital to condemn feminism, which she claims creates 'asexual human beings', to condone National Socialism, and to refer in one breath to 'Turks, Chechens, Asians, Gypsies and the Negroes'.[3] In Denmark the wish of a woman MP to wear a burqa in Parliament caused such public discontent that it led to the adoption of an amendment to law, so as to render impossible such an expression of radical religiousness at the very heart of the democratic system. Pioneering on the issue, Belgium is the first country to take legislative measures against veiling in public space.

The particular case to be analysed here is the French debate over the burqa. It is part of a wider debate on French national identity, which the government of Nicolas Sarkozy initiated in 2009 and which has continued to be actively conducted in 2010.

The chapter addresses the concept of a transition from national towards human security (Chen, 1995; King and Murray, 2001–2002):

> For too long, the concept of security has been shaped by the potential for conflict between states. For too long, security has been equated with the threats to a country's borders. For too long, nations have sought arms to protect their security. For most people today, a feeling of insecurity arises more from worries about daily life than from the dread of a cataclysmic world event. The world can never be at peace unless people have security in their daily lives. Future conflicts may often be within nations rather than between them. Human security is not a concern with weapons – it's a concern with human lives and dignity (UNDP, 1994).

This analysis shares a wider understanding of non-income dimensions of human well-being, of development as expanding people's choices in any relevant way, of Amartya Sen's theory of capabilities focused on expanding individuals' freedoms (Sen, 1999).

The threats to human security differ. Some are more visible and tangible, such as hunger, disease, crime and repression. Others are less tangible, but equally significant: 'sudden and hurtful disruptions in the pattern of our daily lives-whether in our homes, in our jobs, in our communities or in our environment' (UNDP, 1994). This chapter focuses on the second group.

The chapter is predicated on the assumption that the new security concept, centred on the individual rather than on the state, is attacked by new figures of insecurity, referring to the micro rather than to the macro level, to identities rather than to borders. During the period of adoption of the national security concept

---

3 In the *Fakten* magazine, edited by her husband.

the central actor of difference is the terrorist, the one who succeeds in crossing the borders and attacking the very core of the state – the Twin Towers in New York, the commuter trains in Madrid. This actor also has religious and gender characteristics – he is a man and a Muslim.

The new concept of human security offers two different types of rationale: absence of a threat and degrading conditions on the one hand, and the development of human abilities on the other hand. For this concept to be contested, actors are needed who threaten identity rather than life, and cultural and moral, rather than physical, integrity. It is this type of construction, veiled Muslim women in this particular case, in the light of which a more general problem of new threats to human security is to be analysed.

The number of veiled women in France is negligible. Yet the symbolic policy obviously deals with symbols rather than figures. The phenomenon is insignificant and marginal;[4] the political debate renders it central. On January 26 2010, a parliamentary committee[5] chaired by André Gerin published a public report on the practice of wearing a full veil on French territory. This voluminous report contains no less than 658 pages. The draft resolution, a motion by MPs, underscored that the full veil was incompatible with the secular values of the French Republic.

The debate on the burqa has produced insecurity for all parties: for the veiled women who have been overpublicized, stigmatized and marginalized; for French society which feels challenged by this radical public expression of a religious identity; for the political and nongovernmental actors, several of whom do not appreciate the instrumentalization of national identity for electoral reasons, yet are unable to avoid it. The debate will be analysed from four different perspectives: the way it is articulated by the political discourse and the new cleavages not only between the political families, but also, and more importantly, within the same political party; the reactions of the religious communities – Muslim, but also Catholic, Protestant and Jewish; the analytical and critical perspective of intellectuals; the perceptions and attitudes of public opinion; and the international perspective. This chapter will not reify the political perspective transforming the veiled women into an object of political controversy but will give them the floor to voice their motivation and to stand as actors. It is based on a discourse analysis of the French public debate (three newspapers with different political orientations – *Le Figaro*, *Le Monde*, *La Liberation*,[6] as well as weekly magazines such as the *Express*, *Le Point*, *Le Monde Magazine*), on the political analysis of political texts such as the parliamentary report, statements of the minister of immigration

4 The figures vary from a few hundred (*Le Monde Magazine*, No. 20, 30 January 2010) to few thousand (1500–3000).

5 Mission d'information parlementaire.

6 In her comparative analysis of the French and the Greek press, Anastassia Tsoukala in this volume concludes that cultural-threat-focused discourses on immigration exist only in France, that is, a country where such claims are also being put forward by mainstream political elites.

and other governmental officials, as well as interviews with lawyers, historians, sociologists and feminists conducted by the author.

Political reactions: the emergence of new cleavages

The political class has produced many and different voices. What unites them is that there is no political actor shunning the debate. Not that there are no parties or politicians who dislike the debate – in fact, the reverse is true, there is an array of them – but once the debate has been initiated and is resounding throughout the media and in Internet forums, they have been dragged into it.

The *Socialist Party* has reservations, finding the measures proposed in the report too opportunistic, difficult to implement, likely to prove inefficient, and most probably contravening constitutional principles. Marie-George Buffet, leader of the *French Communist Party*, voiced an even stronger stand: 'A law against wearing a burqa would even further stigmatize a community in which women would once again be the ones creating problems for the entire community.'[7] The *Green Party* is far more careful – Cecile Dufflot leaves the issue open: 'All this hurts me both as a human being and as a woman, as a feminist. Stigmatization, however, is not a solution either. Yet I am at a loss as to how to find a solution to this difficulty.'[8] The far-right *National Front* is unsurprisingly uncompromising. Though many accuse the government of too much restrictiveness, Marine Le Pen, the party's vice president, sends precisely the opposite challenge, accusing them of laxity.[9] The *Democratic Movement*, led by centrist Francois Bayrou, openly accuses circles within the orbit of President Sarkozy of using this debate for electoral purposes.

The various political parties produce different discourses, but the interesting thing is that the great differences are not between the parties, but within them. The parties themselves are torn by contradictory visions of solutions to the problem. The socialists are determinedly against the full veil, being at the same time resolutely split over the ways to counter it. The majority of their MPs boycotted the vote of the parliamentary report, yet two Socialist MPs supported the motions of the incumbents. The president of the parliamentary commission, a communist, supports an anti-burqa position, while his own parliamentary group is against any law which would ban it, because, they argue, 'it would be dramatic for national cohesion'.[10]

7 Gabizon, C. 'Ce que dit la mission parlementaire sur la burqa', *Le Figaro*, 26 January 2010.

8 'Le voile intégral, les clés du débat', *La Libération*, 18 June 2009.

9 Marine le Pen, 'La religion musulmane doit faire en sorte de ne pas choquer', *Le Monde*, 2 February 2010.

10 'Le vote sur le port du voile intégral divise la droite et surtout la gauche', *Le Monde*, 8 April 2010.

Disparate attitudes towards Islam are about to shatter the classic political cleavages. The differences between left and right are not relevant in the debate over radicalism: the right find themselves in the same camp as the communists; some of the socialists are against the government, while others support it. In order to visualize this new configuration, the *Le Point* magazine[11] lists on two opposite sides of a page respectively the champions of the anti-burqa law and those who have reservations. Thus Sarkozy finds himself in the same group as socialist Martine Aubry and far-right Jean-Marie Le Pen, while in the opposite end the leader of the parliamentary group of Sarkozy's party is grouped with left politicians.

What is most striking and astonishing is the effect produced by the debate on the far-left *New Anti-Capitalist Party* (NPA), led by Olivier Besancenot. For the regional elections in the spring of 2010, a young activist, Ilham Moussaid was the New Anti-Capitalist Party's candidate, although her low position on the party list made her unlikely to be elected. She comes from a typical background for the common people: she was born to a humble family with many children, who moved to France when she was three, she received French citizenship at the age of 18; at this age she started becoming involved in civic initiatives. In early 2009 she joined the NPA. She dresses in standard clothes, yet she wears a small white headscarf, wrapped around her neck. It is this small cloth that made her a media celebrity. She focused the public controversy. The first criticism of the promotion of a female candidate who is both far-left and veiled is *ideological*, relating to the inconsistency between the proclaimed left feminist and secular ideology, on the one hand, and, on the other, the symbolic weight of the headscarf as an expression of the inequality and subordinate situation of women in Islam, and a 'flag of traditionalist and reactionary Muslims'.[12] The second criticism categorically defines the *public nature of a private choice*. The public perceives the decision to wear a visible sign of one's faith as solely public, not as private. The third criticism is *political* and pertains to the instrumentalization of the candidate in the strategy of the party leader Besancenot: 'Sarkozy bans headscarves? Then, let's include a scarf-wearing candidate!' If the right leader waves the flag of laïcité, the left one is playing with (soft) religiousness. Some of the members and followers of this left party are very disappointed both with the public association with Islam, which does not comply with the secular nature of their ideology, and with the unacceptable concessions to both anti-feminism and religious radicalism. The ambiguous choice of a veiled left wing candidate not only provoked a lot of controversy, but had a visible negative electoral impact – women deserted the vote. Women accounted for only 35% of New Anti-Capitalist Party[13] support, while in previous elections there had been equilibrium between the male and female vote. This negative reaction by members and supporters of the NPA was one of the reasons for its failure at the 2010 local elections.

11 21 January 2010, No. 1949.

12 Fourest, C., 'NPA: nouveau parti antiféministe?', *Le Monde*, 6 February 2010.

13 Zappi, S., 'Le NPA ébranlé par son échec aux élections régionales', *Le Monde*, 3 April 2010.

The political debate brings to light several characteristics:

- There are no left or right positions on religious differences; the cleavages are greater within the parties than between them;
- The cleavages themselves are much more tactical than strategic or ideological, and refer to the ways of screening the public from extreme forms of religious extremism;
- The high pitch of the debate is in contradiction with the quite broad consensus on the secular nature of the French public sphere;
- The debate is initiated by the government rather than by the subjects of the politics of (non)recognition. It is not triggered by the veiled persons to open up the public to their religious difference, but by the political goals of the government. Initiated for electoral and tactical reasons, the debate fails to produce a beneficial effect on improvement of public policies for management of religious differences. The debate is not intended to do so, either. But it also fails to accomplish its Machiavellian purpose of bringing positive electoral results to the ruling party.

Public opinion: new proximities vs. old cleavages

French public opinion is seriously concerned about burqa-wearing, and the majority would vote for an anti-burqa law (57%). A mere 37% hold the opposite opinion.

Table 11.1 In relation to a law imposing a ban on burqa-wearing in France ...

... you fully or partially	Total	Socialist Party	Democratic Movement	Union for a Popular Movement	National Front
support	57%	42%	75%	69%	74%
reject	37%	52%	23%	29%	25%

Source: 'La bataille du voile integral commence', *Le Point*, No. 1949, 21 January 2010, p. 34

A nationwide opinion poll sought the political projections of (non)rejection of the full veil. The almost identical numbers in the far-right and the left are more than indicative: 75% against 25% were the votes respectively for supporting and rejecting the idea of a ban both among the supporters of the far-right National Front and the centre-left Democratic Movement.

A more 'classic' distinction between left and right positions is witnessed among the supporters of the Socialist Party and the UMP. Among the supporters of the SP, the preferences are the opposite of the national ones and even more distanced from those of the right: more oppose (52%) than accept (42%) the idea of an outlawing law. Predictably, Sarkozy's followers share the policy of their party: 69% would vote for the law, but almost one third (29%) hold the opposite opinion.

This analysis shows that, similar to the political parties, public opinion on the burqa and Islam is marked by new proximities between far-right and centre-left on one side, and is divided by classic cleavages between left and right opinions on the other.

Religious reactions: withdrawal and scepticism

The pillar of French identity is *laïcité,* a separation of state and religion, with the latter considered to belong to the private rather than to the public domain.

As for the burqa, different denominations assume different positions. The position of the *Jews* is to have no public position at all. The Grand Rabbi of France believes that it is outside his jurisdiction to take a stand on burqa-wearing as well as on the necessity of a law, as both are concerned with problems of Islam. Catholics express regrets at the fact that the parliamentary committee has failed to consult them. The *Catholic Church* also does not formulate a public stand, but the media voice its position.[14] It is critical both of burqa-wearing, as the latter veils the faces and expressions of women, and to passing a law for three reasons: such a law would treat the symptom rather than the cause; it would play into the hands of the radicals, weakening the position of moderate Islam; the necessity of identification and security does not need such a special law. A similar position is held by the *French Protestant Federation.*[15]

The parliamentary committee has not wanted to take into consideration the interreligious perspective, but has not omitted to ask Muslims about a key issue of the public use of their religion. The *French Council for the Muslim Religion* (FCMR) has a mission to promote moderate Islam. It is this mission which explains the stand of the Council, which does not support the burqa and proclaims itself to be against the law. The Council advances two arguments pertaining to burqa-wearing, a sociological and a theological one. Burqa is a marginal phenomenon and wearing it is not provisioned as a duty by Islam. In relation to the law, their concerns are aroused by fears of a discriminatory effect both on the women wearing burqa and the entire Muslim community: these women, holding themselves aloof, will do so even more; the Muslim community itself is being stigmatized as creating radical trends, which, in its turn, fosters Islamophobia. These arguments lead the FCMR

14 Guenois, J.-M. 'Burqa: les religions ne voient pas l'efficacité d'une loi', *Le Figaro,* 20 January 2010.

15 Ibid.

to a position with a dual message: 'If we are to fight radical practices, we must also fight Islamophobia.'[16]

Intellectual reactions: more human rights than security concerns

In a debate which stirs up, and is intended to stir up, strong political passions, the voice of intellectuals brings critical reflexion and analytical arguments. Lawyers are at the centre of the debate, a number of others enrich the debate with a variety of disciplinary perspectives – psychoanalysts, philosophers, social scientists, feminists.

Lawyers measure the political by the lawful. Most of them agree that there are no legal grounds for a concrete and direct burqa ban. Guy Carcassonne, professor of public law, considers the possible arguments for outlawing the full veil and describes them as unsubstantiated:

> Laïcité cannot be considered as an argument as it concerns the Republic, not individuals. Nor can violation, committed against women, for some of them make this choice freely and intentionally. Neither the principle of human dignity, as it is intended to prevent inhuman and degrading actions, rather than to allow the lawmakers to outlaw everything they dislike, describing it as an affront to dignity, for then dignity would become an enemy of freedom.[17]

Denis de Bechillon, another professor of law, largely concurs, adducing the following argument:

> I do not imagine how the state could be conferred the legitimacy to indicate to the body's owner which uses of the body are dignified. If you ask someone to cut your finger, the one who did it would not be able to cite your request and would be brought to trial. If someone cuts their own finger, there will be no legal consequences.[18]

The arguments of academics interviewed by the author could be summarized in four groups: feminist, historical, philosophical and educational. The first one is clearly formulated by a lawyer who specializes in the human rights of children. She explains how the veil contradicts the aims and the achievements of women's movements from the 60s onwards: the value of equality and empowerment of

16 Declaration of Mohamed Moussaoui, president of FCMR, during the meeting with the parliamentary committee. Le Bars, S., 'Le combat contre la burqa s'accompagne de gages à la communaute musulmane', *Le Monde*, 26 January 2010.

17 Carcassonne, G., 'Oui, on peut l'interdire', *Le Point*, No. 1949, 21 January 2010, p. 36.

18 'Interdiction du voile intégral: l'avis des juristes', *Le Figaro*, 29 January 2010.

women is negated by a symbol of differentiation and subjugation. The second argument is connected with *laicite*, with the secular values of French feminism.

All the other arguments are rather critical of the political debate on burqa. An historian underscores that, over the last century, each time the debate on immigration has intensified, it has not been connected with its quantitative growth, but rather with the wish of the government to make up for the want of a mobilizing project through exclusion policies.

The philosophical argument expresses the critical vision of most of the academics, that the very term 'national identity' introduces an essentialist and exclusionary vision, liable to seek radical otherness in order to define itself in relation to it. This argument is voiced more strongly by French academics of foreign origin.

A more pragmatic and promising perspective is formulated by a lawyer, deputy director of a research institute, who stresses that the question of veiled women has to be posed in radically different terms: of education, so that informed and responsible citizens are formed, capable of making decisions both in their own and the public interest.

The external perspective of Tunisian writer Fawzia Zouari, author of the book *The Veil Which Tears France*, underscores an aspect unnoticed by any of the French commentators. Some of the veiled women, newly converted from Christianity to Islam, still keep their Christian reflections, willing 'to reach the end of Christ's road, in order to gain a better insight into themselves than the others. But they err: Islam is not a religion of self-torture, but of life.'[19]

The political discourse mixes gender, Islam and radicalism to forge the message that national identity is under threat and that insecurity is rising. The intellectual discourse distinguishes between these components, showing that they are not analytically connected to make the phenomenon intelligible, but are instrumentally assembled to blur distinctions and create security scares.[20]

International perspective: national policies rather than a European model

The international perspective is present in the French debate in two different ways: as a response to French politics by popular foreign media and as French interest in

19 Golliau, C., 'Ce que dit le Coran', *Le Point*, No. 1949, 21 January 2010, p. 44.

20 Very indicative of the constructed relation to security are the opinions of security experts. Police experts introduce the subject of the applicability of a possible law outlawing the burqa. They are unanimous in their negative answer. Arguments are adduced of a different scope, I will highlight two here. The most important is that it would distort their activity, making it devoid of its major focus – defence of security, fight against criminality. The second one is that it would alter the geography of their activity: instead of concentrating on dangerous and problematic regions, they would have to concentrate around sensitive and symbolically charged places such as mosques and schools ('Et si cette loi était impossible a appliquer?', *Le Point*, No. 1949, 21 January 2010, p. 36).

the politics of other countries relating to the full veil. The international comparison demonstrates that the French public debate is stronger, more passionate and lasting, but the issue of the burqa is a political priority for other European countries also. Belgium has been much faster and efficient in banning the veil.

Most audible among the French public is the view of the *New York Times*.[21] In an editorial, the headline of which sets out both the tone and the attitude, '*The Taliban Would Applaud*', the popular US daily says:

> It is easy to see that a woman's human rights are violated when a government requires her to wrap her body and face in an all-concealing veil, as the Taliban used to do when it ran Afghanistan. It should be just as easy to see the violation when a French parliamentary panel recommends barring women who wear such veils – the burqa and the niqab – from using public services, including schools, hospitals and public transportation.[22]

Then a moralizing lesson in liberalism follows:

> People must be free to make these decisions for themselves, not have them imposed by governments or enforced by the police.[23]

Sarkozy repeatedly explains that 'the problem of the burqa is not a religious problem, it's a problem of liberty and women's dignity. It is not a religious symbol, but a sign of subservience and debasement.'[24] Whilst the President of France argues using the high rhetoric of human rights, the *New York Times* accuses him of the open political instrumentalization of a religious subject:

> With regional elections scheduled for March, Mr. Sarkozy and his allies are desperately looking for ways to deflect public anger over high unemployment. It is hard to produce jobs and far too easy to fan anti-Muslim prejudices.[25]

Paradoxically, the two opposing positions, the French and the American, proceed from an identical logic, i.e. the full denaturalization of religion, depriving it of all its spiritual, social and cultural dimensions and reducing it to 'pure' politics.

The second way in which the international perspective is present is through the policies and experience of other European countries in regulating the full-veil

21 'Burqa: le New York Times accuse Sarkozy d'attiser la haine', *Le Monde*, 27 January 2010.

22 'The talibans would applaud', Editorial, *New York Times*, 26 January 2010.

23 'The talibans would applaud', Editorial, *New York Times*, 26 January 2010.

24 'Sarkozy Backs Drive to Eliminate the Burqa', *New York Times*, 22 June 2009.

25 'The talibans would applaud', Editorial, *New York Times*, 26 January 2010.

issue in the public domain.[26] These vary widely, from totally ignoring the issue on the part of Germany, because it is marginal, to the ban on veiled faces in public space in Italy. Belgium has also banned the veil in public, and is the first European country to regulate the issue by law.[27]

The debate in Denmark is conducted in the context of the passions sparked by the cartoons of Mohammed, which explains their heatedness and intensity. It is provocation that proves to be the pillar of public controversy. In the same way as many Muslims across the world refuse to accept that the cartoons of Mohammed are an expression of freedom of speech, equally a number of Danes fail to understand the position of a woman MP of Palestinian extraction who wishes to sit in Parliament in a niqab, which is not explicitly forbidden. Not every political act which is allowed is legitimate and accepted. A wave of discontent both on the part of politicians and of public opinion shows undeniably that veiling on the parliamentary rostrum is perceived as an unacceptable and inadmissible political provocation. The responses have been so numerous and strong that the government has taken steps to eliminate this option. By a recently adopted (28 January 2010) decree, the government gave administrations, schools and even privately run companies the opportunity to impose a ban on veiling.[28]

An analysis of the (non)openness of the public domain explicitly shows that there is no 'European' model, but a mixed mosaic of practices. Several conclusions summarize the representation of this multicoloured picture:

- A number of countries avoid passing special laws and giving institutions and local authorities full leeway in the matter, while Belgium passed[29] the first law in Europe which bans veiling in public space;
- Public policies relating to ostentatious religious symbols such as the niqab and the burqa range widely from full acceptance to lesser or greater degrees of restriction;
- The changes in public policies result from political, not from quantitative factors, not when the number of veiled women soars, but when certain action is perceived by the society as provocative, a good example of which is the will of the Danish MP to perform her parliamentary duties veiled.

26 'Et chez nos voisins?', *Le Point*, No. 1949, 21 January 2010, p. 37; Fournier A., 'Débat sur le voile intégral: ou en sont nos voisins?', *Le Monde*, 28 January 2010.

27 Voted on April 29 2010.

28 Fournier, A. 'Débat sur le voile intégral: ou en sont nos voisins?', *Le Monde*, 28 January 2010.

29 April 29 2010.

The reactions of the subjects of reactions: the Veiled Self

So many reactions, positions, interpretations inevitably reduce the burqa-wearing women to objects of policies, discourses and attitudes. The niqab and the burqa are the visible symbol of women's subjection. The gender aspect is crucial to the criticisms of the burqa, unlike other religious symbols such as the Christian cross. If the latter involves universal, gender-neutral practice, the burqa is a female item. Treating public visibility of the female body in a radically different way than that of the male allows gender inequality.

The topic of subjection is even more fundamental, as it pertains to the very nature of the social actor and concerns the key issue of the authorship of the decision. Who makes the decision: the women or others? Regardless of the fact that these others are personalized as the husband or the parents, or depersonalized as tradition, the critics of the burqa emphasize that the source of veiled women's behaviour is external – not the women themselves, but the family, the community, the tradition.

Objectification and domination – these are the two focuses of the view from outside, which turn the veiled women into an object of discourses and public policies, with the latter being argued and legitimized on the basis of the inequality and subjection of women. Analysing the views of the burqa-wearing women themselves reverses the perspective: the women become the subject of their own discourse, authors of their own actions, and several present these actions as freedom and self-development rather than as coercion, subjection and conformism.

The key to this analysis of the speech of the women themselves is the turning over of the focus: from objects of a discourse they become the subject of their own activity, responsible for their positions, practices, achievements and exaggerations.

The essential question is: who is the social actor, who stands behind the veil? Who takes the decision? On some women the veil is imposed. At a time when the parliamentary report on the burqa is being actively discussed, France's immigration minister Eric Besson has made public his decision to refuse to grant citizenship to a Moroccan who makes his French wife wear a burqa. In this as well as in many similar cases, women are objects of domination. In other cases the women are the ones who decide to wear the veil. This latter case is the most interesting as it runs counter to other public mindsets.

A conditional typology would distinguish three groups of veiled women. The first comprises comparatively young women. This is their way to show that they are zealous Muslims. This is a cultural 'omission' in the new family environment (such is the case with the Vietnamese woman married to a Maghrebian husband) or a discrete aspiration for the type of husband who would be expected and appreciated.

Some of these Muslim women are flexible in building a relationship between religious difference and professional career, as is the case with Hayette[30], who is about to receive a Master's degree in banking law and opts for a 'French' version of the veil, wearing it as a turban as an explicit strategy to avoid confrontation: 'I agreed to this compromise between my beliefs and French society.'[31]

The second group comprises women of more mature years, who often seek relief and support in religion from disappointments in life, existential problems, great love or great drama, loss of physical or affective stability: Sihab has lost her job, being a single mother of two daughters; Nijat has had a difficult childhood, she is a divorced mother of a 17-year-old daughter; after her parents divorced, Nadia admits she does not feel well, she has left school and has started drinking and smoking pot; Oum Aymen has severely suffered from a marital infidelity in her family, while her boyfriend got involved with criminal gangs until Islam 'soothed' them all.[32]

Islam performs different identity functions. To some, religion makes up for a social, professional or family deficiency. It is the last that restructures existence. If before the change it was filled with activity, now it is filled with spirituality. Salafism is strongly restrictive with regard to presence in the world, it requires withdrawal, restricts freedom, but this is not experienced as a want or a vacuum, rather as a fulfilled completeness, as a life 'each moment of which is devoted to Allah'.[33] On others, it has a psychotherapeutic effect:

> Before the meeting with the religion I was lost, I sought myself. I had problems with aggression, rebelled against everything. Islam soothed me, cultivated and built me, I learned how to love again, to approach the other and respect the other, however different he or she might be. [34]

The third group consists of older women who have devoted themselves to religion. They dedicatedly hand it down to other young Muslim women. One of them gathers girls and young women at her home on Sundays and teaches them the Koran. Others give advice and reinforce belonging to the religious community. They play a very important role, and the new converts often mention the '*sisters*' in their stories, one of them telling precisely how they took her from the pub into the mosque.

Most of the burqa-wearing women are the first veiled generation. They are not bearers of a tradition carried on from their mothers, rather they are at the beginning of a tradition in which they involve their children. Each beginning means severing,

30 The names quoted in the article are the ones published in the analysed corpus of media publications.

31 Le Bars, S., 'Petits arrangements avec le voile', *Le Monde*, 17 February 2010.

32 *Le Point*, No. 1949, 21 October 2009, p. 41.

33 'Le Nikaq: un autre regard', *Le Monde Magazine*, N 20, 30 January 2010, p. 18.

34 Najat, 37 years old, *Le Point*, N 1949, 21 October 2009, p. 40.

crossing a border, a change or a move. To some, the move is from moderate to more radical Islam, to others it is from Christianity to Islam. A number of them are not born Muslim, they have converted. Oum Aymen was born Julie and baptized, she converted to Islam at the age of 19.[35] Sandra converted at around the same age. Unlike most of the converts she keeps her French name.[36] Caroline is of Cameroonian origin, born in the French province of Picardy. All her family are atheists, while she has shown an interest in religion since childhood. Her parents were sure she would become a Buddhist. She converted to Islam.[37] The ethnic diversity of the converts is impressing – Hawwa Hue Trinh Nguyen is a French woman of Vietnamese extraction, a convert in order to 'feel secure as regards the family of my husband and men.' [38]

Marginalization and insecurity

The situation is felt as disturbing by both sides: by the state, which sees its fundamental republican values betrayed; by the veiled Muslim women, who feel themselves betrayed by the French state for two different reasons. The first one is the contradiction between the ideology of freedom of choice and denomination and the restrictive practice:

> I feel betrayed by the France I love. I grew up to be a free woman thanks to France and now France wants to shut me up for not approving of my choices.[39]

The second reason is because the state closes its public sphere even further to the veiled Muslim women and inevitably marginalizes them. Long before the passing of the legislative restrictions, Siham almost never left her neighbourhood. In her multicultural suburb, there are many Muslims, she can buy halal meat, and people do not treat her as a broken rule. She feels a strained attitude when she leaves her neighbourhood, blaming it predominantly on the politicians, as since they initiated the debate on national identity, immigration and Islam, people have become less tolerant:

> When I have to go away from the neighbourhood, I feel distressed. They look at me as though I am a monster. At times they insult me or are rude to me … Since the politicians started this dispute, people have become less and less tolerant.[40]

35 Oum Aymen, 30 years old, *Le Point*, No. 1949, 21 October 2009, p. 41.
36 Sandra, 23 years old, *Le Point*, No. 1949, 21 October 2009, p. 40.
37 Caroline, 26 years old, *Le Point*, No. 1949, 21 October 2009, p. 43.
38 Le Bars, S., 'Petits arrangements avec le voile', *Le Monde*, 17 February 2010.
39 Siham, 50 years old, *Le Point*, No. 1949, 21 October 2009, p. 38–9.
40 Siham, 50 years old, *Le Point*, No. 1949, 21 October 2009, p. 39.

Chaima makes an equally negative evaluation of the debate on national identity:

> Since the full-veil debate has evolved into a nationwide issue, I rarely leave my home, I am afraid.[41]

The third reason is distrust in the public institutions and discourses. The veiled women do not accept the image of themselves which society builds:

> No, I do not want to talk to the media, they manipulate us very much, our words are always distorted. We do not find ourselves in the portraits they make of the women wearing the niqab.[42]

The privatization of existence, withdrawal from the public into the private domain is the social price paid for radicalization:

> There are no firms willing to hire a niqab-wearing woman. So much the better, I do not wish to work for any company.[43]

So, Umm Aymene, despite her language competency and her fluent Arabic and English, works at home as a tailoress.

The privatization of religious life is one strategy. Another is the building of publicity, with which to shelter with dignity the faith they want to practise and show off. New convert Sandra formulates the new project:

> We are a new generation of young Muslims, who will build our own hospitals, schools, businesses to enable Islam to live adequately.[44]

The Freudian concept of 'narcissism of minor differences'[45] is helpful in understanding the strong focus on veiled women in French public debates: a number of the veiled women are not unemancipated immigrants; on the contrary, they are French women converts. The image of radical foreignness is not produced by someone radically foreign, but by people close to us who have turned into strangers.

---

41 Caroline, 26 years old, *Le Point*, No. 1949, 21 October 2009, p. 43.

42 De Feo, A., 'Le Nikaq: un autre regard', *Le Monde Magazine*, No. 20, 30 January 2010, p. 17.

43 Oum Aymen, 30 years old, *Le Point*, No. 1949, 21 October 2009, p. 41.

44 Sandra, 23 years old, *Le Point*, No. 1949, 21 October 2009, p. 42.

45 Freud in *Civilization and its discontents* coined the phrase to emphasize that often our hatred, fear, and contempt are directed at people who resemble us, while our pride is attached to the small markers that distinguish us from them ('An Interview With Freud Biographer Peter D. Kramer', accessed at: http://calitreview.com/72).

Invisible women, visible threat

The visibilization of the invisible takes place on different levels. Veiled women protect themselves from being seen, only to be put in the limelight. They withdraw from the public sphere, only to be made central to it by the political discourse. They are invisible, their number is negligible,[46] whereas the public debate makes a central problem of them.

The sphere in which the public debate is conducted is narrow and problematic. On the one hand is the right to religious freedom, on the other, the will to counteract the public expression of extreme forms of religious beliefs. More important are two other questions, namely, which is the religion and who are the bearers of radicalism? The point is not just in a religious extremity, but in a religious extremity within a certain religion – Islam. It is symptomatic which figure is chosen to denote its radical trends – the female.

This debate is symptomatic of a transition from national to human security. The figure of radical otherness in the first case is the terrorist. Borders are guarded with weapons; the terrorist turns himself into a weapon. This figure has clear gender and religious characteristics: a male and Muslim.

The new stage in the focus on human security needs a new figure of otherness and one such has been immediately constructed: the veiled woman. She also has clear gender and religious characteristics: a female and Muslim.

Three decades ago Margaret Thatcher remarkably formulated shortly and to the point the post-Cold War doctrine: '*Islamism is the new bolshevism*'. In this new political strategy, the threat to security is defined not any more ideologically, but culturally and religiously.

Religious radicalism is inevitably seen as Muslim. Unlike a first initial stage, when the state was at the centre of the security concept and the male figure of the terrorist was the one to pose a threat, now individual security is in the centre and a female figure is destined to be the right one to attack the private space.

This chapter has considered the particular case of how French political discourse constructs as significant and scary the figure of a religious practice, which is marginal and almost not represented in society. The conclusion has a more general validity and pertains to the very specifics of the securitarian paradigm. At this new stage it shows two contrary trends. On the one hand, it shows a positive trend towards a transition from the state to the individual, as well as towards a broader notion of human security as that of a long-term lack of threats and development of abilities. On the other hand, it shows a negative trend towards increasing securitization as a ground for a new form of governing. Maguire in his chapter in this book examines one of its expressions, the major developments in biometric security over the past decade, the transition from 'smart' borders to the securitization of access to spaces.

---

46 One of the experts I interviewed, a 45 year old French citizen, stresses he has never met a veiled woman.

The present chapter has analysed another trend, the construction of new figures to be presented as a threat on the micro level of identities. The mechanism is complex. It starts by identifying a real problem. In this case study it is that of the veiled women. Then the problem is overemphasized and overpublicized. French society is largely consensual in its support for the secular values of *laïcité*. In contrast with this overwhelming consensus, the debate has been going on for months and has been elevated to a top priority by the political discourse. Belgium proceeded in an opposite way – without any major public debate, a radical legal ban has been introduced. The Belgian case serves to illustrate that what is at stake is not a difficulty in regulating a marginal problem which concerns 1000–2000 citizens. What is at stake is the need to create figures of otherness, which challenge the republican political culture and threaten the collective identity. This new threat, reified by the political discourse, is a favourable ground for a further securitization of policy. Eventually, the ultimate end, that of human security, is not strengthened, rather a very securitarian vision and policy is reproduced and enhanced.

References

Castles, S. and Miller, M. (2003) *The Age of Migration: International Population Movements in the Modern World*, New York: Palgrave Macmillan

Chen, L. C. (1995) 'Human Security: Concepts and Approaches' in Matsumae, T. and Chen, L. C. (eds) *Common Security in Asia: New Concepts of Human Security*, Tokyo: Tokai University Press

Grant, R. (1991) 'The Sources of Gender in International Relations Theory' in Grant, R. and Newland, K. (eds) *Gender and International Relations*, London: Routledge

King, G. and Murray, C. (2001–2002) 'Rethinking Human Security', in *Political Science Quarterly*, Vol. 115, No. 4, pp. 585–610

Sen, A. K. (1999) *Development as Freedom*, New York: Knopf

Smyth, I. (1995) 'Gender, Change and Insecurity', in *Gender and Development*, Vol. 3, No. 3: 6–15

UNDP (1994) *Human Development Report: New Dimensions of Human Security*, accessed at: http://hdr.undp.org/en/reports/global/hdr1994/

Chapter 12
Convergence of Securitization of Migration and 'New Racism' in Europe: Rise of Culturalism and Disappearance of Politics

Burcu Toğral

In the last decades, we have witnessed a heated academic and political debate over the emergence of so-called 'Borderless Europe'. The debate came on the scene with the introduction of the Single Market and Schengen Area which envisaged the free movement of people within the European Union (EU). Undeniably, today nationals of EU member states and of some privileged countries belonging to the 'developed west' enjoy this freedom without much of an interruption and without being imprisoned within any kind of borders (either territorial or non-territorial). On the other hand, we have experienced the strengthening of existing borders and the creation of new ones for *certain groups of people*, especially for those immigrants coming from poor and Muslim countries, asylum seekers and undocumented immigrants. The borders have taken the form of restrictive and discriminatory migration polices. Even if they were born in, or have been living for many decades in, one of the member states, there are still other borders for them in the form of racialized and securitarian discourses and policies.

These new forms of borders, operating within and outside the EU, have rendered certain economic, social, ethnic and religious differences, by definition, as decisive dividing lines between 'them' and 'Europeans' and within this rebordering process, the mobility and presence of these groups have been easily fabricated into a security issue.

This securitization process, namely constructing or framing the issue of migration as a security threat, has been strengthened in the wake of September 11. The 'war on terror' has not only linked migration to terrorism, it has also consolidated the place of migration as a threat to cultural identity. Particularly, the 'cultural differences' of certain group of migrants (mainly Muslims and Arabs) have been reified as a striking destabilizing factor, a security threat to national identities, so-called cultural homogeneity of the member states, and European 'values'. In this sense, this securitization of migration through appeal to culture bears the hallmarks of the so-called 'new racism', or, as Ibrahim contends eloquently, it has emerged as 'racism's most modern form' (Ibrahim, 2005); a racism which capitalizes on 'cultural differences' rather than on biological ones to discriminate and subordinate certain group of people. It is new in the sense that it has transcended classical definitions and no longer operates by reference to superiority/inferiority dichotomies or through physical elimination of racialized populations. Today's racism is much more hidden, 'respectable', it has been built upon more 'innocent' configurations, such as preservation of one's identity, own way of life and values in the face of the 'destabilizing' and 'damaging' effects of other cultures.

This convergence between the securitization of migration and new racism should not be seen just as some kind of linguistic practice, 'speech act'[1] or as a matter of representation of the issue through the utilization of security language or racially coded messages. Of course, language has a performative power in the sense that it shapes our perceptions of the world. But this approach does not explain the complex relationship between the performativity of language, power and institutional structure. As Foucault puts it, discourses, which come into being through the exercise of power, create a specific economy of 'truth' or 'knowledge' (Foucault, 1980: 93). They authorize certain types of individuals to speak the truth, thus making them 'experts', speakers of the dominant discourse holding power over what can be legitimately said (Foucault, 1972). Hence, their position and power are not only structured through the discourses; they are also engaging in the reproduction and perpetuation of these discourses.

Yet, the story does not end here. These discourses of truth structure and inform policies and laws, which in turn reproduce and consolidate the dominant discourses. So the exploration of today's securitization and new racist discourses necessitates also the analysis of policies and laws through which discourses become materialized. Last but not least, the analysis of discourses can not be detached from the historical and social context in which they emerge. Discourses regulating a certain issue may change over time and they are reshaped or challenged by other discourses in specific historical and social contexts. These insights shed light on how discourses on migration and racism have shifted over time, changed their forms and created new truths or *regimes of justification*,[2] and how these discourses

1 The conceptual tool of securitization was first brought into the agenda of security studies by the so-called Copenhagen School of Security Studies. More specifically, their approach signifies that securitization refers to a 'speech act' through which 'the issue is presented as an existential threat, requiring emergency measures and justifying actions outside the normal bounds of political procedure' (Buzan *et al.*, 1998: 23–4; See also Wæver, 1995, 2000). Even though their theory of securitization has provided important revenue for research, the exclusive focus on linguistic practices, without paying much attention to the role of context, power and the institutionalization of securitization, limits the sociological applicability of their approach.

2 I borrow the term 'regimes of justification' from the works of Boltanski and Thevenot (1999). Their approach concerns how the actors justify their stance at 'critical moments', namely when a dispute emerges among actors 'who are doing things together … and who have to coordinate their actions … realize that something is going wrong' (Boltanski and Thevenot, 1999: 359). Most importantly, they point to the plurality of regimes of justification (ibid.: 360). These regimes of justification, each with their 'own criteria of validity and

inform policies in different ways in line with historical and social changes. They are also important in exploring how securitization of migration has emerged as a dominant discourse alongside the marginalized, subjugated, but at the same time, still challenging desecuritization discourses.

On the basis of these concerns, drawing much on the Foucaultian approach of discourse, the main aim of this chapter is to discover how the securitization of migration and 'new racism' have been converged under the 'auspice' of the cultural difference-as-a-threat narrative. To achieve its purpose, this chapter first traces what 'new racism' stands for in contemporary debate. The objective here is to briefly give some theoretical and conceptual explanations of the shift in the racist discourse. This will reveal how securitization of migration has internalized the 'new racist' discourses through the same appeal to the cultural-difference-as a threat narrative. To make the issue more tangible, it shall focus on how the discussions turning around the issue of integration have embedded these securitization discourses, particularly the culturalist ones, and the 'new racism' throughout Europe, particularly in the post-September 11 era. Also, the products of these discourses on integration policies across Europe shall be scrutinized.

Here, from the outset, some points should be clarified. The discourse on integration should not be taken for granted and included in the analysis without critical reflection. Integration discourse is not as sinless as the general public thinks or as politicians assert, because the emphasis on the necessity to integrate migrants may 'directly or indirectly confirm a nationalist desire for a culturally homogeneous society and identifying immigrants as the obstacle to a successful realization of this desire' (Blommaert and Verschueren, 1992, 1998 cited in Huysmans, 2000: 765). This is a crucial point. As Huysmans observed, integration attempts assume that previously there was a homogenous society, which was disrupted by the arrival of migrants (ibid.). Indeed, in certain cases the underlying logic of integration is the same as that of exclusionary policies. Both see migrants' 'difference' as a threat to the stability of the 'host society', and call for the elimination or disciplining of these differences. Likewise, as detailed later, multiculturalism, represented for a long time as a favorable strategy to deal with these differences, also needs careful reflection. The real issues can be veiled, as culture or cultural difference is framed either as the source of the 'migrant problem' or as the remedy to solve the same problem. Keeping in mind these critical points, finally, this analysis will draw attention to how these discursive practices and policies suffocate politics by means of culturalism and ethnicism, and bring identity concerns and cultural differences

---

internal consistency', provide a framework for the 'situated actors to engage in disputes with others on the common good' (Diken, 2002: 2). Diken, through drawing on the work of Boltanski and Thevenot, shows how the justifications and critiques regarding the issue of migration operate in the framework of different regimes and provide different 'truths' and 'realities' on the issue. For example, disputes over the economic utility of migration operate in a different regime with different frames of justifications and rationalization than discussions on migrants' effects on the hosting societies' cultural identity.

to the fore both in the political debate and in the course of anti-racist struggles. Thereby, it will demonstrate how real social and economic problems associated with structural factors have been depoliticized and hidden.

What is new in 'new racism'?

'New Racism' as baptized by Barker (Barker, 1981), defined also as 'neo-racism' (Doty, 2003, Spears, 1999, Balibar, 1991), 'differentiated racism' (Taguieff, 1990), 'Euro-racism' (Sivanandan, 1990; Webber, 1991), 'xeno-racism' (Sivanandan, 2001; Fekete, 2001) 'cultural racism' (Fanon, 1967; Gilroy, 1987) and lately, in the context of the 'war on terror', 'anti-Muslim racism' (Kundnani, 2007) does not mean the banishment of older forms of racist practices; rather 'new racism' implies that current racist discourses are being dominated by the so-called 'insurmountability' of cultural differences. These discourses work as a pretext or an excuse to exclude, discriminate, subordinate and classify certain group of people, unlike the old or classical racism, or so-called 'scientific racism' that used biological discourses for the same aims. Even though there is no reference to 'race' in the classical sense, 'new racism', as Spears puts it, 'is still racism in that it functions to maintain racial hierarchies of oppression' (Spears, 1999) and as Balibar posits, despite the changes in the 'doctrines' or 'discourses' of justification, it still sustains the 'same old acts' (Balibar, 1991: 18).

However, in spite of the unchanging and enduring purposes of racism, it is still possible to define today's racist discourses as novel by referring to certain points. First of all, migrants constitute the most apparent target of the discourses of 'new racism', unlike those of previous periods which were exclusively directed towards blacks and Jews. As Balibar (1991: 222) contends, this 'new racism' is 'racism without race' and:

> Immigration has become, par excellence, the name of race, a new name, but one that is functionally equivalent to the old appellation, just as the term 'immigrant' is the chief characteristic which enables individuals to be classified in a racist typology.

Likewise, as Sivanandan (2001: 87) argues, today's racism is not only a problem of color, rather

> It is a racism that is not just directed at those with darker skins, from the former colonial territories, but at the newer categories of the displaced, the dispossessed and the uprooted, who are beating at Western Europe's doors, the Europe that helped to displace them in the first place. It is a racism, that is, that cannot be colour-coded, directed as it is at poor whites as well, and is therefore passed off as xenophobia, a `natural' fear of strangers. But in the way it denigrates and reifies people before segregating and/or expelling them, it is a xenophobia that bears all

> the marks of the old racism. It is racism in substance, but `xeno' in form. It is a racism that is meted out to impoverished strangers even if they are white.

Second, it is argued that 'cultural relativism', which was initially proposed by UNESCO in the aftermath of the Second World War in the fight against 'old racism', and which put emphasis on cultural differences to define difference between human groups, has paradoxically contributed to the justification of the 'new racist' discourses[3]. As noted by Taguieff (1990), in the framework of 'differentiated racism', this is related to one of the underlying arguments of the 'new racism', that even though there is reference to the equality of cultures, 'different cultures' should remain separate for the preservation of cultural identities. Yet, neither the lack of reference to 'race' in the 'new racism' nor the emphasis on culture render it less dangerous; on the contrary, it facilitates justification of racist practices and discourses. Perpetuators of these discourses can claim innocently that they are not racist, but concerned only with the preservation of each cultural identity and particular 'way of life', without reference to 'colour or phenotype' (Solomos and Back, 1996: 27). Given this 'innocence', these culturalist discourses have placed far too much power in the hands of racists, as they offer a 'respectable racialized world view; a popular, politically legitimate means of social differentiation and exclusion that works through the immediacy of its truth claims and easy communication of its core ideas'[4] (Duffield, 2006: 71).

Furthermore, as some scholars have explored, this 'innocent' appeal to culture has proceeded alongside the rise in emphasis on national identity; an emphasis which paves the way for a 'post-modern racism' or 'corporate national populism' constructing itself as against both racism and multiculturalism (Žižek, 1993; Goldberg, 1994). Here, essentialism is the driving force – that is cultural difference and national identity are reified and objectified as if they were unified and homogenous entities. These arguments authorize the belief that people who are part of the same culture or nation are devoid of differences of any kind; so different cultures can disturb this 'unity' or 'homogeneity' of the nation and society. This essentialism and homogenization has made it possible for conservative and racist groups to build their arguments around the defence of 'our values' or 'our way of life'. As illustrated by Campani's focus on the Italian case and Krasteva's on the French case in this book, these essentialist projections onto migrants have strengthened the hand of populist rhetoric and racist framing of certain group of immigrants. Especially through looking at the security-migration nexus from a gender perspective, they show how essentialist, populist and gendered arguments, coupled with 'national identity' discourses, have reinforced the criminalization, exclusion and marginalization of immigrants, especially female ones.

So, racism 'is in constant movement and constantly mutates itself' (Diken, 2001: 8). Today, it has increasingly become culturally coded and migrants, with

3 See Alana Lentin (2004) *Racism and Anti-Racism in Europe*. Pluto Press: London.

4 See Barker, 1981 and Balibar, 1991.

different cultures have become the principle victims, 'dangerous' outsiders disturbing the 'cultural harmony' of the 'hosting societies'. Today, the group of people to be considered 'white' has been changed and, for the 'new racism', migrants coming from the poor, Muslim and/or Arab world, clearly remain outside this 'whiteness', and are constructed as dangers to the mythical national and cultural homogeneity due to their distinct values, life styles and religious practices. This 'new racism' culminating in fears and antagonism, is significant in understanding current exclusionary practices and how the substitution of 'racial differences' by 'cultural' ones has reproduced new regimes of justification for the subordination and exclusion of certain groups of people, and 'has modernized racism and made it respectable' (Duffield, 1996: 175).

A culturalist convergence between securitization of migration and 'new racism'

In the last decades, there has been an increasing shift in the discourses and policies governing the issue of migration. Previously seen as an innocent economic activity, blessed with a vital role in the construction of European economies after World War Two, welcomed and encouraged by European states, dealt with through economic and humanitarian discourses, now migration has been transformed into a security matter in that it 'constitutes' a threat to European societies. It has been increasingly linked to criminality, socio-economic problems, cultural deprivation, and lately to terrorism. This transformation started mainly in the mid-1970s, but accelerated with the redefinition of security in the wake of the Cold War. To be more precise, the traditional security thinking of the Cold War years, which confined itself to state-centric and military-oriented conceptualizations, has shifted to include various issues, ranging from the environment and poverty to population movement. This widening of the security agenda to different areas, and utilization of security language for nonmilitary issues, has had important ramifications for the field of migration. In this transformation of security, migration has come to be increasingly framed, analysed and governed through security lenses. A new discourse emerged, a new 'truth' which has become a dominant one, overwhelming the economic and humanitarian discourses of previous years, that of the securitization of migration. As detailed by Karyotis in his chapter, the securitizing actors (politicians, security experts, media, NGOs, etc.) have played a decisive role in this process, increasingly connecting migration with various security threats, especially societal ones.

Of course, the rise of the securitization discourse should be also understood in light of international and regional developments, especially in the context of Europe. The oil crisis of the 1970s, with its traumatic effect on the world capitalist market, the rise of neo-liberal policies alongside the demise of the welfare state, the emergence of the Single Market, and associated calls for the strengthening of Europe's external borders, have all intensified the securitization of migration. September 11 and the subsequent attacks in Madrid and London have contributed much to the justification of claims regarding the migration-security nexus in Europe, and this 'security knowledge' has become an important determinant in devising responses to the issue of migration (Huysmans, 2006: 21).

In line with these developments, although Huysmans eloquently contends that migration has been securitized in three thematic fields, namely the internal market, welfare state and cultural identity[5], the securitization of migration via culture has become the most prominent discourse, with the rise of culturalist discourses, identity politics and the depoliticization of structural problems. Since the 1980s, Thatcher/Reagan style neo-liberalism has pushed this process further, and migrants have increasingly become the scapegoats for socio-economic problems, alongside the 'threat' arising from their 'insurmountable' cultural differences. Culturalism has been cemented with a rampant populism to mask the socio-economic problems of 'hosting' countries, and secured simple answers to complex questions. This discourse has thereby provided an important political remedy, legitimation in the face of rising economic and social unrest among the 'native' population. In this vein, as a mechanism of power which serves the economic and political interest of the situated actors, it is not difficult to understand why the securitization of migration has become a dominant discourse today and why it is so 'effectively incorporated into the social whole' (Foucault, 1976: 39).

This securitization of migration discourse via culture operates and has been 'incorporated into the social whole' in a similar way to the production of 'new racist' discourses. As with the 'new racist' discourse, this securitarization of migration through the fabrication of cultural differences into a threat has taken much of its power from a kind of homogenization and essentialization of culture and identity. The popular rhetoric utilizing these culturalist discourses tries to establish a dialectical and antagonistic relationship between the 'Europeans/us' and the 'immigrants/them' and locates this relation in a Schmittian approach of friend vs. enemy.[6] Right wing politicians as well as the conservative sections of European societies began to build their arguments on this binary grouping of

5 See Huysmans, 2000, 2006.

6 Schmitt (1996) defines the friend vs. enemy distinction as the essence of politics. He argues that 'Every religious, moral, economic, or other antithesis transforms into political one if it is sufficiently strong to group human beings effectively according to friend and enemy' (Schmitt, 1996: 37). To be more precise, in his formulation of this grouping, the enemy is 'the other, the stranger; and it is sufficient for his nature that he is in a specially intense way, existentially something different and alien, so that in the extreme case conflicts with him are possible' (ibid.: 27). Paradoxically, this enemy is necessery, constitutive for the identity of political community, which should be, at the same time, protected against this enemy through exceptional measures. Indeed via the fabrication of an enemy and existential threat, a political community defines itself and the political authority secures its legitimacy. Huysmans (1998) demonstrates that this Schmittian realist conceptualization of political community built on the friend/enemy distinction is the crucial force informing the securitization process by framing and administering the issue of migration as a threat to the political community.

people, and talk about the erosion of national identities, values and tradition as well as morality, because of increasing migration. In this context, migrants have become the 'ideal public enemy for all identity claims' (Tsoukala, 2005: 167). The 'invasion', 'inflow' of migrants, with highly 'differentiated' cultural and religious traits have become popular and dominant themes in the representation of migration both in public and political discourse; a discourse which has established a certain 'truth' through political speeches, expert opinions, security analysis and media reports. This process is exemplified very well by Tsoukala's chapter, which focuses on the representation of migrants by the media in France and Greece. She illustrates that despite the mobilization of desecuritization moves in line with changing domestic needs and interests within these countries, framing of the migrants (in this case by the media) as the kernel of societal problems has added fuel to the dominant securitarian discourses.

Indeed this enemy/friend, us vs. them construction with its political and economic utility was already in place. Many times in history, Europe and the West in general have ascribed to themselves the mission of civilizing 'others', who have been framed as backward and primitive. Whether in the context of the imperial or colonial period, or of today's 'bona fide' policies for the democratization of the so-called third world, the self-identification of the West and justification of these policies have dwelled upon an alleged divide between the civilized 'West' and barbaric 'Other'. Indeed, this 'civilizing mission [has become the] moral defence of imperialism ... to force others to conform to the norms prescribed by universal values' (Wallerstein, 2000). Dividing lines shift constantly and take different forms, but are always there, located in a dialectical relationship (such as the past's so-called East-West and today's South-North divisions), and play a constitutive role in sustaining racialized and hierarchical relations for the preservation of the economic and political interests of the dominant class.

In the contemporary security landscape, especially after the disappearance of the 'communist threat', and the 'successful integration' of the Central and East European countries into the *liberal, democratic West*, the constitutive role of immigrants (specifically of those non-Western ones) in reproducing these divisions, has gained utmost importance, as it substituted very well for the 'communist enemy' of the Cold War.

Bigo (2002) asserts that the securitization of migration, or the replacement of the 'communist threat' with the alleged threat from migrants, is structured by the interest of security professionals not only in the foreigner but in the immigrant, as they try to increase their power, legitimacy and budget, as well as affirm their role as providers of security in the context of the re-definition of security risks following the end of the Cold War. In this conjecture, certain categories of immigrant, with their 'distinct' cultural, religious threats, have been easily constructed as the new 'other' or new 'threat', and the securitization process has provided a way of managing the issue of migration in accordance with economic and political concerns, and justifying policies under cover of a kind of concealed racism working through the discourse of cultural difference.

As the case study below illustrates, discourses and policies utilizing 'cultural difference' as a pretext to discriminate, exclude and subordinate the 'others' have intensified following September 11, and manifested themselves in the discussions around the integration of certain groups of migrants, exclusively Muslims and of Arab origin, and their conformity with the 'values' of Europe.

Integration as a means of suffocating 'threatening' differences

Following the September 11 and subsequent attacks, which have brought the alleged 'radicalization' of Muslim communities to the fore, the integration of Muslim people has become the focal point of a much larger sense of anxiety. Especially after the discovery of the September 11 hijackers' link to Germany, the emergence of terrorist cells in Spain, and, more importantly, the subsequent attacks in Madrid and London, eyes have turned to those 'enemies within'. However, it may be misleading to attribute the intensification of integration debates solely to the terrorist attacks. The eruption of urban 'riots', such as those that occurred in Britain and France, as well as dramatic events such as the murder of Theo van Gogh in the Netherlands and the clashes on the Mohammed caricatures, the headscarf issue or, lately, debates surrounding the construction of minarets, all contributed to the mobilization of racialized and securitarian integration discourses and policies in recent years. Multiculturalism has come under attack throughout Europe, and it has been portrayed as the reason for integration problems, for the emergence of 'ghettos' or 'parallel societies,' 'sleeper cells' or, in more extreme form, for the multiplication of 'potential' terrorists. As Carrera (2006a: 89) describes, in the post September 11 period, we have witnessed the securitization of integration as well, which involves taking it for granted that 'lack of integration' constitutes a risk, and is detrimental for the receiving countries' social and cultural unity. This securitization, which circumscribes the integration 'problem' presented by certain group of migrants (generally referring to Muslim people) has been driven by certain motifs in national agendas: 'in the Netherlands, "standards and values"; in Sweden and Norway, cultural barriers to inclusion; in the UK, "community cohesion"; in France, the principle of laicité (state secularism); in Germany, the primacy of "Leitkultur" (leading culture); in Denmark, the "intolerant culture" amongst immigrants that prevents integration; in Spain, public safety and crime' (Fekete, 2004: 18). Concomitantly, the 'clash of civilizations' thesis has become much more popular, as used by various politicians across Europe, ranging from Aznar and Berlusconi to Westenthaler. To be more precise, the Danish politician Pia Kjaersgaard, known for her anti-Islam stance, stated that 'September 11 was not the start of the clash of civilizations as a clash would indicate that there are two civilizations, when there is only one civilization, and that's ours' (cited in Fekete, 2006: 9). Similarly Kristian Thulesen Dahl from the Danish People Party- commented that 'Islam is the biggest threat to world peace since the fall of Communism… It is eating us [from within and] destabilizing [our societies]' (ibid.). In Netherlands, Geert Wilders, leader of the Dutch Party for Freedom, has also been playing on this securitization as one of the best known figures invoking anti-immigrant and anti-Muslim sentiments. His film, Fitna, represents Islam and Muslims in a way which reinforces the already established image of Islam as a fundamentalist and backward religion. He argued that 'Fitna is the last warning to the West. We can choose to pass freedom on to our children or allow our freedom to sink into a multicultural swamp' (Spiegel, 2008a). Moreover he warns Europe against Islamization, which he argues, is an ongoing process due to the high rate of immigration from Turkey, the Middle East and Africa. In one interview, he further explained his stance:

> It's true, [the Muslim youth] don't carry the Koran under their arms. But it's at home. And their fathers go to the mosque. They don't tell their children that beating women or believers of other religions is not allowed. That's why we have to push harder for a kind of *Leitkultur*, a guiding culture. Not a monoculture but a culture that draws on our Christian, Jewish, humanistic traditions and that poses a challenge to the Islamic problem. This is patriotism, not nationalism; this is pride in our own culture (Spiegel, 2008b).

These words clearly prove, what Fekete (2004: 4) argues, that 'under the guise of patriotism, a wholesale anti-Islamic racism has been unleashed which itself threatens to destroy the fabric of the multicultural society' by way of homogenizing Muslims and Islam. However, these anti-Islam voices fail to recognize, consciously or unconsciously, the fact that like Christianity and any other religion or ideology, Islam is not monolithic, but is subject to multiple interpretations. The general approach within Europe is to homogenize these groups, fixing them under the same label. As Mamdani (2004: 13) argues, this framing of Islam as an unitary religion, and Muslim people as an homogenous group, facilitates the equation of Islam and terrorism with each other, or the differentiation between 'good' and 'bad' Muslims through 'ethnocentric and polemical interpretation of Islam'. He further posits that these discussions clearly stem from 'cultural explanations of political outcomes' which ignore the importance of '*history* and issues' (ibid.: 45, emphasis original).

Such types of discourse are not limited to Wilders or to other anti-immigrant or conservative political groups. The centre left and feminists, namely so-called 'white' or 'bourgeois' ones, are also in danger of falling into the trap. This is evident when we look at their arguments regarding genital mutation, forced marriages or violence against Muslim women. They link these problems directly to Islam, without reference to the legacy of patriarchy and sexism embedded in all societies, or to socio-economic developments structuring these problems. They do this through 'reducing and immobilizing [cultures] to one element and condemning that whole community as backward or even "uncivilized"' (San Juan, Jr, 2007: 142). Some feminists have gone so far as to call for a halt to immigration from Muslim countries or to supporting a ban on headscarves, as they consider that Islam and headscarves are symbols of women's oppression and detrimental for the hard-won rights of western women as well.[7]

Yet these voices again tend to universalize the Eurocentric notion of modernity, or dictate what Hesse calls a 'racialized modernity' (Hesse, 2007), equating the headscarf with backwardness, and, at the same time, claiming the same monopoly over the body of Muslim women that men have been doing for a long time. The notable fallacy in this approach is the logic which foresees the possibility of emancipating Muslim women through the ban on headscarves, and which at the same time ignores the fact that wearing headscarves is not necessarily an involuntary act. So rather than prompting forceful 'emancipatory' projects by the way of repressive and racialized discourses and policies, it is necessary to follow a more historical, political and just approach, and 'to offer Muslim women better protection and equal access to justice in the host country, but not to deny them rights of residence or settlement via stricter immigration controls' (Fekete, 2006: 7).

These discourses clearly reflect the fact that culture has become the first and foremost explanation for problems, while structural deficiencies such as racial discrimination in employment, education, housing and other social services, the lack of political rights for long-term residents, and patriarchy have been left untouched. In the context of the dominant discourse purported by politicians, right or left wing, media reports, or even by certain academic works, we see overemphasis on the link between culture and the problems of migrants. Ghettoization has been presented as if it was a voluntary choice or resulted from self-segregation and signified a certain kind of anomie. High unemployment among immigrant communities is asserted to be the result of cultural deficiency which prevents them gaining certain skills. Even their poverty is seen and presented as culturally inherited. As the analysis of integration measures below shall demonstrate, invocation of culture both as an explanation of and as a remedy for the problems of integration has not remained a purely linguistic practice. Instead it has infiltrated into policies as well.

Forced 'integration' programmes

The above delineated discourses on integration have become much more tangible and institutionalized through the integration measures adopted across Europe. These measures reaffirm the very fact that culture (or cultural difference) is the key issue to be dealt with in the course of integration. As cultural diversity is seen as a threat to the harmony and values of European societies, and as migrants are presented as if they were resisting integration, aggressive and forceful integration measures have become the common policy strategy across the EU. Member states differ in how they manufacture dangerous 'others' under the guise of integration debates, but the proposed and implemented measures have taken similar forms,

7 See Fekete, 2009.

which signify a move to a more assimilationist model, and which enforce the European 'way of life' and 'values' as a prerequisite for living in Europe.

The most crucial change, starting from the late 1990s and becoming much more apparent in the wake of September 11, is the character of civic integration policies. Civic integration policies are the integration programmes through which newcomers as well as long-term residents are obliged to learn the *language, history, culture and values of the hosting country*. Previously, participation in these programmes was not mandatory. Now, all over Europe, and especially in Germany, Netherlands, Denmark, Austria, and Belgium (Flanders), it has become compulsory. Non-compliance with the 'obligation' to attend, or failure to pass these programmes and courses, leads to harsh sanctions, such as financial penalties, denial of extending resident permits and citizenship applications, withdrawal of welfare supports or even deportation[8]. So learning the language or culture-- or conforming to the 'values' of 'hosting' countries has become key to acquiring a 'secure judicial status'[9] (Carrera, 2006a:87). Another common feature is that, by means of 'remote integration', family unification has also been restricted. With the 'reforms' of integration policies, those who applied for family unification are obliged to take language and integration tests or to prove at least a basic language proficiency in their country of origin before their departure. Lastly, laws have been tightened so that grants of citizenship are subject to tests or courses through which applicants' conformity with the country's culture can be 'evaluated'.

Outwardly, there seems to be no direct link between the language and/or integration courses, securitization and 'new racism'; these policies are justified in the name of more 'integration' into the labour market and society, and said to be for the benefits of migrants. Indeed, one is tempted to say, language courses in particular may help newcomers in their search for jobs or in their social contact. But some points with regard to these measures need to be critically assessed. The mandatory character of these courses and the sanctions, as well as the emphasis on knowledge of *'our' culture, values, and respect for the way of life of the hosting societies*, which are presented as the prerequisite for the successful integration, are very controversial. They highlight how integration discourses and policies are colonized by power relations, and how they have been encapsulated within a hierarchical structure of subordination and domination. They are also perceived as othering migrants in negative terms, while at the same time glorifying the culture of 'hosting' societies. This reflects, as at many times in history, the idea

8 Germany's New Immigration Act of 2004, Austria's New Settlement and Residence Act of 2006, Denmark's new versions of the Act on Integration of Aliens from 1 January 2004, Netherland's Civic Integration of 1 January 2007, Civic Integration Decree of the Flemish Government of 2003, all, highlight the importance of a knowledge of language, history and culture as well as of civic and social aspects of the country for better integration.

9 Here, as stated by Carrera (2006b:1), 'secure judicial status' refers to 'having access to the rights and freedoms provided in the legislation on immigration and citizenship, and to be treated equally and in non-discriminatory manner'.

of the so-called supremacy of Europe, which 'stands for universal values – let us rehearse them again; democracy, human rights and the rule of law' (Lentin, 2008) in the face of the allegedly less worthy, backward values of the non-Western. The logic behind these measures seems to be to transfigure the disturbing 'aliens' and their 'values' in accordance with a European model of community, discipline them, make those 'abnormal' 'normal', and assimilate them into the presumed homogenous societies, or if this can't be done, exclude them.

Secondly, these policies appear to address only non-Western people, exempting the nationals of 'developed' countries from their scope. For instance, in the Netherlands, the citizens of US, Canada, Australia, New Zealand and Japan are exempted from the integration contracts and tests on the condition that they have a certificate which proves 'they are officially "integrated"' (Fekete, 2004: 22). On the other hand, people of Turkish and Moroccan origin are discriminately subjected to these measures. For instance, in the Integration Abroad Act of 2006, these conditions are applied only to family members from these countries, and make family unification for them much harder through the imposition of financial and language requirements. This is becoming the case in other European countries, such as in Germany, Denmark, France and the UK; indeed the Netherlands has become the guiding model for other European countries, and provides a good example of how assimilationism has started to override multicultural approaches.

Another common aspect among member states is that the mandatory integration programmes are not applicable to EU citizens. Either they belong to the exempted groups or have the right to participate in these programmes on a voluntary basis, e.g. in Germany.[10] As pointed out by Guild, they experience measures of 'convergence avoiding the subordinate/dominant difference implicit in integration' (Guild, 2005: 105). Discourses of 'convergence' demonstrate that:

> The member states indices are converging one upon the other in a consistent manner. The discourse of convergence is a voluntary one where there is a programme within which there is a natural movement. It does not contain the coercive idea of the movement of lesser valued towards more valuable norms … . [On the other hand] the construction of mandatory integration programmes in some countries usually indicates a fear of diversity no matter how much the proponents seek to justify these projects as for the benefits of the 'immigrants'. The right of immigrants, nationals of EU member states, not to integrate, but to

10 Even though the exempted groups differ from country to country, mainly EU citizens, EEA nationals, highly skilled workers, those workes in possession of short-term work permits, researchers, scientists and asylum seekers are the groups who are not obliged to participate in integration courses. Given the scope of integration programmes, some schoalrs rightfully draw attention to the linkage between integration and poverty. Carrera and Guild argue that as long as immigrant groups are not a financial burden on the state, they should not be obliged to attend integration programmes (Guild, 2005; Carrera, 2006a and 2006b).

> remain distinct, is constructed as a benefit to the whole of the Union, evidence of diversity (ibid.: 105–10).

Furthermore integration policies tend to link the reasons for social unrest and the problems of immigrant communities to their so-called uncivilized character, and turn a blind eye to structural problems such as long-standing racial discrimination, poverty and the marginalization of immigrants in Europe. For instance, 'Britishness' has become the driving theme in the course of integration policies, especially since the 2001 summer riots, the cause of which was represented as being a very 'liberal' multicultural settlement, resulting in cultural/ethnic enclaves in the UK. The remedy has again taken the form of an assimilationist model, the Cantle Report, entitled *Community Cohesion*, under the guidance of Home Secretary Blunkett. Despite the fact that, those ethnic communities involved in riots had been living in Britain for decades or were even born there, their alleged lack of English proficiency or of knowledge about the 'democracy', 'culture' and 'values' of British society were emphasized and presented as the source of polarization in these towns. The 'remedies' proposed to deal with the causes of the riots have invoked criticisms from community leaders. Manwar Jan Khan, a community leader in Bradford, stated that: 'I'd like to know what David Blunkett means by integration. Like learning English – many of us do speak English, many are quite westernized … but let's have the real debate, let's talk about racism, let's talk about why white employees won't give Asian kids in Bradford any jobs. Let's talk about the impact of Islamophobia' (BBC, 2001).

Another example reflecting such an approach is the latest French Law on Immigration and Integration, which was drafted by Sarkozy and came into force in 2006. In particular, regulations concerning family reunification illustrate how culturalism has blurred the debate. The main aspects of this are demonstrated by three aims: 'to ensure that immigrants respect French values, to promote their integration into French society, and to fight forced or polygamous marriages' (Murphy, 2006). This bill again led to accusations of racism by human rights organizations and leftist groups as well as Muslim and Christian communities. Sarkozy reacted strongly to these criticisms, and stated that recent riots proved the failure of past immigration and integration policies. The debate went even further. In supporting the new bill, the riots were described 'as a sign of increasing Islamic radicalism and the start of a so-called clash of civilizations between the Muslim East and the Christian West' (Al-Jadda, 2005). The perpetuators of this approach, who misuse the religious background of these youths, ignored the fact that their concerns were not about religion, but prolonged injustice and exclusion from the social fabric. They united not on the basis of their religious and cultural background or common identity, which were already annihilated by France, but because of their common 'misery' (Murray, 2006: 30). Again, riots were viewed in culturalist terms, linked to the incompatible 'culture' and 'values' of 'minority' groups. Again the voices which highlighted institutional racism, and discrimination in the labour market and social fabric, as well as stereotyping of African and Arab people, have been overshadowed by simplistic and culturalist arguments.

The growing emphasis on immigrants' alleged lack of knowledge of democratic values and human rights has been manifested across Europe in the form of integration and loyalty contracts as well. The most controversial one is the Danish loyalty declaration that must be signed by those applying for citizenship. The web site of the Ministry of Integration made the 'prescription' to the citizenship applicants clear; 'work, pay tax, don't hit your children and show respect for equal rights between the sexes' (Baltic and Nordic Headlines, 2005) .Various criticisms were leveled against the contents of this loyalty declaration. UNHCR stated that 'the way the declaration is formulated creates anticipation that foreigners will not respect certain values that are fundamental in a democratic society, and that they will commit the type of crimes mentioned in the declaration' (Migration News Sheet, 2006: 4). These trends prove that liberal democracies in Europe use immigration policy to discipline non-white and Muslim communities, and consequently link 'race' and 'immigration' to each other (Kundnani, 2002: 67).

Concluding remarks: rise of culturalism and disappearance of politics

The discourses and policies surrounding the integration debate simply exemplify how migration has been constructed as a security threat, how this discourse has been converged with 'new racism', and how the cultural difference-as-a-threat narrative has been increasingly gaining the status of 'truth', and structures migration policies. This chapter does not delve into the 'real' reasons for this convergence, but is intended to demonstrate that the articulation of culture or cultural difference as one of the main problems or as a remedy through which problems can be solved has become a powerful method of masking real issues. Even some sections of the anti-racist movement, and some desecuritization voices, have fallen into the trap of 'cultural talk' and encapsulated their struggles in issues of identity, culturalism and ethnicism. They overemphasize culturalist strategies such as.support for inter-cultural dialogue, multiculturalism, or more 'integration' of migrants, etc., but they overlook the persisting structural problems, such as racial discrimination, and the subordination and domination of migrants in the economic, political and social fabric of European societies. In this way, they reproduce the same discourses of 'new racism' and securitization of migration or, as Gilroy posits, they put their activities at risk of 'simply reversing the terms of racist discourses under the label of anti-racism' (1990 cited in Solomos and Back, 1996: 115) This is closely related to the fact that racism today is viewed as just a matter of cultural prejudices, something personal, and, most significantly, is still defined and viewed 'in terms of superiority/inferiority'; a definition which 'has helped conceal how common is a form of racism that does not need to make such assertions – indeed can make a positive virtue out of not making them' (Barker, 1981: 4). This, in turn, clouds the political character of the issue and mobilizes apolitical solutions. In fact, with the help of this seductive strategy, we bear witness to the under-representation of immigrants as a political issue, but overrepresentation of their culture, their ghettos and their 'differences', which allegedly disturb the homogeneity of European societies (Diken, 1998: 2).

In this context, multiculturalism, integration or intercultural dialogue (which are the main focus and departure point of many governmental and non-governmental initiatives), lead to depoliticization of the struggle of migrants and other 'strangers'. Or, as Huysmans describes it, they politicize them 'via their cultural identity' (Huysmans, 2006: 77).

So we need to be critical of the idealized representation of multiculturalism and other culturalist strategies. 'Multiculturalism [and similar culturalist policies] as long as [they are] conceived within the existing framework of the racial polity of a hegemonic order founded on class inequality, can not offer the means to realize justice, fairness and recognition of people's singular identities and worth' (San Juan Jr., 2002: 8). But they are retained as an 'ideology of conservatism, of preserving the status quo', which grants migrants only the means of 'survival' within an existing system contaminated by structural inequalities (Kundnani, 2002: 68). Multiculturalism as an approach respecting differences and challenging monoculturalist discourses needs to be encouraged, but as a policy of settlement limiting the problems of migrants to their cultural or identity aspects risks concealing the role of state practices and the economic structure, and the resultant unjust, hierarchical social relations.

To conclude, there is a need for a more radical break with these existing conservative culturalist, racialized, securitarian discourses and policies in order to disturb the current regimes of justification. Undoubtedly, this is a hard process which calls for universal justice and which privileges human mobility over capitalist mobility.

References

Al-Jadda, S. (2005) 'In French Riots, a Lesson for Europe', in *USA Today,* accessed: 23 April 2006 at: http://www.usatoday.com/news/opinion/editorials/2005-11-08-french-riots-edit_x.htm

Balibar, E. (1991) 'Is There a "Neo-Racism?"' in Balibar, E. and Wallerstein, I (eds) *Race, Nation, Class: Ambiguous Identities*, London: Verso

Baltic and Nordic Headlines (2005) 'Loyalty Oath Demand', accessed: 7 April 2008 at: http://www.unhcr.se/se/news/pdf/November_2005.pdf

Barker, M. (1981) *The New Racism,* London: Junction Books

BBC Online News (2001) 'Immigrants Should Try to Feel British', accessed: 10 January 2010 at: http://news.bbc.co.uk/2/hi/uk_news/politics/1699847.stm

Bigo, D. (2002) 'Security and Immigration: Toward a Critique of the Governmentality of Unease', in *Alternatives,* Vol. 27, No. 1, pp. 63–92

Blommaert, J and Verschueren, J. (1992) *Het Belgische Migrantendebat*, Antwerp: International Pragmatics Association, cited in Huysmans, J. (2000) 'The European Union and the Securitization of Migration', in *Journal of Common Market Studies*, Vol. 38, No. 5, pp.751–77

Blommaert, J and Verschueren, J. (1998) *Debating Diversity: Analyzing the Rhetoric of Tolerance*, New York: Routledge, cited in Huysmans, J. (2000) 'The European Union and the Securitization of Migration', in *Journal of Common Market Studies*, Vol. 38, No. 5, pp. 751–77

Boltanski, L. and Thévenot, L. (1999) 'The Sociology of Critical Capacity', in *European Journal of Social Theory*, Vol. 2, No. 3, pp. 359–77

Buzan, B., Waever, O. and de Wilde, J. (1998) *Security: A New Framework for Analysis*, London: Lynne Rienner Publishers

Carrera, S. (2006a) 'Integration of Immigrants Versus Social Cohesion: A Typology of Integration Programmes in the EU' in Balzacq, T and Carrere, S. (eds), *Security versus Freedom? A Challenge for Europe's Future*, Hampshire: Ashgate

Carrera, S. (2006b) 'A Comparison of Integration Programmes in the EU: Trends and Weaknesses', *Challenge Papers*, No.1/March, accessed: 13 February 2009 at: http://aei.pitt.edu/6773/01/1310_01.pdf

Diken, B. (1998) 'Immigrants on the Margin: Introduction to Strangers, Ambivalence and Social Theory', *Online-Papers*, Department of Sociology: Lancaster University, accessed: 25 May 2008 at: http://www.lancs.ac.uk/fass/sociology/papers/diken-immigrants-on-the-margin.pdf

Diken, B. (2001) 'Immigration, Multiculturalism and Post Politics after 'Nine Eleven'', in *Third Text*, Vol. 15, No. 57, pp. 3–12

Diken, B. (2002) 'Justification and Immigration in the Network Society: A New Ambivalence?' *AMID Working Papers Series*, Department of Sociology: Lancaster University, accessed: 25 May 2008 at: http://www.amid.dk/pub/papers/AMID_4-2002_Diken.pdf

Doty, L. R. (2003) *Anti-Immigrantism in Western Democracies*, London and New York: Routledge

Duffield, M. (1996) 'The Symphony of the Damned: Racial Discourse, Complex Political Emergencies and Humanitarian Aid', in *Disasters,* Vol. 2. No. 3, pp. 173–93

Duffield, M. (2006) 'Racism, Migration and Development: The Foundations of Planetary Order', in *Progress in Development Studies*, Vol. 6, No. 1, pp. 68–79

Fanon, F. (1967) *Towards the African Revolution*, New York: Monthly Review Press

Fekete, L. (2001) 'The Emergence of Xeno-racism', in *Race and Class*, Vol. 43, No. 2, pp. 23–40

Fekete, L. (2004) 'Anti-Muslim Racism and European Security State', in *Race and Class*, Vol. 46, No. 1, pp. 3–29

Fekete, L. (2006) 'Enlightened Fundamentalism? Immigration, Feminism and the Rights', in *Race and Class*, Vol. 48, No. 2, pp. 1–22

Fekete, L. (2009) *A Suitable Enemy: Racism, Migration and Islamophobia in Europe*, New York: Pluto Press

Foucault, M. (1972) *Archeology of Knowledge*, trans. A. M. Sheridan Smith, New York: Pantheon Books

Foucault, M. (1976), 'Two Lectures' in Kelly, M. (ed.) (1994) *Critique and Power: Recasting the Foucault / Habermas Debate*, Cambridge: MIT Press

Foucault, M. (1980) *Power/Knowledge: Selected Interviews and Other Writings, 1972–1977*, New York: Pantheon

Gilroy, P. (1987) *There Ain't No Black in the Union Jack: The Cultural Politics of Race and Nation,* London: Hutchinson

Gilroy, P. (1990) 'The End of Anti-Racism', in *New Community*, Vol. 17, No. 1, pp. 71–83, cited in Solomon, J. and Back, L. (1996) *Racism and Society*, London: Macmillan Distribution

Goldberg, D. (1994), 'Introduction: Multicultural Conditions' in Goldberg, D. (ed.), *Multiculturalism: A Critical Reader*, Oxford: Blackwell

Guild, E. (2005), 'Cultural and Identity Security' in Guild, E. and van Selm, J. (eds) *International Migration and Security: Opportunities and Challenges*. London and New York: Routledge

Hesse, B. (2007) 'Racialized Modernity: An Analytics of White Mythologies', in *Ethnic and Racial Studies,* Vol. 30, No. 4, pp. 643–63

Human Rights Watch (2008) 'The Netherlands: Discrimination in the Name of Integration. Migrants' Rights under Integration Abroad Act.', accessed: 2 May 2008 at: http://www.hrw.org/backgrounder/2008/netherlands0508/netherlands0508web.pdf.

Huysmans, J. (1998) 'The Question of the Limit: Desecuritization and the Aesthetics of Horror in Political Realism', in *Millennium – Journal of International Studies*, Vol. 27, No. 3, pp. 569–89

Huysmans, J. (2000) 'The European Union and the Securitization of Migration', in *Journal of Common Market Studies*, Vol. 38, No. 5, pp.751–77

Huysmans, J. (2006) *The Politics of Insecurity: Fear, Migration and Asylum in the EU*, London and New York: Routledge

Ibrahim, M. (2005) 'The Securitization of Migration: A Racial Discourse', in *International Migration*, Vol. 43, No. 5, pp. 163–87

Kelly, M (1994) *Critique and Power: Recasting the Foucault/Habermas Debate*, Cambridge: MIT Press

Kundnani, A. (2002) 'The Death of Multiculturalism', in *Race and Class*, Vol. 43, No. 4, pp. 67–72

Kundnani, A. (2007) 'Integrationism: The Politics of Anti-Muslim Racism', in *Race and Class*, Vol. 48, No. 4, pp. 24–44

Lentin, A. (2004) *Racism and Anti-Racism in Europe*, London: Pluto Press

Lentin, A. (2008) 'Europe and the Silence about Race', in *European Journal of Social Theory*, Vol. 11, No. 4, pp. 487–503

Mamdani, M. (2004) *Good Muslim, Bad Muslim: America, the Cold War and the Roots of Terror*, New York: Pantheon

Migration News Sheet (January 2006)

Murphy, K. (2006) 'France's New Law: Control Immigration Flows, Court the Highly Skilled', *Migration Policy Institute*, accessed: 15 April 2008 at: http://www.migrationinformation.org/feature/display.cfm?ID=486

Murray, G. (2006) 'France: the Riots and the Republic', in *Race and Class*, Vol. 47, No. 4, pp. 26–45

San Juan JR., E. (2002) *Racism and Cultural Studies: Critiques of Multiculturalist Ideology and the Politics of Difference*, Durham and London: Duke University Press

San Juan JR., E. (2006) *In the Wake of the Terror: Class, Race, Nation, Ethnicity in the Postmodern World*, Plymouth, KY: Lexington Books

Schmitt, C. (1996) *The Concept of the Political*, Chicago: University of Chicago Press

Sivanandan, A. (1990) *Communities of Resistance: Writings on Black Struggles for Socialism*, London: Verso

Sivanandan, A. (2001) 'Refugees from Globalism', in *Race and Class*, Vol. 42, No. 3, pp. 87–100

Solomos, J. and Back, L. (1996) *Racism and Society*, London: Macmillan Distribution

Spears, A. (1999), 'Race and Ideology: An Introduction' in Spears, A. (ed.) *Race and Ideology: Language, Symbolism, and Popular Culture*, Detroit: Wayne State Press

Spiegel (2008a) 'A Missionary with Dark Visions', accessed: 4 April 2008 at: http://www.spiegel.de/international/europe/0,1518,543627,00.html

Spiegel (2008b) 'Interview with Dutch Populist Geert Wilders: Moderate Islam is a Contradiction', accessed: 4 April 2008 at: http://www.spiegel.de/international/europe/0,1518,544347,00.html

Taguieff, P.-A.(1990) 'The New Cultural Racism', in *Telos*, Vol. 83, pp. 109–22

Tsoukala, A. (2005), 'Looking at Migrants as Enemies' in Bigo, D. and Guild, E. (eds) *Controlling Frontiers: Free Movement Into and Within Europe*, Burlington: Ashgate

Waever, O. (1995), 'Securitization and Desecuritization' in Lipschutz, R. (ed.) *On Security*, New York: Columbia University Press

Wallerstein, I. (2000) 'Cultures in Conflict? Who are We? Who are the Others?', accessed: 16 May 2007 at: http://fbc.binghamton.edu/iw-hk-pao.htm

Webber, F. (1991) 'From Ethnocentrism to Euro-racism', in *Race and Class*, Vol. 38, No. 2, pp. 77–81

Žižek, S. (1993) *Tarrying with the Negative: Kant, Hegel, and the Critique of Ideology*, Durham: Duke University Press

Chapter 13
Impossibilities of Social Citizenship: On the Statelessness of Migrants

Mojca Pajnik

Contemporary trends of job market deregulation, increasing precarization, and the decline of social services particularly aggravate the situation of the socially and economically most marginalized groups. The demand for hard physical work and non-skilled labour in the EU states is satisfied mainly by migrants. According to the Employment Service's data across the EU, the greatest demand for migrant labour force is generated by the sectors traditionally characterized by poor working conditions: the construction sector, the care sector, manufacturing, seasonal work, sales and cleaning services. Research shows[1] that employment and migration policies systematically relegate migrants to low-paid sectors and non-skilled jobs associated with low levels of social security and safety at work frequently combined with a high risk of health hazards, as well as limited labour rights.

Contemporary EU social policies are based on the idea of the welfare state that originated in the 1950s and is related to the notion of social citizenship (Marshall, 1950) which added social rights to the corpus of political and civil rights. This chapter explores the notion of citizenship while critically challenging its (a)political conceptualizations (Soysal, 1994; Lister, 1997; Dean, 2003). I first define citizenship as membership of a political community which enables action within the public sphere. Next, I explore the disappearance of the political dimension of citizenship, with one of its consequences being the exclusion of certain groups from a political community. Various approaches to social citizenship are examined and the validity of this concept is assessed using the example of labour and social policies considered with respect to migrants.[2] I refer to migrants

1 See the findings of the Primts project (Prospects for integration of migrants from 'third countries' and their labour market situation: towards policies and action), accessed: at http://www.primts-mirovni-institut.si., Trimikliniotis (forthcoming) for specific cases of Cyprus, Germany, Italy and Slovenia.

2 The concept of social citizenship widely praised after Marshall's introduction in the 1950s is based on the idea that granting people social rights increases their security. Some later understandings of citizenship (Soysal, 1994; Dean, 2003) that take into account the weaknesses of the welfare state model and explore citizenship in relation to the marginalized groups, migrants among them, are more critical of Marshall's model. One of the main arguments of the critique is that social citizenship devalued people as political beings;

in this chapter and refrain from further categorization of migrating populations. However, the analysis presented is also valid for asylum seekers and refugees whose statuses are presented in more detail in chapters by Schuster, Naagtegall and Parusel in this book.

National states see citizenship as a status, and membership of a national community; consequently migrants are regarded as non-citizens and non-members, with the rights arising from social citizenship applying to them only to a limited extent. In this chapter, I use the example of current migration polices in Slovenia to problematize the circular condition employed in granting social rights/citizenship. Since I advocate the need to rehabilitate the political aspect of citizenship, and since citizenship is understood here as a full participation (of migrants) in the activities of a political community, I am not concerned solely with the institutional dimensions of citizenship (labour market policies, migration policies etc.) but also make room for migrants' narratives. Accordingly, in illustrating the (non)functioning of social citizenship, the chapter draws on the narratives of women migrants in Slovenia obtained through biographical interviews.[3]

On the political aspect of citizenship and its disappearance

Interpretations of citizenship most often draw on specific theoretical traditions in which debates usually revolve around differences between metatheories, i.e. the liberal and the republican model of citizenship. Further classifications frequently include communitarian, economic, discursive and multicultural citizenship. Over the past two decades, the changing role of the national state, the tendency towards its greater inclusion in globalization trends, and larger migration flows gave rise to theories about post-national, trans-national and cosmopolitan citizenship. These problematize national state-based conceptualizations and discuss redefinitions of

---

instead of promoting political activity it introduced social services that are by definition always biased when it comes to the security of vulnerable groups.

3 During the period 2006–2007 we conducted 26 biographical interviews with migrant women who had moved to Slovenia during the previous 15 years. Country of origin was determined by birthplace: 7 interview partners came from the former Yugoslav republics, 4 from Kazakhstan, 3 from Ukraine, 2 from China, 2 from Moldova, and 1 each from Russia, Uzbekistan, Japan, Peru, Colombia, Thailand, Lithuania, Turkey, Slovakia and the Czech Republic. The women were between 22 and 48 years of age, they had various socio-economic and educational backgrounds, they came from differing geopolitical contexts, and lived and worked in differing social situations. The interviews were part of the 6FP Femipol project (Integration of female immigrants in labour market and society: policy assessment and policy recommendations) that aimed at analysing the impact of integration policies on the position of migrant women in EU countries. More on the project is available at http://www.femipol.uni-frankfurt.de. See also Slany, K., M. Kontos and M. Liapi, eds. (2010) and Pajnik and Bajt (forthcoming).

citizenship at the transnational level. In this chapter, I avoid well-known definitions.[4] I will not draw on any metatheory but will define citizenship as a political activity, or as the 'right to have rights', to borrow from Arendt (1948/1976); as membership of a political community and activity in the public sphere (Somers, 2008).

For Arendt, citizenship is a precondition that enables life and activity within a community. Central to the ethos of citizenship is the principle that the right to have rights is not dependent on individual circumstances, meaning that membership of a political community cannot be denied because one holds citizenship of another country, or on the grounds of ethnic origin, language, religion and the like. In this sense, citizenship prevents statelessness as a consequence of someone not being what he/she should be, for example, a citizen of a specific country or a member of a certain (dominant) ethnic group. However, migrants, as shown in the empirical part of this chapter, alert us to the fact that membership of a political community is not self-evident; rather than being something that naturally belongs to a person, it requires the creation of conditions and extensive political will to make membership possible. The accounts of migrant women presented below point to the worrying lack of a will that would contribute to the security of migrating populations.

As such, citizenship constitutes the worldliness of people (Arendt, 1967), their membership of the world, their political value, and the expression of rights in the public sphere. It denotes 'citizenship of the world', meaning rights and activity in the world as a political community (Balibar, 2008: 534). It also reflects the fact that people share many things on a global level. Diseases, unemployment, aging etc. are collective rather than individual circumstances, and the ethics of citizenship exact collective answers to collective issues: unemployment is not the 'fault' of an individual, but a structural problem that calls for structural measures and protection against the ideologeme of the invisible hand of the market. Defined in political terms, citizenship is thus not a contractual relationship, but an association based on solidarity and mutual responsibility (Arendt, 1967; Somers, 2008).

Some authors (Arendt, 1967; Habermas, 1972; Somers, 2008) argue that the theoretical basis for the depoliticization of citizenship in the eighteenth and nineteenth centuries was provided by social contract theories, which, by analogy with Hobbes's *politike techné*, defined citizenship as a contractual relationship and necessity enabling legitimization of the government. Somers (2008) termed this 'contractualization of citizenship', as citizenship is reduced to contractual relations. According to Hobbes (1651/1985), citizenship defines isolated individuals who come together accidentally and sign a *quid pro quo* social contract as an exclusively instrumental mechanism in the function of social management. Contrary to this are understandings of citizenship as constituting membership of a political community and activity within it. *Polity* presupposes the autonomous, intersubjective action of individual members of the community. It is plural by nature, and arises from the network of activities that are carried out on an equal footing. It is not a private matter, nor is it aimed at satisfying individual interests

4 For citizenship models see Delanty (2000).

through a contractual relationship. The nation and national identity, based on ethnic and cultural premises and delimited by a country's borders, 'devoured' the idea of a political community at some point in time. Depoliticization (Habermas 1972) meant that the national has become the 'political' and the nation has become a 'political community', a change that resulted in the exclusion of certain populations, such as migrants.

Citizenship, originally denoting membership of a political community, became a status and primarily an administrative criterion, implying membership of a national community either by virtue of birth (*ius sanguinis*) or residence (*ius soli*); it has become identity ascertained by the passport that functions as proof that one belongs to a specific 'political' (=national) community. In this definition of citizenship, an individual is a member of a national community. Such a definition serves the purpose of territorial demarcation, the legitimization of the borders of a national state and the power of its elites (Habermas, 1994: 24–5) rather than implying citizens' participation in and co-creation of political community. This is visible, for example, in citizenship and naturalization policies which often prevent dual or multiple citizenships, and require the renouncement of original citizenship before granting a specific citizenship, i.e. a specific national affiliation.

Habermas (2001: 53) linked the apolitical understanding of citizenship with not only the processes of political community nationalization but also with the flourishing of entrepreneurship in state practices. The state has become entrepreneurial, based on the contractual principles of market operation. The nationalization processes and uncritical adoption of market principles in administering the nation states of the nineteenth and twentieth century established citizenship as necessary in enabling the existence of national communities. The 'status' nature of citizenship has become a necessity, preconditioning the existence of a 'political' (=national) community; citizenship has become inclusion in or exclusion from the privileged membership of (western) national communities. These shifts are a source of insecurity for migrants who are excluded from membership. Migrants in EU countries also compete unequally with the 'native' population in national labour markets since, as shown below, policies are adopted that aim at 'securing' the national labour market (against migrants).

Statelessness and the right to have rights

The *locus classicus* of modern debates about citizenship is held to be Marshall's (1950) conceptualization of citizenship based on political, civil and social rights. The triad of rights addresses citizenship primarily as a status: only a citizen has the right to vote and only a citizen has the right to a pension. Theorists of post-national citizenship point to the problematic hypothesis here (Soysal, 1994; Habermas, 2001), that citizenship encompasses all members of a community. In practice, this supposed all-inclusiveness and equality maintains and reproduces inequalities with respect to migrants. The problem is that this type of citizenship does not recognize actual status differences. On the contrary, it presupposes status equality (citizenship as national membership) and treats it as a social fact. Such a presupposition also implicitly defines what is left outside this order. Citizenship (as a status) becomes a privilege of 'primordial' members of a national community. Understood in this way, it produces its opposite, i.e. statelessness as *de facto* 'otherness' and inequality. The simple fact of not belonging to a specific national community generates inequality of non-citizens. Statelessness arises from the fact that a person has found him/herself outside the community into which he/she was born (Arendt, 1948/1976). The moment we encounter an instance of unequal citizenship (as in the case of migrants) or statelessness (Arendt's thematization of the case of Jews during the Nazi era), the principle that everyone is a member of a political community is betrayed.

Arendt (1948/1976) argues that the right to have rights that enable citizenship is much more fundamental than human rights. It implies equal membership of a community, and this actually enables one to refer to human rights. To make human rights possible, it is first necessary to ensure the right to have rights. While Arendt problematized the factual statelessness of Jews, Somers (2008), in actualizing her theory, has established that today statelessness does not mean only non-membership of a national community. It is also applicable to the situation of migrants who are regarded as 'inappropriate' members of a community, those who allegedly threaten its security. Xenophobic and racist rhetoric that is increasingly adopted not only by extremist organizations but also by some political parties across the EU promotes migrants as a threat to European security. Migrants are presented as insecure 'elements' that endanger the well-being of 'first world' countries. Such discourses usually propose policy solutions that aim at the closure of borders, the expulsion of migrants without documents, the limiting or banning of migrants' labour etc.

At the same time, contemporary statelessness is also the ever-growing 'statelessness of *citizens*'. According to Somers, statelessness in contemporary times should be problematized as the exclusion or absence of certain populations (e.g. the unemployed, the poor) from the public sphere. Membership of a national community is therefore not a sufficient condition for citizenship; *de jure* citizens, too, can be non-citizens (Somers, 2008: 26–7). To illustrate that *de jure* citizenship does not automatically imply *de facto* citizenship, Somers (2008) uses the example of New Orleans residents affected by the hurricane Katrina. They lost the right to have rights in a situation of 'internal apartheid'. The unresponsiveness of political elites and inefficiency of the system led to them not only losing their homes, but becoming refugees or undesired migrants in their own country, seeking shelter while fleeing from catastrophe (ibid.: 63–117).

Social citizenship in the shackles of market fundamentalism

In Marshall's view, social rights would lead to social protection programs, and consequently full citizenship, through state regulation of the market and economic development policies. The welfare state was conceptualized as a project aimed at reducing the conflicts arising from labour market inequalities. It was expected to bring 'reconciliation', or at least mitigate the conflict between capitalism and political citizenship. It was meant to decommodify individuals, and reduce their dependence on the logic of the market. Social citizenship was based on the ethical principle of 'equal social worth' and was expected to equalize life chances (Handler, 2004: 88).

According to Marshall social citizenship includes minimal standards of housing, and citizens' participation in economic and social welfare through the system of social protection (which is often erroneously interpreted as a system of 'benefits'). It includes the right to education and health protection, reimbursement in the case of unemployment, the right to pensions, sick leave and social assistance. Social rights are therefore meant to ensure social safety, and their effect, in Marshall's opinion, should be the overcoming of the conflicts of capitalist society and the reinforcement of the feeling of belonging to a political community.

The paradox of social rights manifests itself through the conflict between their ostensible (distributive) justice (they should ensure a just distribution of social protection) and the assumption regarding their peaceful coexistence with the existing social and economic system. The opportunism of this concept seems to lie in the fact that social rights could be exercised in practically any socio-economic system (reformist, authoritarian, liberal and so on), since they arbitrarily adapt to various governing strategies. Ever since the introduction of social rights, it has been clear that they do not have the real power to eliminate the inequalities of the capitalist system but only have corrective potential. Therefore they can be seen as thin rights that actually maintain structures of inequality. If political and civil rights are, at least professedly, ensured once and for all, and believed to be above national interests, social rights are heavily dependent on national economic policies and daily developments on the labour market. These policies are directed towards protecting the national, so social rights are always limited for migrants.

Likic Brboric in her contribution to this book speaks about the 'liberal paradox' contained in the incompatibility of the rights-based perspective and the economic perspective on migration. While the first advocates transnational citizenship rights, the second promotes policies of labour market closure and limitations to freedom of movement. This paradox can help to explain the fallacies of the European social model that doesn't work because of its subordination to the interests of global capital. The author claims that while the model frames migration along the lines of partnership and solidarity, EU migration policies, specifically the border regime, prioritize policies of return and security of borders over migrants' rights. Social rights favour the members of a national community: the more one is integrated into society and the labour market, the easier one's access to social services.

The welfare state foresaw fully employed citizens; for them, access to social services is indeed facilitated, but not for those who most need it. The greatest number of social rights is thus enjoyed by highly productive workers with regular wages (Handler, 2004). Normatively, social citizenship is based on the principle of solidarity, but it seems that this does not function in practice. It is clear that social 'benefits' are enabled by workers who pay contributions. What, then, is the true meaning of the universality of social rights and solidarity? In practice, social rights are dependent on participation in the labour market; it is the labour market that enables the exercising of social rights. Criticism that the welfare state creates 'social dualism' seems valid (ibid.: 91). As such, the welfare state may be seen as an actor which deepens rather than resolves stratification, and reproduces divisions between deserving and undeserving receivers of social 'benefits', while circumventing structural relations which determine the situation of so-called undeserving claimants.

The principle of separating deserving claimants from undeserving ones is embedded in the welfare state system: equal access to social services is available solely to citizens, meaning those who contribute to the state budget by paying taxes. Migrants as non-members do not have access to these services (moreover, even the equality of citizens themselves is questionable). For migrants, this access is invariably conditional (Layton-Henry, 1991: 112–13), with those conditions including legal entry into the country, the right type of work or residence permit, adoption of 'right' citizenship (naturalization) and renouncement of 'inappropriate' citizenship, and so on. Distributive justice has often proved to be inaccessible to non-citizens. In this context, social citizenship can be critically viewed as a strategy for maintaining the superiority of the original community, and for constructing the Other as an outsider. Naturalization policies are often proof that this is so: before one is granted new citizenship, he/she is required to renounce original citizenship. With various narrowly defined policies, ideas about citizenship of the world, nomadic or diasporic citizenship (Balibar, 2004, 2008; Habermas, 2001) have gained urgency. These new understandings of citizenship recognize the transgressive aspects of contemporary migration and urge us to think about human relations in a transnational perspective, beyond the boundaries of the national state. Migrants' transnationalisms (various patterns of transnational family life, patterns of work-related migration, diasporic realities etc.) point to new fashions of interconnectedness on a global level and to new kinds of mobility across spaces. Classical understandings of citizenship that are tied to the national state and that have produced the equation of citizenship with nationality don't have the potential to grasp these realities, and thus seem inadequate. Ideas of world citizenship or of nomadic citizenship in contrast to classical understandings are not based in a national context; they put at the forefront the agency, the migrating subject and his/her mobility and aspirations towards greater security in a globalizing world. The ideas of world citizenship, of nomadic citizenship, as an alternative to classical conceptualizations urges us, as Balibar (2004: 177) would say, to 'attack the obsessive question of collective insecurity by beginning precisely with the situation of the most 'insecure', the nomadic populations.

During the 1990s, discourse on active work policy replaced discourse on the welfare state (*workfare* replaced *welfare)*. Under the new regime, citizens have both rights and duties. In return for the 'benefits' they must increasingly be ready for work; social rights are increasingly tied to work contracts (Handler, 2004: 140, 142) and lack of these for migrants also means lack of rights. Even if we accept the thesis about minimal differences between welfare and workfare as regards the active population, we cannot overlook the fact that the negative consequences of the workfare ideology, as Handler (ibid.: 209, 211) argues, most affect the most vulnerable groups, including long-term unemployed people and migrants. Their position is explicitly unequal to that of their employers and they have practically no negotiating options, which further aggravates their marginalized starting position.

The strategy of circular condition: social (non)citizenship and migrants' experiences in Slovenia

The lives of migrants living in contemporary national communities show that the disappearance of the political aspect of citizenship, i.e. the right to membership of a political community, and it's replacement by citizenship based on activity within the public sphere, has led to the reproduction of the statelessness of migrants as a group. The strategy of circular condition and selective admission into citizenship begins when a migrant enters the country. To be admitted to the margin of rights, entry must be legal. Migrants who enter a country undocumented are considered not only non-members, but 'illegal' non-members, i.e. twice non-citizens, those who threaten the so-called first-comers and pose a serious danger to the safety and security of their country. Migrants who enter a national community legally are confronted with what I call 'the strategy of circular condition' when they want to access the labour market, when undergoing the naturalization process etc., and this is an obstacle to their security.

In the next section I check the validity of social citizenship using the example of the labour market and migration policies in Slovenia, while examining various condition-setting mechanisms – work permits and the quota system, residence permit, access to social services, and conditions of naturalization. These place limitations on social citizenship. In problematizing these mechanisms, their practical repercussions for migrants will be examined, while their institutional aspect will be presented along with the migrants' narratives. The purpose of this analysis is to draw attention to the need for a critical deliberation about current policies that regulate migrants' citizenship. The need to rehabilitate citizenship is evidenced by the narratives of migrants and their experience with the current policies of selective admission to citizenship.

'Work your fingers to the bone and don't complain': labour market policies

Admission to citizenship is to a large extent dependent on the conditions of admission to the labour market. The analysis of laws and policies in Slovenia shows that the first condition for migrants' admission to the labour market is a work permit. There are three types of work permit: a) personal work permit, b) employment permit, and c) permit for work.[5] A personal work permit may be issued to migrants who have a permanent residence permit in Slovenia, refugees, asylum seekers, and self-employed individuals. In principle, it should entitle migrants to the same rights as those enjoyed by Slovenian citizens. The personal work permit is either renewable or permanent, and it enables a migrant to work for any employer or to be self-employed. It is also the only type of permit that allows migrants to register with the employment service as job seekers.

The second type of permit, the employment permit, is usually issued at the time of a migrant's first employment and is tied to the employer's need for labour. In contrast to the holders of personal work permits, who can switch to another employer, a migrant holding the employment permit is tied to the employer who applied for the permit on his/her behalf. While personal work permits are issued for an unlimited period of time, employment permits are generally issued for up to one year and may be renewed for a further period of up to one year. According to Employment Service data, most permits issued to migrants are of this type (69% in 2008).[6]

The third type of permit, permit for work, is based on demand for specifically defined migrant labour. It entitles a migrant to temporary employment undertaking work for which the permit has been issued. This permit has a fixed expiration date, and is issued to foreign representatives or foreigners who have been referred for a specific job, attend training or education courses in Slovenia, offer individual services or do seasonal work. According to Employment Service data this permit is often issued for work in the construction sector, agriculture, hotel, catering and tourism industries.

Work options for migrants are further limited by current demand on the labour market and by the quota system. The number of employment permits and permits for work must not exceed the annual quota of 5 per cent of the active population (according to the Employment and Work of Aliens Act 2000, personal work permits are not included in this quota). The quota system is usually described as the mechanism which enables the employment of migrants, contributing to an open society. The bigger the quota, the greater the number of working migrants, and the more open the society. What is not frequently mentioned is that the migrant work

5 Employment and Work of Aliens Act (2000).

6 Data collected as part of the Primts project in June 2009.

quotas usually imply so-called 3D jobs (dirty, dangerous, demanding) shunned by local workers. In practice this means that the quotas further limit the already limited options available to migrants.

Next we look into how policy regulating access to the labour market and the quota system affects the lives of migrants. We base our observations on interviews conducted with migrant women from various social and economic backgrounds and patterns of family life. Some of our interview partners have completed secondary or vocational school, others have university degrees, some are married with children, others single, some have left families in their country of birth, while all have experienced precarious labour such as working on the black market as a cook, cleaner, kitchen aid, waitress etc. Most of our respondents obtained a permit for work prior to their arrival in Slovenia. The interviews revealed that the existing permit system enables employers to exert self-will and reduces the negotiating options for migrants. Some interviewees told us that working conditions proved to be worse than those promised to them prior to their arrival. It often turned out that they had to work for lower wages with worse working conditions. Fearing that they would lose their jobs and consequently their permit for work, they had no other option but to endure the exploitation, for example, performing tasks beyond those previously agreed, working longer hours for low pay, being always available for work and unable to take sick leave. Some employers refused to extend their permits despite previous agreement, or tied the permit extension to additional duties.

> The first job in Slovenia was a catastrophe. I found myself somewhere where I don't belong. I worked in a bar, as a barmaid. I worked practically for nothing for two years. When I saw that they were going bankrupt, I wanted to go away. And then I realized that he did not register me [with the authorities]. Actually, that I was registered for half a year, but I worked for two years … And now I've been promoted, except that I don't known how much I'll be paid. It's irresponsible. I ask: 'How much will I get?' And he says: 'What d'you think, that you know everything, that you're smart …?' Is it the way to talk? Is it a boss, talking like that? … When I see that, I don't feel like going to work. … I live like a robot. I work Monday to Friday, seven to three, or eight to four, I do eight hours. I go straight to the school to fetch my daughter – she has extended hours at school – I come home, grab the vacuum cleaner, tidy up, do the linen, ironing, cooking, dinner. Then my daughter goes to bed … And on Saturday it's house cleaning, on Sunday it's cooking. I never have time to spare (Marija, 30, Moldova).[7]

> I came to Slovenia. All of us girls lived together, and all of us had temporary address there. It was tied to the contract, its duration, half a year. Other girls came for, say, 14 days, one month. And he does not register them. I had a half-year contract and I could go home whenever I wanted. That is to say, I had to

7 All names have been changed for ethical reasons.

leave 200 euros with my boss, as a warranty that I'll come back. If I didn't do that, if I stayed at home and he had to terminate the contract … it needs to be paid (Natalija, 24, Czech Republic).

I got a physical job. And that was thanks to myself, not someone else. I'm not satisfied with my work; in the beginning it was not that bad, let's say that for a young woman who didn't even know how to hold scissors, who couldn't speak Slovene, it was quite ok. I came and I had to try hard. In the literary sense of the word, I tore myself apart. I first tore myself apart for the sake of people, to establish contacts with them, then I tore myself apart for the sake of the system, because I needed it in order to be able to do anything at all. Then I tore myself apart at work. Then to get the money I earned … I work an awful lot. I sweated and I still do. And conditions are unacceptable. Every migrant has to adapt to them. Work your fingers to the bone, indeed, and you must not complain. Yeah, you must not complain … The state does not help. As to me, nobody helps me. Quite the opposite, the state wants something from you all the time. But when you need something, then no way, they'd never help you, never give in. And all that bureaucracy, the system, I really don't like it (Kizimka, 27, Central Asia).

After several attempts to find a job corresponding to their work experience and education, migrants are forced to accept work below their skills (*de-skilling*) and education level. To keep their employment permit and remain in Slovenia, they are willing to accept any kind of work; they become completely dependent on the employer and are forced to moonlight or accept black market jobs. They talked about losing the courage to grapple with long bureaucratic procedures for approving education certificates. After working for several years as unregistered cleaners or kitchen helps, they cease to hope that they will ever be able to perform the work for which they are skilled. Hard working conditions preclude further education. Our respondents also referred to their experience of discrimination.

So they said [during a job interview]: 'Tell us something about yourself.' So I started: 'I'm so-many years old, I come from Bosnia.' And they said: 'Stop.' I said: 'Why stop?' They were a man and a woman. The man said: 'Let her speak. See how well she speaks French.' She said: 'No, she's from Bosnia, she cannot speak.' I said: 'But why?' She said: 'Well, we don't take Bosnians.' So I said, *jebela cesta*, that was straight into your face. … And at the Employment Service, where I went for interviews I don't know how many times, they said: 'Why nobody takes you?' That's what I'd like to know, too. Then she said: 'You know what? Why don't you change your surname? You'll never get a job with that surname of yours.' … So I said: 'I won't, out of spite.' And even if I did. If I had, I don't know, some Slovenian surname, it'd be still obvious that I'm not a Slovene, from how I speak, so it's pointless (Ada, 39, Bosnia-Herzegovina).

> I got the job, although it required a lower level (of education). I have level 7 education, I graduated in economy and I have my certificates nostrificated in Slovenia. There was no problem getting a job – although that position required level 5 education, but I wanted to start from below, to gather experience slowly. At that time I had nine years of work experience in Russia, but I wanted to see how and what, how things were done, business procedures in general. That's why I went for a lower level position, an ordinary administrative position in a company that dealt with finances (Vera, 29, Russia).

> I danced for a while, then I went home and then came back – at that time it was not a problem. But it's different now – dancing is not what it used to be. In the past there were shows, professional ballerinas, but it's no longer like that. And when you do it for such a long time, it's no longer a pleasure … And you cannot find another job immediately. You can't. As simple as that. You have to look, to make an effort, find someone who is ready to help you as a foreigner, to employ you. It is also difficult to find a regular job. Well, I found it, I was lucky. Lucky indeed. To change job. To move from that bar to a normal company, to work as a cleaner. I worked for six years. I immediately said yes. I didn't care whether it was a cleaning job or whatever, all that I wanted was to move away from that. Then I worked as a cleaner, I stayed in Ljubljana for six years, cleaning offices. I cleaned offices in a bank, it was a regular job (Nika, 35, Ukraine).

'Citizen and still a foreigner': social services and naturalization

The migrants' work permit is tied to the permanent residence permit, and this is one of the strongest selective mechanisms for admission into Slovenian citizenship. The residence permit may be issued to migrants after they have been living in Slovenia for five years continually on the basis of a temporary residence permit.[8] As already mentioned, permanent residence is a precondition for obtaining a personal work permit, without which migrants are left to the good will of their employers. The EU directive concerning the status of third country migrants who are long term residents allows member states to require migrants requesting a permanent residence permit to prove that they have health insurance, adequate resources to support themselves and their families, and that they observe integration conditions in accordance with national legislation. Before the law on foreigners in Slovenia was amended, migrants seeking a permanent residence permit in Slovenia were asked to provide evidence of adequate resources to support themselves. The narratives below show the practical effects of such a policy:

> There was no way for me to register with the Employment Service, because I only had temporary residence [permit]. I went there [to the Employment Service]

8 Article 41 of the Aliens Act (2009).

to find something, some job … They told me I couldn't register because I didn't have the permanent residence permit. It's my great wish to work in my field, as a teacher, but I don't know how likely it is, because I have to speak Slovene perfectly, so that I can be sure when I write down something, or say something, that it's correct, to teach others, small children, teach them properly. I really don't know when it will be possible. And I'd like to ask, if there is a possibility, to get a 50% help from the Employment Service, or I don't know, I'd take a course, and then go on. I have a strong will! But small resources (Melanija, 42, Croatia).

I lived in Ljubljana for a year or so without documents so I couldn't work. I had simple problems, for example, I couldn't find any work, nobody wanted me because of that and I decided to do something about it. So I married, it was fake marriage, to obtain status and some basic rights without which you cannot, you cannot live in another country. I examined the Foreigners Act and I realized that it was the only option. So I found a husband … After the wedding the search started, applications for temporary residence, for permanent residence, all that. That process lasted about four years in my case and … the biggest problem was that the man I married didn't have a regular job, he was a free-lance artist, and that was not good enough for the state [administration] because they didn't have proof that he could take care of me in the material sense. That was quite a big problem. Because of that I couldn't get permanent residence status. As for obtaining documents, it was a wearisome process, I didn't know if I'd ever be through with it … (Ana, 35, Croatia).

The milestone was when we married, here in Slovenia. That was one such point that simplified a lot of things for me, if I compare myself to other women, regarding various documents, papers and such things. So I started to get things done, temporary residence, medical insurance, all that was necessary in order to live here; actually, the most important thing was that I got temporary residence, that I registered, obtained medical insurance and that I had a doctor (Anuška, 31, Slovakia).

The type of work and residence permit determines access to social services: employed migrants are included in the obligatory insurance scheme for unemployment,[9] but they can register with the Employment Service as job seekers only if they hold a personal work permit. Or, for example, employed migrants are included in the compulsory health insurance scheme but only if they have a permanent residence permit, otherwise they are left to their own ingenuity and depend on the self-sacrifice of doctors and employers. Migrants' rights to parental protection are also limited – they are not entitled to parental bonuses, while financial assistance at the birth of a child is tied to at least one parent having a permanent

9 Employment and Insurance Against Unemployment Act (2007).

residence permit, and the right to child bonuses to the child's residence.[10] The circumstances within the housing field are similar; migrants cannot rent a non-profit apartment and cannot own real estate.[11]

Speaking about poorly paid jobs they perform, our interview partners emphasized the difficulties they had because of high rents. Since they are excluded from financial schemes and since they have difficulties obtaining housing loans, they are forced to accept hard physical work, and jobs offered on the grey market. They also talked about discrimination; for example, some landlords refused to rent out their apartments or charged higher rents on the grounds that they were not of Slovenian nationality. Migrants who enter Slovenia on the basis of an employment permit arranged by their prospective employers often find themselves living in unbearable conditions. Some interviewees told us about living with other migrants in a small room deprived of privacy.

> Only a few years ago it was very difficult to find an apartment, especially for foreign women from Eastern Europe, perhaps because of bad reputation, because of things like women selling themselves. And because I come from Eastern Europe I too came across such obstacles when I was socializing … I tried to work legally, find a regular job, but it simply did not work, I still ran out of it. Even the social worker told me: 'Yes, your situation is difficult because you don't have your own apartment.' I cannot get a non-profit apartment, because I still don't have Slovenian citizenship. Even if I had it, it is a question how quickly I'd get it. There are many young families here, single moms, Slovenians, and they cannot get it, apart from some communal apartments in basements. And the social worker said: 'All you can do is moonlight and receive social assistance', because you simply cannot make it. Maternal home … I turned to them for accommodation, but it is fully occupied and that's only a makeshift solution for six months or so and then you have to look for it again. You don't get anywhere (Vera, 29, Russia).

> I teach English and I work on contract. You have a contract for one school year. Mainly for nine or ten months when schools are open. That means that you do not have your annual leave paid and you are less protected in general. You don't have health insurance, you have to arrange it on your own, and so on. That worries me a lot. What will happen in the future, what shall I do, where shall I work and what sort of job will it be? Of course I want to get a regular job, you can't always work on contract. It's not a job if you don't have insurance, it's the main problem. You don't have your annual vacation paid, no health and pension insurance (Tereza, 34, Lithuania).

10 Parental Protection and Family Benefit Act (2008).

11 Housing Act (2003) and the National housing program (2000).

Speaking about health issues, our respondents emphasized the inadequacy of legislation. Persons without regulated status have access to emergency health services only. This policy should be redefined and expanded to other medical services. We noted situations in which migrants who were unfamiliar with the procedures or did not have a good command of Slovene were denied access to medical services covered by additional insurance because, for example, they unknowingly paid less than the required additional insurance fee. Some migrants discovered that their employers did not pay the agreed health insurance contributions only when they needed medical assistance in a concrete situation. They are frequently denied their right to sick leave or maternity leave.

> When I was in hospital they found that I had problems of a specific nature and that's why I couldn't move my hand. Then they put me in plaster. I had to pay for the therapy from my own pocket. I had to pay myself. I didn't have the money. And if E. did not send it! E. sent the money for the therapy. They didn't pay for it, the ministry pays only for urgent cases. If you're dying they'll help you, if you are not, then … (Valbona, 41, Kosovo).

> I had three or four days to go before giving birth and I was without health insurance … The lady at the ministry told my husband to go to Trieste with my passport and get the tourist visa for me until I get papers in order. The residence permit. So my husband did it, he went to Trieste and got the tourist visa without problem. On the 15th in the morning I had labour pains and I still did not have health insurance. I went to hospital intending to pay from my own pocket. Because if you are a foreigner and you don't have medical insurance, you pay for everything. On the 15th we arranged everything, I don't know what happened but we did, so I got medical insurance through my husband who worked in a restaurant … That month was really, really difficult, so now when I think of it I can't believe that I made it through all of that. After I gave birth I had an allergy. Many doctors came to look at it, and they all said that they never saw something like that, so in the end their diagnosis was that it was a sort of, how to put it, a post-birth shock. That it was a trauma, all that stress I went through (Sandra, 29, Serbia).

> When I first went to the clinic she wrote a prescription for insulin, which I needed; at that time I was in Slovenia for, I don't know for how long, one week, and of course I couldn't speak Slovene. Because I only just came, I started speaking in Bosnian, apologizing for disturbing her, saying that I had a big request, I only just arrived. She simply shut the door in my face and said: 'Don't come back until you learn Slovene' (Ada, 39, Bosnia-Herzegovina).

Slovenia has one of the more restrictive citizenship policies: the condition for granting citizenship is ten years of residence in Slovenia.[12] Applicants are required to demonstrate their communication skills in Slovene, to prove that they have adequate financial resources, that they have never been sentenced to more than one year in prison, and to renounce their existing citizenship.[13]

Of the 26 migrants we interviewed, six had been granted Slovenian citizenship. All of them recounted long procedures that governed their lives for ten years. The granting of citizenship was a milestone that facilitated their daily lives. At the same time, however, the majority of those who obtained citizenship stated that their lives did not change radically and that the acquiring of citizenship did not solve all problems. They had difficulties with finding work even after they had acquired citizenship, were still forced to take jobs on the grey market, or perform unskilled work below their qualifications level.

> Now when I am a citizen it's easier, but no, although I'm a citizen, I'm always a stranger! I'm still a stranger, because when people watch me on the street they surely say 'that woman is a stranger', they will never say that I'm a Slovene … People here keep to themselves, they don't care about people they don't know, especially strangers, few people really want to help you (Katarina, 44, South America).

> For me citizenship has no meaning, citizenship does not give me anything. I can travel with the Croatian passport … I have the work visa, I have health insurance, I have all that I need to live here. The passport is one step forward but I actually don't need it now. But some people do. The Slovenian passport enables you to travel freely, to move around. I don't have problems with that. I think nobody is worried about citizenship if he has the passport to move around. Because it is the main reason why you want citizenship of a country. Because it

---

12 Similar conditions are in place in Austria, Greece, Spain, Portugal and Luxembourg, and in Italy where this condition does not apply to the citizens of EU member states.

13 The conditions for regular naturalization stipulated in the Citizenship Act (2003) are as follows: 1) 18 years of age or over; 2) a release from current citizenship or proof that such a release will be granted if the person acquires citizenship of the Republic of Slovenia; 3) residence in the Republic of Slovenia for the period of 10 years, of which at least five years prior to the application for citizenship must be without interruption; 4) guaranteed residence and a guaranteed permanent source of income sufficient to enable material and social welfare; 5) active command of the Slovenian language demonstrated in an obligatory written and oral examination; 6) the applicant must not have been sentenced to a prison term longer than one year in the state of which the person was a citizen or in the Republic of Slovenia; 7) no ban on the person's residence in the Republic of Slovenia; 8) the person's admission to citizenship of the Republic of Slovenia must not pose a threat to public order or the security and defence of the state; 9) no outstanding tax obligations; 10) the person should take an oath that he/she will respect the free democratic constitutional order based on the Constitution of the Republic of Slovenia.

is well organized and its passport is recognized so you can travel around (Ana, 35, Croatia).

Redefining citizenship

We have defined the policies governing employment, residence, access to social services and acquisition of citizenship as the strategies of circular condition. One permit is tied to another one. Without a permit for work it is not possible to obtain a residence permit; without a personal work permit it is not possible to obtain a permanent residence permit; without a permanent residence permit it is not possible to obtain citizenship; without a personal work permit it is not possible to register with the Employment Service; without a permanent residence permit it is not possible to join the medical insurance scheme; without citizenship it is not possible to apply for non-profit rents, and so on.

The circular condition policies create the 'statelessness of migrants'. The right to have rights is denied through the circular condition processes and selective admission to citizenship. In Balibar's words (2008: 530), although migrants are conditionally inside the system, 'they are located 'outside' the system'. We have also shown that *de jure* acquisition of citizenship does not solve migrants' statelessness. Migrants are excluded from the political community and left at its margins. When after ten years of residing in Slovenia they prove that they have a good command of Slovene and that they can earn for their living, and even when they *de jure* acquire citizenship, they continue to be 'non-citizens within the country'; they are in the position of internal exclusion (ibid.). Migrants always live in conditional relations, 'robbed of their humanity' (Arendt, 1948/1976); being invisible and unheard within the public space, they have no other option but to 'work their fingers to the bone without complaining', as one interviewee put it.

Contemporary strategies that create statelessness call for the reconceptualization of citizenship in the direction of politicizing it. *De jure* citizenship is not necessarily a solution for migrants' statelessness. The latter can be avoided only by recognizing migrants as equal members of a political community and by opening the public space towards post-national deliberation in which migrants would have a visible place. An inclusive public sphere resting on the egalitarian principles of political culture can only emerge within the context of plurality. Discursive, deliberated citizenship presumes the openness of institutions and their sensitivity to the topics, values and directions arising from the lifeworld (Habermas's *Lebenswelt*). Citizenship is more than simply an aggregate of individual interests and the passive exercising of rights paternalistically imposed on citizens from above. It requires interactivity, 'an interplay between institutionalized processes of communication and informal networks of public communication' (Habermas, 1994: 32).

Empirically, the rehabilitation of citizenship entails (at least) the redefinition of the structural framework in the direction of participation. If debates on citizenship are dominated by talk about rights and duties, then the normative framework in postmodern society needs at least structural changes in the direction of participation, a 'social and economic structure of chances' (Adriaansens, 1994: 67), redefinition of the status-based nature of citizenship towards a 'more solidaristic form' (Fraser and Gordon, 1994: 105), and towards citizenship as global political co-responsibility (Falk, 1994). Balibar (2008: 536) advocates the politicization of social citizenship in such a way that it would not operate as incessant conditioning that consistently keeps non-citizens on the margins of rights. (Social) citizenship calls for a reform that would move it away from economic fundamentalism and ambitions to maintain (fictitious) 'original ethnicity'. It calls for the type of imagination that would see and realize citizenship as an equal belonging to a political community.

References

Adriaansens, H. (1994) 'Citizenship, Work and Welfare' in Van Steenbergen, B. (ed.) *The Condition of Citizenship*, London: Sage

Arendt, H. (1967) *Vita Activa oder Vom Tätigen Leben*, München: R. Piper & Co.

Arendt, H. (1948/1976) *The Origins of Totalitarianism*, New York: Harcourt

Balibar, É. (2004) *We, The People of Europe? Reflections on Transnational Citizenship*, Princeton, Oxford: Princeton University Press

Balibar, É. (2008) 'Historical Dilemmas of Democracy and their Contemporary Relevance for Citizenship', in *Rethinking Marxism*, Vol. 20, No. 4, pp. 522–38

Dean, K. (2003) *Capitalism and Citizenship: The Impossible Partnership*, London: Routledge

Delanty, G. (2000) *Citizenship in a Global Age: Society, Culture, Politics*, Buckingham: Open University Press

Falk, R. (1994) 'The Making of Global Citizenship' in Van Steenbergen, B. (ed.) *The Condition of Citizenship*, London: Sage

Fraser, N. and Gordon, L. (1994) 'Civil Citizenship against Social Citizenship?' in Van Steenbergen, B. (ed.) *The Condition of Citizenship*, London: Sage

Habermas, J. (1972) *Theorie und Praxis: Sozialphilosophische Studien*, Frankfurt am Main: Suhrkamp

Habermas, J. (1994) 'Citizenship and National Identity' in Van Steenbergen, B. (ed.) *The Condition of Citizenship*, London: Sage

Habermas, J. (2001) *The Postnational Constellation: Political Essays*, Cambridge, MA: MIT Press

Handler, J. F. (2004) *Social Citizenship and Welfare in the United States and Western Europe: The Paradox of Inclusion*, Cambridge: Cambridge University Press

Hobbes, T. (1651/1985) *Leviathan*, London: Penguin Books

Layton-Henry, Z. (1991) 'Citizenship and Migrant Workers in Western Europe' in Vogel, U., Moran, M. (eds) *The Frontiers of Citizenship*, 107–24, Hampshire: Macmillan

Lister, R. (1997) *Citizenship: Feminist Perspectives*, New York: New York University Press

Marshall T. H. (1950) *Citizenship and Social Class*, London: Pluto

Pajnik, M. and Bajt, V. 'Migrant Women's Transnational Experiences: Family Patterns and Policies', in *International Migration*, Vol. 48, No. 3 (forthcoming)

Slany, K., Kontos, M. and Liapi, M. (eds) (2010) *Women in New Migrations: Current Debates in European Societies*, Krakow: Jagiellonian University Press

Somers, M. R. (2008) *Genealogies of Citizenship: Markets, Statelessness, and the Right to have Rights*, Cambridge: Cambridge University Press

Soysal, Y. N. (1994) *Limits of Citizenship: Migrants and Postnational Membership in Europe*, Chicago: The University of Chicago

Trimikliniotis, N. (ed.) (forthcoming) *Contested Integration, Migration and Societal Transformation: European Dilemmas and Challenges*, Nicosia: University of Nicosia Press

Chapter 14

Plastic Citizenship, (In)securities and Processes of Abjectification: The Case of Albanian Migrant Women in Greece

Gabriella Lazaridis and Anna Maria Konsta

Introduction[1]

Relations between Albania and Greece today are shaped by the contradictory and ambivalent coexistence of discourses sustaining a feeling of belonging to a disowned and disowning nationalist project, which was embedded at the core of the communist experience in Albania, and at the core of the expansionist doctrine known as the 'great idea' in Greece, which aimed to bring under its jurisdiction all Orthodox Christian areas in the Balkans, including part of South East Albania known as 'Northern Epirus'. In this part of Albania there is a relatively large Greek-Orthodox minority, the size of which is estimated by Albanian authorities to be around 59,000 whereas the Greek government puts it at around 260,000, justifying the disparity in terms of attempts made by the Albanians to weaken the Greek minority and violate its human rights (Lazaridis, 1999: 109).

This chapter focuses on the experiences of Albanian migrant women, making a distinction between 'ethnic-Greek-Albanians' (originating from Northern Epirus) and 'other' Albanians, looking at the differential treatment of these 'strangers' by the host country and at ways in which they try to carve out spaces of control. As we will show, different members of the Albanian migrant population in Greece are positioned differently in relation to different dimensions of citizenship[2], according to their gender, ethnicity, legal residence status etc. We claim that there is a *plastic citizenship* emerging in relation to these women where boundaries are blurred and processes of becoming or not are fluid, changing over time and influenced by notions of who should belong and who should not, who is entitled to what rights and who is not; at the same time, we argue that this *plastic citizenship* leads to a

---

1 This chapter is based on the findings of a research project by Gabriella Lazaridis, funded by the Leverhulme Trust (Grant No: F/00230/D, 2001-3), in collaboration with the University of Sussex.

2 As Erel (2003: 263) writes, 'there are different levels of citizenship (i.e. legal, social, political …), different aspects of citizenship, i.e. active/passive and public/private … as well as different tiers of citizenship (local, regional, national, transnational)'.

transformation of their subjectivities into *plastic subjectivities*, which can resist and can have all kinds of possibilities to wiggle and escape from the rigidity of the societal structure.

The chapter also deals with the dislocated and uprooted female legal subject, who shares time, space and bonds of affection with the country of origin and the country of destination. New transnational forms of female subjectivity are emerging in Europe today, which are regulated by immigration, family and labour laws. Greek and European legal regulations prove to be insufficient to cope with the needs of the transcultural female legal subject. National and EU citizenship rights, if interpreted in a narrow sense, do not apply to these migrant women. The transcultural female legal subject finds herself excluded from the host society. In this respect migrant women are transformed by law and through law from legal *subjects* into 'legal *abjects*'. As we will show below, the process of *abjectification* is not linear and differs according to one's differential inclusion and/or exclusion in the host country.

Theoretical Context

A contemporary differentiated legal subjecthood in a Member State (MS) and the EU context opens the doors to political, economic, social and legal relations of great complexity amongst individuals whose legal status may become more and more differentiated from the so-called 'neutral', but predominately 'male', legal subject of positive law. In this context, recent EU legislation on equality and non-discrimination goes beyond the boundaries of traditional citizenship rights, which are reserved only to MS nationals and takes into account the transcultural legal subject, the bearer of rights and obligations deriving from various legal systems and legal cultures (see Ong, 1999 on mobile subjects who respond to dynamic borderless market conditions). Migrant women enjoy a transnational subjectivity and a multiple belonging. For instance, migrant women may be transnational mothers, dividing their time between one country in which they work, and another in which they share time and space with family members. This mobility may be possible and at the same time limited by legal regulation.Women are often victims of multiple discrimination on the grounds of sex, race, ethnicity, religion, beliefs, legal, economic and social status. The predominant role of economic migration in Europe today requires the reconception of citizenship rights and a proper institionalization of so-called European social citizenship (see papers by Pajnik and Likic Brboric in this volume). Contemporary reality requires citizenship not to be restricted to groups which claim to be ethnically and culturally homogeneous. Greek citizenship and immigration legislation, which views citizenship merely as 'formal membership' of the Greek polity, does not leave room for accommodating ethnic and cultural diversity within the nation-state and for the mutual recognition of rights and diversity of both the majority and the minorities (Kymlicka, 1995).

Arendt (1951) defines citizenship as the right to have rights. Thus, citizenship is a prerequisite for the enjoyment of human rights. This broader conception of citizenship allows us to reflect on the concept of subjectivity. The migrant is by definition the unprivileged legal subject in Europe today who is constantly shaped, changed, eventually altered; rights are conferred to and taken away from migrants according to the interests of supranational or national entities. According to Malabou (2004), the subject is plastic, not elastic, it never returns to its original form, it may be shaped, but in the process of being shaped it undergoes a sort of transformation into something new. A process of trans-subjectivation of the subject occurs in Foucaultian terms (Konsta and Lazaridis, 2010; Foucault, 2004: 214), which can produce a new self. This new self however retains elements from the old self. Malabou (2004) explains the concept of *plasticity* as follows: 'The subject is not supple and soft, and it is not rigid either; it is something in between. The subject is "plastic". Plastic, if you look in the dictionary means the quality of a matter, which at the same time is fluid but also resisting. Once formed, it cannot go back to its original state. For example, when the sculptor is working on the marble, the marble, once sculpted, cannot be brought back to its original state. So, plasticity is a very interesting concept, because it means, at once, both openness to all kinds of influences, and resistance' (Vahanian, 2008: 6). *Plasticity* might be the name of this trans-subjectivation (Vahanian, 2008: 5). Migrant women undergo a process of trans-subjectivation when they are granted limited citizenship rights, through the regularization process, for example (see Konsta and Lazaridis, 2010). They are transformed into new subjects of law; they lose their old subjectivity, and new subjectivities emerge, differentiated from one another, which, even though one may think they are shaped by public authorities and that they remain static, unable to escape their newly created form, incorporate sperms of resistance which eventually will be revealed. Foucault argues that power, the power over subjects, always creates resistance (Foucault, 1978: 96).

People may be the subjects of law, the bearers of rights and obligations, but they can be transformed by law and through law into objects, or, in Kristeva's terms, into *abjects* (Kristeva, 1992; Konsta, 2009). The concept of *abject*, which originated in the works of Kristeva (1992), exists in between the concept of a subject and the concept of an object. The *abject* is something which used once to be a subject, and which can provoke a traumatic experience in one who comes into contact with it. A corpse, for example, used once to be a subject; it is something that should be alive, but isn't. The corpse especially exemplifies Kristeva's concept since it literalizes the breakdown of the distinction between subject and object that is crucial for the establishment of identity and for our entrance into the symbolic order. What we are confronted with when we experience the trauma of seeing a human corpse (particularly the corpse of a friend or family member) is our own eventual death made real. As Kristeva (1992) put it, 'the corpse, seen without God and outside of science, is the utmost of abjection. It is death infecting life'.

On the level of our individual psychosexual development, the *abject* marks the moment when we separated ourselves from the mother, when we began to recognize a boundary between 'me' and 'other', between 'me' and '(m)other'. According to Kristeva, the *abject* has to do with 'what disturbs identity, system, order. What does not respect borders, positions, rules' and, so, can also include crimes such as Auschwitz. Such crimes are *abject* precisely because they draw attention to the 'fragility of the law'. Thus, 'legal *abject*' is connected to 'legal subject', whereas legal subject means the bearer of rights and duties. The contrary, i.e. the one with no rights and duties at all, could be characterized as a legal object, a *res* (a thing). The legal *abject* refers to people and not to things. So, a legal *abject* represents the form of utmost exclusion in society and *abjection* describes the state of often marginalized groups, excluded bodies by law and through law who eventually become repulsive to society at large, such as the drug addict, the homeless, the convict, the poor, the aged or the disabled, and, in the case of this chapter, migrant women (Konsta, 2009).

Here we apply the notion of *plasticity* in citizenship, implying that it emerges when boundaries are blurred and processes of becoming or not are fluid, changing over time and influenced by notions of who should belong and who should not, who is entitled to what rights and who is not; we argue that this *plastic citizenship* leads to a transformation of women's subjectivities into *plastic subjectivities*, which can resist and can have all kinds of possibilities to wiggle and escape from the rigidity of the societal structure (Konsta and Lazaridis, 2010). At the same time, we claim that these women may be transformed by and through EU and Greek immigration law into legal *abjects*. The *abjectification* process undergone by these women leads towards exclusion and utmost insecurity. The *abject* is a metaphor for the ultimate form of exclusion. In this chapter, we embrace the 'human security' approach which is calling for the pursuit of the security of individuals, not just the security of their nations and of territory. In other words, it is about the security of people. As Karyotis states in this volume, 'this is based on a core, moral belief that the only appropriate referent object of security is the individual, whose welfare, wellbeing and freedom cannot be compromised in order to safeguard the state as a political unit'. So here we examine migration not as a perceived threat to state security but through a humanitarian perspective in terms of protection of the human rights of the 'other'. In order to achieve this, citizenship cannot be viewed as a contractual relationship between the State and the individual, but as an association based on solidarity and mutual responsibility.

The different types of citizenship emerging in the Greek legal order have different security-insecurity implications for the state and for the migrant. The irregulars are in danger of immediate deportation as opposed to the regularized ones. Ethnic-Greek-Albanian women face less discrimination but run the risk of unemployment in times of recession and due to their engagement in the informal economy (which is nowadays estimated to be around 40% of GDP), encounter difficulties in obtaining social security and other benefits from the employers. In reality, the current economic collapse in Greece has resulted to a sort of pseudo-security for all migrants, illegal or regularized ones, since they eventually end up working in the informal economy or joining the queues of the unemployed. The choice of the EU and Greek public authorities to grant differentiated citizenship rights in a context of security considerations has resulted in human insecurity for these immigrants, which may eventually jeopardize state security. As De Genova (2007: 440) argues, '... during an era when the abjection of non-citizens has become scandalously routine and the insecurity of citizens has been rendered a political resource of onerous gravity, the gathering revolt of the denizens may yet signal a stringent clarification of our universal political predicament – as always-already susceptible to suspicion, always-already potentially culprits' (De Genova, 2007: 442). He adds: 'Within the global regime of capital accumulation, the more flagrant abjection of the world of denizens only shows, to the more properly domesticated citizens, the image of their own future' (ibid.). One way or another, we are all marginalized, susceptible to a possible process of *abjectification.*

Methods

Extensive fieldwork and data collection took place in Athens. Athens was chosen as it contains the main concentration of Albanians in Greece, being a metropolis where low-cost migrant labour is much in demand in various sectors of its economy. A large number of in depth, semi-structured interviews with migrants [41 in Athens (22 with 'other' Albanians and 19 with ethnic-Greek-Albanians, 19 with women and 22 with men)], 69 with migrants' associations, anti-racist organizations, NGOs and state agencies, employers and other key informants were conducted in Greece. Given the known high level of suspicion amongst Albanian migrants, the interview subjects were selected carefully and approached with sensitivity. The role of intermediaries was critical here, combined with carefully controlled snowball techniques. Although the migrant population surveyed was unknown as far as its strict demographic and social parameters are concerned and hence not subjectable to a rigorous sampling frame, we tried to achieve intuitive representativeness by interviewing people across age groups and socio-occupational backgrounds.

The EU Context

In the EU context, different kinds of citizenship rights apply: regular European Union nationals' citizenship rights, rights to non EU nationals married to an EU citizen, rights to non-EU nationals. The two EU non-discrimination directives, the Race Directive 2000/43/EC and the Employment Directive 2000/78/EC, grant non-discrimination rights to non-EU citizens as well. A kind of a *quasi-social citizenship* on behalf of migrant women emerges in the context of these two directives, which is constantly undermined by immigration legislation, which is restrictive of the rights of third country migrants. The predominant role of economic migration in Europe today requires the re-conception of citizenship rights and a proper institutionalization of so-called European social citizenship, which would entail more consistency between non-discrimination and immigration legislation aimed at the protection of rights of all people living and working in EU territory (see also chapter by Pajnik in this volume).

Gender equality and non-discrimination law in the EU context is the most developed law in the world today, and has elaborated concepts that go beyond formal equality and direct discrimination, incorporating, for example, indirect discrimination and affirmative action. However, this law is somehow not meant to apply to its full extent to migrant women. Women who are EU nationals enjoy a full set of rights and privileges, while migrant women face restrictions to these rights imposed by severe immigration law. Gender studies in this context need to shift the focus onto the 'unprivileged' legal subject, and migrant women and their children are by definition **the** unprivileged legal subject in Europe today (ibid.).

The issue of differentiated subjectivities for women arises in this context as well. Migrant women who are mothers of EU citizens enjoy more social citizenship rights than migrant women who are not mothers of EU citizens. The right of residence in a MS involves a hierarchy amongst migrant women, since it is often connected with their own and their families' biological survival, and rights in this respect are not equal for all migrant women. Law shapes migrant women's subjectivities in this respect. At the same time law connects social citizenship rights for migrant women to motherhood, perpetuating the public/private sphere division.

The Directives adopted under article 13, i.e. The Race Directive (2000/43/EC), The Framework Employment Directive (2000/78/EC) and The Goods and Services Gender Directive (2004/113/EC), grant a sort of *plastic citizenship* to women by including certain rights on their behalf, and at the same time excluding others. Non-discrimination in relation to a woman's race and equality of treatment in employment are well protected by formal EU social law. However, the limited application of the Framework Employment Directive only to employment, leads to lack of protection for women when multiple discrimination occurs on other grounds besides gender outside employment. Also, The Goods and Services Gender Directive's material scope is very limited, excluding areas such as education, domestic violence, taxation and the media (areas which were initially inserted in the Draft Directive by the Commission, but were removed after it became clear that no consensus could be reached inside the Council on these matters (Masselot, 2007: 153). 'The limitation of the scope of anti-discrimination law on the basis of sex therefore sends the message that some forms of gender inequality and discrimination are acceptable, while others are not allowed as far as racial discrimination is concerned' (ibid.).

Of course, one should not neglect the fact that gender equality in the EU has taken the form of a fundamental right (especially after the ratification of the Lisbon Treaty, which makes article 23 of the EU Charter of Fundamental Rights legally binding for the MS). Article 23 requires equality between the sexes to be ensured in all areas. As Kravaritou (2003) states, 'That equality "must be ensured in all areas" means as much that men must assimilate to the social roles undertaken by women as vice versa. The emphasis on the global dimension of equality in article

23 – in the public, work and private spheres – indicates awareness of the breadth of problematic gender relations in European society'.

Background context

Country of origin

The gender regime underpinning social relations in Albania is embedded within an authoritarian and personalistic model of social relations prevalent within an agro-pastoral mode of production, an extended patriarchal family characterized by the supremacy of the adult male as the moral authority over all other members, and rooted in the 600 years old 'Kanun of Lek', a specific version of the Albanian customary law. It is more common in the north part of the country (especially areas which did not benefit from Hoxha's cultural revolution) and less in the south where other local versions of customary law are practised.

In the south, these values and traditions were partially challenged by the Greek Orthodox cultural tradition, and its patriarchal values and institutions. Unlike the honour-shame value system (Lazaridis, 1995), the Kanun bases its strength not on individual responsibility but on the institution of the extended family clan, with authority the exclusive prerogative of men. Unlike the honour-shame value system, under the Kanun, males dominate private as well as social life. According to Article 29, 'a woman is known as a sack, made to endure as long as she lives in her husband's house'. According to customary law a husband 'is entitled to beat his wife and to tie her up in chains when she defies his word and orders ... The wife is obliged to kneel in obeisance to her husband' (quoted in Ash, 1974: 233 and in Hall, 1994: 82). The husband is entitled to kill his wife for adultery or refusal to attend to some sacred duties, such as honouring guests. The parents of the bride used to give the husband a bullet during the ceremony of marriage to express their giving away of their daughter's life. The woman's purpose in life is to bear children, work in the house and renounce marriage, and to become the man in the house if all men are killed in a *vendetta* (blood feud).

Similar to the Kanun, the honour-shame value system prevalent in the southern parts of Albania demanded honour (*besa* or *philotimo*), hospitality (*filoxenia*) and pursuit of retribution through the blood feud (*vendetta*). This acted as a means of social control by which individuals and families had the right and duty to protect their and their women's honour. In the Muslim region they are confined to the home and have to wear the veil. In Albanian culture the birth of a girl was seen as a misfortune and, as an old Albanian saying put it, 'even the beams of the house shed tears when a girl was born' (Prifti, 1975: 112). Arranged marriages have been the norm.

The communist regime of Enver Hoxha who ruled Albanian for over 40 years until his death in 1985, reinforced the isolation of Albania from the rest of the world. Under the communist regime these values never completely lost their influence in the areas in which they operated, and Albanian contemporary society is still characterized by the overwhelming supremacy of the institution of the family and its patriarchal values and practices. Within this context, according to Mai (2001: 268) 'violence and abuse against women have easily found legitimation in Albania'. Under the communist regime, although formal equality between men and women was legally sanctioned, women were gradually granted entitlement to education and employment, and the concept of the head of the family was undermined, hegemonic patriarchal mentality remained. This continued even after the institution of the patriarchal extended family clan was attacked during the Cultural Revolution of the 1960s as a perceived major impediment to the creation of a communist state. Women had to work, subsidized day care nurseries were established, and canteens were set up at neighbourhoods and workplaces with labour saving devices aimed at lightening the women's load and assisting them in their role as workers, mothers and wives. 'Communist laws and propaganda were fairly effective in convincing many people that the *Kanun* in northern Albania was no longer respected' (Young, 2000: 133).

Until the late 1980s, strict bureaucratic, border and social controls (social surveillance via an extensive network of informers operating at every level of social life) were rendering Albanian migration impossible. Albania's exposure to the outside world was next to nil. Telephone conversations were checked, foreign television was forbidden, contemporary foreign literature was not available, the press was state-run and controlled, and internal migration was made difficult by the bad road network connecting the north to the south. People wore uniform-like clothes, and jewellery and anything that was seen as not strictly functional was banned.

Due to the economic crisis that hit Albania at the end of the 1980s, the collapse of the communist state in 1991 and the politico-economic changes that followed, and 'the ensuing lack of any effective policing and judicial system, the revival of the *Kanun* laws soon became apparent in its early area of control (northern Albania)' (Young, 2000: 133). They form the basis of Albanian morality and govern rules of behaviour on every level (ibid.: 51). In addition, the resulting chaotic rapidly urbanizing market economy and change from extreme scarcity of goods to exposure to new values and ways of life once equated with moral depravity meant that both men and women experienced insecurity. Acquisition of foreign consumer goods and 'way of life' became the 'dream to be fulfilled', and migration became a strategy through which the dream was to be realized.

By and large Albanian women migrate *en famille*; however, there are instances where women try to escape economic hardship and break free from the submission and violence they experience at home by migrating solo. Stimulated by the world of 'freedom and wealth' seen in the media, and/or deceived by promises of a good job, they decide to migrate abroad underestimating the level of insecurity, exploitation and exclusion they are likely to experience. Yet others are traded by their own family members for cash to those involved in organized trafficking. So past values which legitimized the disposal of women/brides for cash have now been re-adapted to legitimize and serve the world of trafficking for prostitution.

Women have experienced a process of reification in traditional Albanian society according to the *Kanun* laws. They were not considered to be legal subjects, but legal objects. The communist regime intervened only superficially to improve their position in society, and the *Kanun* laws have been conserved to some extent until today. Albanian migrant women in Greece, which is an EU member state with a modernized gender equality law, could have escaped the burden of the *kanun* tradition, and become legal subjects, enjoying rights similar to those of all women in EU countries. The hostile EU and Greek immigration laws however, did not allow them to overturn their degraded status in society or to champion their emancipation from the past. The host society's legal framework on immigration antagonized the gender equality framework, and contributed instead to the creation of all the necessary conditions that lead to a process of *abjectification,* which may eventually lead to their exclusion from society. Albanian migrant women in Greece could have regained their subjectivity, but Greek immigration policy and law instead of giving them the chance to become legal subjects transformed them in many instances into legal *abjects*.

Destination: policy/legal framework and society

In the early 1990s Greece, once a country of emigration, experienced a mass influx of migrants, the majority of whom came from Albania. The Greek government, unprepared, reacted by introducing in 1991 Law 1975/1991, which left the majority of migrants under a clandestine status. In 1997 a bilateral agreement was signed between the Albanian and Greek governments regulating the 'invitation' of Albanian workers on a seasonal contract basis. After a lot of deliberation and lobbying from various pressure groups the government introduced a series of regularization programmes (Lazaridis and Poyago-Theotoky, 1999), but these were inundated with implementation problems at all levels, mainly institutional bureaucratic ones, characterized by racist attitudes and corruption. But most importantly, although third country migrants who are regularized are entitled to equal rights at work (that is equal payment for the same job as Greek workers, and welfare and social insurance coverage rights), the majority still get jobs in the informal sector; moreover, regularization has granted them temporary, transient status, resulting in their entrapment in a 'regularization cycle'. For example, most work permits were issued for one year and then had to be renewed, but, due to bureaucratic delays, immediately after one received the permit one had to start the procedure of applying for its renewal all over again. To compensate for delays, the government introduced during the 2001 and 2005 regularizations a 'semi-regularized' status by issuing 'certification of having submitted an application for a residence permit'; this document protects the applicant from deportation without constituting a valid residence permit. One is entrapped in an in-limbo status, deprived of entitlement to social benefits and of freedom of movement. '*Ke na thelo na pao stin Alvania, then moro*' ('Even if I wanted to go to Albania, I can't'), a woman said.

As we have shown in our paper *'Plastic citizenship and plastic subjectivities in Greek immigration policy'* (Konsta and Lazaridis, 2010), the Greek state has opted to grant *differentiated, stratified statuses* to three groups of immigrants, that is residence permits to Albanians, quasi-citizenship to ethnic Greek Albanians, and formal citizenship rights to Pontians from former Soviet Union countries. The implementation of the process resulted in structuring: a quasi-documented status and in-limbo, marginalized belongings in the case of the Albanians; a degraded, devalued, vulnerable, insecure lower class status in the case of ethnic Greek Albanians; and finally the administrative shortcomings of the Greek civil service were illustrated in the implementation process in the case of the Pontians too, especially when they need to prove their Greek origin. Albanian migrant women fall under the first two categories as they, as well, have been granted limited citizenship rights which have led in their case to different degrees of (in)security.

The notion of *plasticity* in citizenship is also observed in the national context. Currently, Greece's major legal instrument concerning immigration is Law 3386/2005 of 18 August 2005 regarding entry, residence and social integration of third country nationals into the Greek territory. The Ministry of Interior has already issued two circulars on the application of this law: number 26 (1 September 2005) and number 38 (23 December 2005). Law 3386/2005 has been revised by Law 3536/2007 of 22 February 2007 with regard to determination of matters of migration policy and other issues falling within the competence of the Ministry of Interior, Public Administration and Decentralization.

The most important provision of the 2007 Law was the one regarding the procedures for the regularization of undocumented immigrants. Under the 2005 Law, hundreds of thousands of illegal immigrants were not able to obtain a residence permit. Aside from visas and stamped passports, an immigrant could be legalized provided that public documents certifying that one of the following had occurred before the end of 2004: a) the birth of a child in the country, b) marriage to a citizen of Greece or another EU country, c) a child's enrolment in primary or secondary school.

The 2007 Law abolished the obligation to supply an employment contract for the renewal of residence permits for construction workers, private nurses and domestic workers, with more than one non-permanent employer. Moreover, immigrants in paid employment could officially buy insurance stamps at 20% of the number needed annually, in order to renew their residence permits and not to be forced, as in the past, to buy them at extremely high prices on the black market. They could also be employed in regions other than the one in which their residence permit has been issued, one year after this permit has been granted.

The 2007 Law also established for the first time a national committee on the social inclusion of immigrants, including representatives of local government, parliamentary parties, social partners and the academic community.

Article 53 of Law 3386/2005, as supplemented by Presidential Decree 131/2006, provided that third country nationals who have lawfully resided in Greece for a period of two years or longer are permitted to request the entry and residence of their family members in Greece. 'Family members' according to Greek Law are considered to include spouses who are over 18 years old, their common children (including adopted children), or any other children of the person applying for family reunification or of his/her spouse, who are unmarried and are younger than 18 years old. Parents and grandparents are not considered by Greek Law, and Common Law partners are not recognized under Greek Family Law. In the case of polygamy, if applicants have already a spouse residing with them in Greece, family reunification with any other spouse is prohibited.

Presidential Decree 106/2007 and circular number 10 (28 February 2008), which dealt with free movement and residence of EU citizens and members of their family, stipulated that same-sex partners who are registered as such, or who have performed a legal wedding in another EU member state, are not considered family members.

In December 2008, Law 3731/2008 was adopted. This facilitated the regularization but not the naturalization of migrants' children born in Greece; they could acquire the status of long-term residents when they reached the age of 18, and only if their parents continued to reside legally in Greece. Greek law in this case envisaged the creation of a new category of migrants with special rights, excluding from this category children who were not born in Greece, but who had moved into the country at a very young age, children whose parents decided to return to their country of origin and left their child behind, and children who were born in Greece, but whose parents were never regularized.

The recent Law 3838/2010, however, tried to cure this hostile legal treatment towards migrants' children, and provided for the acquisition of Greek citizenship of these children by birth, in cases where both parents reside legally for five consecutive years in Greece and if they submit a joint declaration and request for registration of the child at the register of the municipality of their residence within three years after the child's birth. Also, according to this new law a migrant's child who has successfully completed at least six years of schooling in Greece and resides permanently and legally in the country may acquire Greek citizenship. Again, the parents need to submit a joint declaration and registration form at the municipality of their residence within three years of the completion of this time. Law 3838/2010 also provides for voting rights in municipal elections for legally residing immigrants.

The new legal citizenship framework in Greece has undoubtedly ameliorated the position of regularized immigrants. It has marked a shift from the doctrine of 'ius sanguinis' (the law of the blood) to 'ius soli' (the law of the place of birth); it has not granted, however, any rights to second generation immigrants whose parents are not legally residing in Greece. It is estimated that thousands of illegal immigrants are currently residing in the country, the exact number of which is unknown. Thus, a large number of migrants' children will never enjoy the privileged status of being a citizen.

The policy towards ethnic-Greek-Albanians differs. These are viewed as 'temporary guests' rather than 'intruding strangers'. Up until 1998, they entered Greece with a visa issued by the Greek consulates. Upon their arrival in Greece some were issued with residence permits by the police authorities (Law 1975/1991). Special legislation was introduced in 1998 to regulate the employment and residence of those who overstayed when their visas or permits expired, or who entered the country illegally. Although the Greek state did not 'encourage' their permanent settlement in Greece so as not to weaken the Greek presence in southern Albania, which is regarded by many as 'traditionally Greek territory', it introduced a temporary privileged status to accommodate this group, taking into account the long term 'national interest' and the short-term needs of the group. They were issued with a '*special identity card for foreigners of Hellenic descent*' (*Ethiko Deltio Taftotita Omogenous*) which has to be renewed every three years and functions as both a residence and work permit. Its temporary duration aims to discourage their permanent settlement in Greece. But at the same time it grants them equal rights to Greeks in the labour market (both public and private sectors), welfare services and other entitlements with the exception of the right to vote and serve in the army.

A shift in the policy towards Northern Epirotes occurred with the adoption of art.41 par.1 of Law 3731/2008. In terms of naturalization, they have to follow the same procedure as all other migrants, but they do not have to fulfil the requirement regarding minimum length of residence in Greece, which is 10 consecutive years in the last 12 years before the submission of their application. Also, the requirement of an oral interview no longer applies for the so called '*omogeneis*' who are considered *a priori* by the Greek administration as having sufficient knowledge of Greek language, history and culture. Greek citizenship law is a law of administrative decisions. The final decision on whether one may become Greek citizen is taken by the Minister for the Interior, who may decline any application for naturalization without any justification (art.8 Law 3284/2004). This means in essence that the immigrant whose application has been declined has no legal right to bring an action before the Greek courts against the decision of the Greek administration. The reason for this change in policy towards Northern Epirotes has been the amendment of the Albanian Constitution which now allows Albanian citizens to have dual citizenship. Thus, the Greek public authorities no longer fear that if they are granted Greek citizenship they will lose Albanian citizenship, and consequently lose their political rights in Albania, thus jeopardizing the very existence of the Greek ethnic minority in the country.

It is worth noting here the difference in status between that granted to Albanian migrant women and to ethnic-Greek-migrant women, that is *residence permit* to the former and *quasi-citizenship* to the latter. Thus boundaries are constructed between 'us', the 'other' (ethnic-Greek-Albanian-migrant women) and the 'other-other' (other Albanians); both the 'other' and the 'other-other' are granted a '*temporary belonging*' via regularization; but whereas the 'other' is temporarily treated as 'one of us' and granted favourable legal status to facilitate their social and labour market inclusion, in the case of the 'other-other' we have a '*functional inclusion*', the creation of a '*gastarbeiter of the new millennium*' for as long as the market needs them. In this respect, it is obvious that a *plastic citizenship* has emerged for Albanian women in Greece. This *plastic citizenship* leads to the creation of *plastic subjectivities* for these women which, together with other conjunctures such as recession, projection of a negative image of these women by the Greek media (see chapter by Tsoukala in this volume), racism and xenophobia, has transformed these groups of women from potential legal subjects to repulsive legal *abjects*. Thus, the EU and Greek legal frameworks created, in the case of the 'other-other' Albanian women, all the conditions for the emergence of legal *abjects*, i.e. women excluded by law and through law.

One should not neglect, however, that migrant men have also been targeted and transformed into legal *abjects,* being viewed as the ones who are responsible for organized crime and illegal trafficking in Greece. This gender specific process of *abjectification* accompanied by the effects of the current global economic crisis has created great human insecurity in contemporary Greek society.

Another important point is that throughout the 1990s many undocumented Albanian migrants lived in fear of deportation. Under the so called '*skoupa*' or 'broom' operations by which migrants were literally 'swept' by the police off the streets and transported to the borders for deportation, men were more vulnerable to police arrests and deportation than women. Migrants that were 'swept' like rubbish off the streets had been transformed by law and through law by the public authorities into legal *abjects*, thus being subjected to extreme forms of insecurity. As Yannis, an Albanian migrant, said:

> I have been deported three times; they put me in a bus full of other Albanians and took us to the borders. A couple of days later we came back. It's easy as long as you have the money to pay someone.

The media contributed to the criminalization of Albanians, their stereotyping as 'cunning, primitive and untrustworthy' (Lazaridis and Romaniszyn, 1998; Lazaridis and Wickens, 1999) and their stigmatization as the 'dangerous other' which has resulted in widespread Albanophobia and expression of racist attitudes and practices in Greek society (see also chapter by Tsoukala in this volume). However, the negative effects of this are gradually beginning to diminish as more and more migrants achieve regularization. Ethnic-Greek Albanians have had more beneficial treatment which, as mentioned above, has recently been institutionalized, provided that they can 'prove their Greekness', that is as long as they can prove that they are part of the in-group. This is important in explaining the differential treatment of 'other Albanian women' and 'ethnic-Greek-Albanian women', those to be 'included' and those to be 'excluded', and the different degrees of citizenship applied by the Greek state which are moulded in terms of plasticity.

Once perceived as a homogeneous entity, Greece is nowadays a multicultural society; this produces challenges to the national identity and paves the way to, at least, facing 'difference' and rethinking our sense of belonging. The first reactions towards the 'other' were a rise in racism and xenophobia but also discussions about 'who we are' and 'where do we belong'. Immigration challenges the exclusionary construction of Greekness on the basis of Orthodox Christian religion, linguistic and genealogical criteria and a historical myth which lies on a historical continuity between ancient and modern Greece. The revival of nationalist ideology in the 1990s which culminated in the 'Macedonian question' together with periodic crises in Greco-Turkish relations, historical disputes between Greece and some of its neighbours (such as Albania), a general perception that those in the Balkans are 'barbarians', 'inferior', and the attitude that 'we' after all can refer to our glorious ancient past, has influenced the population's perception of the 'other' and had a negative impact on the perception of the migrant as the 'dangerous other'. Such perceptions are accentuated by the political and economic crisis the country is currently experiencing and the resulting rise in unemployment (see chapter by Koser in this volume).

Immigration has become part of the political agendas of major parties; the two main parties, New Democracy and PASOK, have developed a managerial approach to migration involving measures encouraging the smooth integration of migrants into the country balanced by an 'effective border control' approach aiming at balancing the national interests of peaceful coexistence within the country with the external pressures towards securitization of the borders. Although a small but strong minority of extreme right-wingers with an anti-immigrant stance exists (e.g. LAOS party), smaller parties on the left of the political spectrum, such as SYRIZA and KKE have been advocating for migrants' rights. Also, there has been a growth in the influence of the activities of NGOs, trade unions, migrants' associations, the Network for Support of Immigrants and Refugees and various anti-racist organizations which have been lobbying and campaigning for migrants' rights. Moreover, although racist attitudes are still widespread in the country, the rhetoric of media discourse has shifted from aphorisms branding Albanian migrants as 'untrustworthy, dangerous and a threat to society' to embracing a debate and giving from time to time a voice to pro-migrant activists and migrant associations. After nearly 20 years of being in the country Albanians are now interacting with civil society organizations (for example their children attend Greek schools), and, as interaction increases, reciprocity between migrants and locals is growing at the informal level and, as the 'second generation' comes of age, hybrid identities and practices are beginning to emerge. At the same time, movements back and forth from country of destination to country of origin encourage and contribute to the emergence of transnational identities. Nevertheless, a mistrust of the 'other' still exists.

Labour market[3]

Most migrant women in Greece, including those originating from Albania, are employed in 'care' services,[4] thus leaving the gendered division of labour unchallenged. The maid, the nurse and the prostitute are the dominant figures of female migrant. They represent the different degrees of exclusion-inclusion in the host society, and they enjoy different degrees of (in)security.

Why the need for 'carers'?

The demand for migrant domestic workers and 'quasi-nurses' is a function of a number of factors: the spatial break up of the extended family and weakening of traditional family relationships and related inter-generational reciprocal arrangements; the growing number of single-parent households; increasing divorce rates which made the balance between reproductive work and employment difficult to achieve (Lazaridis, 2007a); and demographic ageing due to increasing longevity and declining fertility rates. The welfare state characterized *inter alia* by underdeveloped social services and an emphasis on the role of the family as the core unit of social care, has not met the created caring gap by changes in state provisions in childcare, care for the elderly or for people with special needs (see Katrougalos and Lazaridis, 2003). Moreover, putting an elderly person or a person with special needs into a care home is taboo in a society in which men's honour in the eyes of the community depends on, and women derive pride from, the cleanliness and 'care' of the house and the 'good care' of dependent family members. So, although Greek traditional culture did not necessarily value, in a non-numerical sense, reproductive work less that productive work, the performance of these tasks remains highly gendered. Nowadays it has become permissible for a woman or a family to pay for the care to be bought in, to be done by another woman, rather than shaking up the existing social order of the genders or questioning traditional gender identities (Lutz, 2005; Lazaridis, 2007a). This, in turn, has contributed to a 'demand based migration' which enables local women to meet their caring responsibilities without violating cultural norms.

At the same time, infrastructural inefficiencies, unsatisfactory provision of services, and a shortage of humanpower for all areas of nursing care, generate multiple inefficiencies at the level of in-patient care, and has important consequences, one of which is the creation of a niche for *infirmières exclusives*, or '*apoklistikes*'. This takes place right under the nose of the state, and the hospital administration embraces it. Within hospitals the formal and the informal labour markets run side by side. The latter is supported and assisted by the hospital administration; in almost all hospitals, private and state, there is an office on the

3 This section of the chapter relies heavily on Lazaridis, 2007a, 2007b.

4 The word 'care' is used here as a generic term for 'taking care' of the needs of someone.

ground floor where relatives of the patient go to ask for an *infirmière exclusive*. There is fluidity between irregularity and regularity, the public and the private. This paradoxical coalition encourages the further spread of grey work. The risks and financial cost of this type of informality are heaped onto the shoulders of the patients and their families.

Another type of 'carer' is the migrant woman 'looking after' men's sexual needs, needs which are 'natural', nobler and stronger than women's (Loizos and Papataxiarchis, 1991: 222); these need to be served by other than their 'honest decent women'. Visiting a prostitute (and especially an 'exotic other') is a legitimate option for a man in need of expressing himself heterosexually in ways that do not accord with the values of the conjugal relationship (Lazaridis, 1995). Therefore the Greek state tolerates prostitution and informally condones it, provided that the various laws aimed at the protection of public health are adhered to and the woman working as prostitute is registered and 'labelled' as 'prostitute', and her 'profession' appears on her identity card. This involves a moral loss for the 'labelled' woman, because she becomes stigmatized as 'a commodity to be bought', a 'woman with no worth' (see Lazaridis, 2001). The law does not prosecute the clients, even in cases where the woman is a victim of trafficking.[5]

Migrant women's experiences in the host country

Employment trajectories

Restrictive immigration controls, lack of work permits, and an extensive informal economy estimated to be around 40% of the GDP, confine these women in these privatized spheres of work which are bounded by a highly racialized gendered division of labour. Even those migrant women who, during the regularization programmes that have taken place in Greece have acquired a work permit, may experience racial discrimination which confines them to low-paid jobs in the informal sector. However, the acquisition of legal status is important to migrant women's well-being. Regularized migrants feel secure from deportation and can, at least in theory, move towards registered employment. So in theory 'ethnic-Greek-Albanian' women start walking their working path in the host country from a better platform to that of the 'other' Albanians. In practice however, it seems that while they may move towards 'better jobs', these are 'better' in relative terms only, in that they are still confined within the informal sector offering personalized

5 Trafficking has become a big business for many smugglers. 'Complicated networks are involved in forging of travel and other documents and/or providing assistance to the migrant to get inserted into the Greek informal economy' (Lazaridis and Poyago-Theotoky, 1999: 719). Since the early 1990s, it appears that the Greek police have been actively involved in the trafficking of migrants and forced prostitution rackets (*Athens News*, 1 March 1998; 23 October 1998).

services characterized by exploitative working conditions. Working in the public sector is not an option (due to legislation that excludes non-EU nationals from public sector employment) for those without Greek citizenship. What legality gives them is a platform from which to negotiate relatively better payment and perhaps better working conditions, but only marginally.

The overwhelming majority of women concentrate in the 'care' sector, taking up badly paid unskilled or low skilled manual work which contrasts with their educational and professional background; indeed a significant number of our interviewees had qualifications. Some had university degrees, others technical qualifications, but some undertook work which required skills lower than they had acquired. Hence, it seems that migration has had a *deskilling effect*. Gradual improvement has been noted, especially for the ethnic-Greek-Albanian women, in terms of increased wages and a better bargaining position due to regularization and more effective networks of support. Although migrant men seem to experience re-skilling, in the sense of becoming more skilled in the jobs they do and/or becoming self-employed (e.g. plumbers, decorators etc.), some women experience an upgrading of their material conditions in the host country's labour market, but seem entrapped in niches within the service sector, mobility out of which occurs only in isolated cases. So, over time, factors crucial for migrants' well-being (such as regularization and the resulting security migrants feel, greater language competency[6] etc.) may have contributed to pathways to inclusion (a move towards relatively better jobs with less exploitative conditions) but inclusion is still restricted to the informal sector. Interaction between migrants and locals, and the social networks which are formed, function as effective social capital.

Crude stereotypes apply to different nationalities, which in turn affect work chances and arrangements. For example, in relation to domestic workers (nannies and maids) there seems to be a preference for Filipinas over Albanians, and for 'ethnic-Greek-Albanians' over 'other-Albanians'. '*The Albanians are not to be trusted. They can walk out with all your household ... or send in their men to gag you and rob you ...*' an employer said. Such arguments rendered them vulnerable in terms of payment and are used to justify conditions of work characterized by high degrees of exploitation; there are stories of migrants not being paid and being subjected to sexual harassment which resulted in traumatic experiences for some of these women. The live-in domestic workers in particular, hidden in the confines of the household, are at higher risk of being sexually harassed than *les infirmières exclusives* who are 'visible' in the public domain and thus at lower risk of such experience. 'Other' Albanians are more vulnerable due to problems with recognition of qualifications and delays with regularization. The most vulnerable of all are the sex workers; here hierarchies differ: clients have overt preferences for the relatively educated, white, tall and blond Russians and Ukrainians whereas

6 Language skills improve over time with effects on their daily life. For example they give them access to Greek newspapers as well as immigrants' newspapers published in Athens. But it is TV that most use as their basic information and entertainment source.

Albanians are at the bottom of the hierarchy of preferences, occupying a space of their own, away from the 'others'; they are described as rogue peasants with no manners. They often suffer physical, sexual and psychological violence by the clients and the pimps and traffickers, which they never dare to report. Fear and insecurity, as well as knowledge that police officers often turn a blind eye and offer 'protection' to organized prostitution rings, render them silent. Prostitution can provide migrant women with a better living when other unskilled work and the ability to hold other jobs in the informal economy is scarce, and they can also sell sexual favours in order to supplement their income or pay back their bondage to smugglers and traffickers.

Many women work as domestic workers or quasi-nurses during the day and as *infirmières exclusives* or prostitutes during the night. Or one can be a live-in maid or nanny who at the same time offers sexual favours to the employer. For some women there has been a progression from unemployment to maid, from maid to quasi-nurse, from quasi-nurse to *infirmière exclusive*. For others, there is a *trampoline effect* (Lazaridis, 2007a) in operation: from unemployment to maid, from maid to quasi-nurse, from quasi-nurse back to unemployment, up to maid, back to quasi-nurse. Often the boundaries are blurred, and the work is characterized by discontinuity, performed against the backdrop of a complex web of insecurities, uncertainties and violation of basic human rights. There is a *mixing and matching*. For example, one can combine work as a maid or a quasi-nurse during the week and as a prostitute during the weekends, or work as a quasi-nurse in the mornings and a maid in the afternoons. But whatever the work they do within the 'caring sector', the Albanians more often than not find themselves at the bottom of the ladder in terms of remuneration, conditions of employment and treatment by employers. As domestic work is comprised of a number of different tasks, ranging from cleaning to cooking, ironing, caring for children and providing quasi-nursing for the elderly, as Lutz (2005) argues and in line with the German debate, one needs to differentiate between *performance oriented tasks* and *care tasks*. In this case, 'other Albanians' seem to concentrate more on *performance oriented* and less on those *'care' tasks* which require personal care and emotional engagement or on nursing tasks which require skills like administering injections, changing a drip etc. As a Bulgarian woman said: '*The big problem is racism here ... at work, they were asking whether I was an Albanian*'.

But why are Albanians at the bottom of the ladder no matter what job they do? Part of the explanation lies in the Albanophobia and crude stereotypes spread by the media discourse back in the 1990s constructing Albanians as the 'dangerous others' (see above and the chapter by Tsoukala in this volume). It can also be partly explained by Albanians being 'pushed' by poor working conditions or poor prospects of finding a job in the country of origin into accepting any job in the country of destination. This is a *survival strategy*, a decision to sacrifice the present for gaining in the future. It is viewed as a 'path to future well being'; '*doulevo gia to avrio*' ('I work for tomorrow' [meaning her future life]), a woman said. They view themselves as workers. '*I doulia then ine dropi*' ('one shouldn't be ashamed of work'), an Albanian woman told me. This view shapes their identity as 'hard workers' in the host country but also the socially constructed image of the Albanian as someone who 'will do anything for little money'.

In this sense, it seems that there are different degrees in the process of *abjectification* emerging. 'Ethnic-Greek-Albanian' and 'other' Albanian migrant women are exposed to different degrees of security-insecurity. The 'other' Albanian women may be either regularized or illegal migrants. Regularized ones, however, entrapped within consecutive regularization cycles (Konsta and Lazaridis, 2010), and the current hostile economic environment in Greece may still end up working for the informal economy. The same may happen to ethnic Greek Albanian women, but also to women who are Greek citizens. The first to be attacked by Greece's current economic downfall are likely to be women. The economic situation of the country largely affects the process of *abjectification* undergone by these women. The different degrees of *abjectification* may also be observed in relation to employment practices. The process is milder in the case of the self-employed migrant women, the nurse or the maid, and undoubtedly harsher in the case of trafficked women and prostitutes.

The process of *abjectification* can, also, be connected to the different processes and patterns of exclusion-inclusion. Lazaridis and Koumandraki (2007) have examined the exclusion processes and patterns with regard to Albanian migrants in Greece, and have identified three areas of exclusion: a high risk, a relative risk and a non-exclusion area based on migrants' formal/informal employment activities, work conditions, salary, ethnicity, social insurance coverage and legal status.

Albanian migrant women belonging in the *high exclusion/high risk area*, which is associated with manual, unskilled jobs in the informal economy, with high risk, vulnerability, irregularity and informality and lack of legal status, are more vulnerable to undergoing a possible process of *abjectification.* Albanian women in the *relative exclusion/relative risk area*, which encompasses unskilled but more secure jobs in the formal and/or informal sectors, with relatively good payment, and full or partial social insurance coverage, have more possibilities to escape the process of *abjectification.* Women falling under the *inclusion area,* which encompasses well-paid jobs with social insurance coverage, the same pay as their Greek co-workers, and payment for overtime, escape the process of *abjectification.* 'Ethnic-Greek-Albanians' have mainly found professional jobs in this area. Greek ethnicity and favourable legal status played an important role in being included in this privileged area, but not in all cases; also, age and gender shaped exclusion patterns. These areas are not sealed tight spaces of inclusion-exclusion; migrants move in or out, especially in the high risk and relative risk areas. In all three inclusion-exclusion areas, informal networks of support played a crucial role in supporting migrants upon their arrival and in securing access to employment. The following section deals with the role of social capital, and the relational and organizational aspects of migrant women's inclusion in the host society.

Social capital and pathways to inclusion

This section looks at the extent to which social capital can help migrant women create spaces of control in the receiving country, and also at how the conditions they find themselves in (from a political, economic, social and cultural point of view) affect their social capital formation positively or negatively.

Social capital refers to 'norms of reciprocity and networks/associations which can promote co-operative actions and which can be used as social resources for mutual benefit' (Das, 2004: 30; Putnam, 1993; Woolcock, 2000).

Reciprocity, according to Putnam (1993: 172) is of two types: *specific reciprocity*, which refers to simultaneous exchanges of items of equivalent value. For example, the women interviewed exchanged *inter alia* information regarding jobs. As Eleni said:

> When we arrived in Greece we went to stay with my sister who was already in Athens. She was working as a live-in domestic worker for a family in Filothei [a wealthy suburb in Athens]. She found me a job with a friend of her boss. That is how I started working. Later I decided to work as a cleaner for different houses. It is more secure. I mobilized my friends, who spread the word for me … I now have 6 different houses I work for … from Monday to Saturday and sometimes I get ad hoc jobs in the afternoon. The other day one of the women I work for asked me if I could also clean her mother's house; I said no, but I recommended to her my sister-in-law … this is how it works …

This type of reciprocity was abundant amongst the newly arriving migrant women from Albania we interviewed as well as between them and fellow Albanian friends, family, neighbours or acquaintances who have already been in the destination country for sometime and are in possession of such information. The other type of reciprocity is *generalized reciprocity*, that is a continuing relationship of exchange that is at any given time unrequited or imbalanced (e.g. capital etc.). Each act within the generalized reciprocity type is characterized by a combination of short-term altruism and long-term self-interest. 'I help you now in the expectation that you will help me out in the future'. Eva said:

> My brother and his family came to Athens a couple of years ago. I was already established here, had work and somewhere to stay. I offered to put them up for a while, until they got settled. I lend my sister-in-law some money to become self-employed. She is now running a cleaning group – they clean offices. I know that when she starts earning enough she will pay me back in one way or another …

Trust is crucial here. But *trust* takes time to develop through routinized place-based interactions characterized by what Giddens (1987) calls 'co-presence'. In the case of migrant sex workers, for example, the formation of *trust* is made difficult by the '*transferability*' factor[7] (spending one week here, two weeks there) that characterizes their presence in the host country. Therefore *trust* and *reciprocity*, the basic ingredients of social capital, can create a security cushion and cause limited/temporary relief from high risks of migration, but be constrained by the legal vulnerability of migrants or by the formation of negative social capital (e.g. risky dependencies forged between migrant women and their traffickers). We agree with Das's (2004) position that optimists in social capital fail to understand that the factors which influence the ways in which people earn their living (e.g. their class position), and we would add to that their vulnerability, can undermine the conditions of production of social capital.

We also have networks of interpersonal communication and exchange amongst people of equal socio-economic status. Here we have *horizontal associations*, such as migrant associations. These associations facilitate the flow of information, and help their members cooperate for mutual benefit. Overall and in comparison with other migrant groups, the organizational performance of Albanian migrants in Greece has been poor, whereas that of ethnic-Greek-Albanians has been relatively stronger. Collective membership appears limited in scale. Most Albanian migrants we interviewed said that joining an ethnic association is a waste of time. As Floretta said: '*Hasimo chronou ine*' ('it's a waste of time'). In the early 2000s there were four Albanophone newspapers, a magazine published in Athens, and a website covering Albanian and Greek news. There is also a number of ethnic cultural associations such as Northern Epirote House (*Epirotiki Estia*) which is a federation of several associations of ethnic-Greek-Albanians providing assistance with education and access to health and welfare, job finding, and support of ethnic-Greek-Albanians in Albania. This was founded in 1940. The Forum of Albanian Immigrants in Athens collaborates with Trade Unions, NGOs and other migrant associations, and provides its members with legal advice, organizes cultural events, participates in antiracism demonstrations etc. But participation rates are low. The Albanian community is heterogeneous and therefore ties amongst Albanians are weak. Membership is also hindered by the 'lack of trust' of Albanians for collective institutions, the origins of which trace back to the communist regime. Ethnic groups with strong feelings of identity on the other hand, such as the Ethnic-Greek-Albanians and the Vlachs, seem less reluctant to participate in collective organizations, and have formed associations along nationalistic or localist lines. These provide the basis for them to meet, celebrate cultural events, campaign for their rights in the host country and support their compatriots. Café bars in the city provide a space of interaction where information about jobs is exchanged as well as a forum for socialization and entertainment.

We also have *vertical associations*, such as patron-client links. These are usually formed with employers. To facilitate this, often 'other Albanians' try to conceal their difference and to build bonds with the host population based on

7 See Lazaridis, 2001.

the construction of *fictive identities and belonging*.[8] They change their names into Greek, baptize themselves and their children Orthodox Christian, and learn the Greek way of life and customs. This is what Floreta said: '*Last year we all got baptized Greek Orthodox; we are Greek now. We are not different than you are. People cannot treat us differently*'. In this conscious attempt made towards confirmation of self-identity as Greek, their old identities survive as 'dreams or ghosts from the past', there but not really there, kept *in limbo* as the search for and adoption of new identities better suited to the new socio-cultural landscape takes place. Whether a new Grecoalbanian construct (hyphenated or not) comprising of the new fictive identities and the old identities will emerge, remains to be seen (see Lazaridis, 2007).

Bonding social capital refers to strong ties connecting family members sharing similar demographic characteristics (neighbours, friends) (World Bank, 2001: 128). Here norms of reciprocity and mutual help are strong. For example, as Eleni said:

---

8 Albanophobia is widespread. Amongst the coping strategies 'other Albanian' men and women have developed in order to be accepted by the locals is being baptized Orthodox, changing their names into Greek ones, as names are indicative of one's ethnic origin, hiding their religious beliefs, and presenting themselves as ethnic Greek. In a survey by Labrianidis and Liberaki (2001: 174) the percentage of Albanian respondents using Greek names was 34 per cent. Some use one name at work and a different one at home, indicating the quasi natural development of transnational daily realities.

Another strategy for inclusion is baptism of themselves and their children, rather than rediscovering religion or searching for one's ethnic roots. This indicates the perception of ethnicity as a powerful tool for integration. It coincided with the re-emerging importance of the Orthodox religion as part of Greek identity in giving migrants the opportunity to share the privileges of ethnic Greek migrants (e.g. their favourable legal status).

This opportunistic 'voluntary assimilation' can be a strategy for overcoming the overt lack of welcome by the Greeks for Muslims in particular and for other religions (*allohthones*) in general. The paradox here is that the Muslim is a cultural category in the Albanian context more than a religious one, especially because they come from an 'atheist nation', and this is reflected in the high adaptability of Albanians. They do face hostility and there is a lack of places to worship. These attitudes and practices may lead to 'rediscovery' of religion or the use of religion by Greeks to change the inferior 'other' to 'one of us'. With regard to the ethnic Greek Albanians, these seem to be treated by some as Albanians and by others as Greeks, depending on the situation. This identity confusion in their daily interaction with locals renders it imperative to reconfirm every day their Greekness in the country they consider their ancestral home. Elements of both cultures fuse into one when Eleni cooks the traditional pita for New Year, which is not the same as the Greek one, or when she becomes Eliana when work is finished, or when Orthodox Maria celebrates Easter in Athens during Ramadan. These transnational hybrid experiences point to aspects of both cultures being fused together, but being fluid and constantly re-negotiated with some elements lost and others enhanced throughout the process, the Albanian '*besa*' for example.

> When my nephew had a brain haemorrhage, and the doctor had to be paid in cash for the operation, the whole of the Albanian community was mobilized and we raised the cash within two days; when someone is in real need, we all pull together …

However, some of our interviewees argued that Albanians changed in Greece in that each looks after their own affairs and that they do not go out of their way to help their compatriots unless they know them. '*There is a lot of jealousy ... they can't see someone doing better than them ... the Pontians, Russians ...*' an ethnic-Greek-Albanian woman, Maria, said. Another interviewee said: '*Albanians have changed. The reason is money ... work, work, work*'.

The Albanians in general have plenty of *bonding* social capital which offers immediate practical support, especially, but not only, to new arrivals in terms of alleviating the risks of migration. It functions as a sort of informal safety net for the newly arrived in that it provides protection and serves as a source for risk management. Those who lack it are more vulnerable to the risks of migration.

Social capital in this sense, allows Albanian women to counterbalance the negative effects of plastic citizenship, and to resist following the path of absolute exclusion in the host society. The links they establish with the migrant and local populations formulates 'their way out' from being eventually transformed into legal *abjects*. It is a means for formulating their own 'securities' within the host society, which may however prove to be in practice 'pseudo-securities', i.e. a feeling of belonging, which is constantly undermined by hostile immigration laws and policies including ill implemented regularization policies.

Bridging social capital refers to weak ties among members of a civic organization (e.g. association). People with broadly comparable economic status and political power but with different demographic, ethnic and geographical backgrounds are brought together. Although *bridging* social capital is a powerful means by which migrants can address their problems in the host society, and create spaces of control relative to other groups, Albanians have little bridging social capital. It seems that the sentiment of '*we-ness*' amongst them is weak. This is indicated by the identity revision and voluntary assimilation common amongst Albanian migrants in Greece. Why? Mutual-aid is based on an expectation of a certain degree of security. When one is in difficulty, one expects that one's relations with others will act as insurance, and that one can get help. Pressing conditions can prevent migrants from exhibiting cooperative behaviour, including reciprocity. They can constrain the production of social capital or encourage the production of *negative social capital* which instead of alleviating the risks of migration may in the long or short-run accentuate it. For example, prostitutes. A fundamental aspect of their life has been insecurity accentuated by high levels of transferability; they never knew whether or not they will be here tomorrow as they are often transferred from one place to another, one week here, five days there, and from one country to another according to the market necessities. It is a risk limitation strategy on the part of the traffickers in order to avoid detection. The insecurity can undermine the conditions of reproduction of norms of reciprocity which presuppose a sense of security, a sense of guarantee if you like, that you will be here tomorrow to return the favours when needed. So informal rules of sharing are difficult to sustain for long, and this has acted as a constraining force on social capital, in that it limits its beneficial role.

Migrant women are conscious of their irregularity within the informal sector, that they are vulnerable, facing great risks and insecurity. In addition, the fact that they are scattered in the city means that no forms of reciprocity based on neighbourhood lines can be formed and this can render migrants vulnerable, although one should acknowledge that some migrants are more vulnerable than others (e.g. prostitutes).

In this sense, it seems that there are different forms and degrees of *abjection* becoming apparent in our observation of the different forms and degrees of inclusion-exclusion experienced by migrants. Consequently, migrant women from Northern Epiros and 'other' Albanian migrant women are exposed to different degrees of security-insecurity. Also, within the same group of migrant women, the degree of (in)security that these women experience is directly connected to their professional status. The maid or the quasi-nurse experiences more security than the prostitute.

Concluding remarks

Initially third-country migrant women formed part of what Guarnizo and Smith (1998: 18) described as a highly mobile 'transnational working class', or as Lazaridis and Romaniszyn (1998) described as an 'undocumented underclass', their hard work and flexibility making them 'borrowed women' on 'borrowed time' (Smart, 1997, referring to agricultural workers in Alberta, Canada) being where they are in order to satisfy employers' needs. With changes in state and border control regulations this situation is, at least for some, changing, as more and more migrants are in Greece to stay. Migrant women from Albania form a 'dynamic section' within the 'care' sector; the migratory route may for some lead to an improvement of their condition over time, but this is not accompanied by a shift out of the 'care' sector and towards employment within the mainstream sectors of the economy.

Pathways to inclusion are carved as exclusionary processes and practices are experienced in parallel to one another and intersecting one another. The whole kaleidoscope of experiences of migrant women is embedded within wider processes of economic restructuring, political developments and broader social mechanisms. These have to do with the transformation of the Balkans into a transnational space within the wider globalization process; it has also to do with a complex web of structural, institutional and cultural contradictions which shape the inclusion-exclusion dialectic and the multifaceted grid within which the 'us' becomes the 'other' and the 'other' becomes the 'us'.

In this context, a plastic citizenship has emerged, a citizenship that is fluid and flexible, easily shaped by public authorities. This citizenship has led to the creation of plastic subjectivities for these women, and eventually to a process of their *abjectification*, a process during which they are gradually transformed from legal subjects to repulsive legal *abjects* excluded by the host society. This process also leads to different degrees of exclusion-inclusion for migrant women in the host society, and formulates different forms of human (in)securities. Social capital has been the means for resisting this process of *abjectification*, in the sense that it creates securities or, in practice, pseudo-securities for these women.

References

Arendt, H. (1951) *The Origins of Totalitarianism*, New York: Harcourt

Ash, W. (1974) *Pickaxe and Rifle*, Wimbledon: Howard Baker

Das, R. (2004) 'Social Capital and Poverty of the Wage-Labour Class: Problems with the Social Capital Theory', in *Transactions*, Vol. 29, pp. 27–45

De Genova, N. (2007) 'The Production of Culprits: From Deportability to Detainability in the Aftermath of "Homeland Security"', in *Citizenship Studies*, Vol. 11, No. 5: pp. 421–48

Erel, U. (2003) 'Skilled Migrant Women and Citizenship' in Morokvašić-Muller, M., Erel, U. and Shinozaki, K. (eds) (2000) *Crossing Borders and Shifting Boundaries: Gender on the Move*, Opladen: International Women's University

Foucault, M. (2004) *The Hermeneutics of the Subject*, New York: Picador

Foucault, M. (1978) *The History of Sexuality*, Robert Hurley (trans) New York: Random House

Giddens, A. (1987) *Social Theory and Modern Sociology*, Cambridge: Policy Press.

Guarnizo, L. E. and Smith, M. P. (1998) 'The Locations of Transnationalism' in Smith, M. P. and Guarnizo, L. E. (eds) *Transnationalism from Below: Comparing Urban and Community Research*, New Brunswick: Transaction Publishers

Hall, D. (1994) *Albania and the Albanians*, London: Pinter

Katrougalos, G. and Lazaridis, G. (2003) *Southern European Welfare States: Problems, Challenges and Prospects*, Basingstoke: Palgrave Macmillan

Konsta, A. M. and Lazaridis, G. (2010) 'Greek Immigration Policy: civic stratification, "plastic" citizenship and "plastic subjectivities"', in *Journal of International Migration and Integration (JIMI)*, Vol. 11, No. 4, pp. 365–82

Konsta, A. M. (2009) 'Some Thoughts on American Legal Culture: The Legal Abject in Arthur Miller's "The Crucible" and in William Gaddis' "A Frolic of his Own"', in *Intellectum Magazine*, issue 6, December: 21–32 (in Greek). English version of the paper accessed at: http://www.intellectum.org/gb_home.htm

Kravaritou, Y. (2003) 'Equality between Men and Women (article 23)' in Bercusson, B. (ed.) (2003) *European Labour Law and the EU Charter of Fundamental Rights (summary version)*, Brussels: ETUI

Kristeva J. (1992) 'Approaching Abjection', *Powers of Horror: An Essay on Abjection*, Columbia University Press

Kymlicka, W. (1995) *Multicultural Citizenship*, Oxford: Oxford University Press

Malabou, C. (2004) *The Future of Hegel: Plasticity, Temporality and Dialectic*, London: Routledge

Labrianidis, L. and Liberaki, A. (2001) *Albanian Immigrants in Thessaloniki: Paths of Prosperity and Overviews of their Public Image*, Thessaloniki: Paratiritis

Lazaridis, G. (1995) 'Sexuality and its Cultural Construction in Rural Greece', in *Journal of Gender Studies*, Vol. 4, No. 3, pp. 281–95

Lazaridis, G. and Romaniszyn, C. (1998) 'Albanian and Polish Undocumented Workers in Greece: A Comparative Analysis', in *Journal of European Social Policy*, Vol. 8, No. 1, pp. 5–22

Lazaridis, G. (1999) 'The Helots of the New Milennium: Ethnic-Greek Albanians and 'Other' Albanians in Greece' in Anthias, F. and Lazaridis, G. (eds) *Into the Margins: Migration and Exclusion in Southern Europe*, Aldershot: Ashgate

Lazaridis, G. and Poyago-Theotoky, J. (1999) 'Undocumented Migrants in Greece: Issues of Regularization', in *International Migration*, Vol. 37, No. 4, pp. 715–38

Lazaridis, G. and Wickens, E. (1999) 'Us and the Others: Ethnic Minorities in Greece', in *Annals of Tourism Research*, Vol. 26, No. 3, pp. 623–55

Lazaridis, G. (2001) 'Trafficking and Prostitution: the Growing Exploitation of Migrant Women in Greece', in *European Journal of Women's Studies*, Vol. 8, No. 1, pp. 67–102

Lazaridis, G. (2007a) 'Irregular Migration and the Trampoline Effect: '*Les Infirmières Exclusives*', 'Quasi-nurses', Nannies, Maids and Sex Workers in Greece', (forthcoming)

Lazaridis, G. (2007b) 'Les Infirmières Exclusives and Migrant Quasi-nurses in Greece', in *European Journal of Women's Studies*, Vol. 14, No. 3, pp. 227–45 (forthcoming)

Lazaridis, G. and Koumandraki, M. (2007) 'Albanian Migration to Greece: Patterns and Processes of Inclusion and Exclusion in the Labour Market', in *European Societies*, Vol. 9, No. 1, pp. 91–111

Loizos, P. and Papataxiarchis, E. (1991) 'Gender, Sexuality, and the Person in Greek Culture' in Loizos, P. and Papataxiarchis, E. (eds) *Contested Identities: Gender and Kinship in Modern Greece*, NJ: Princeton University Press

Lutz, H. (2005) 'The "Intimate Others": Migrant Domestic Workers in Europe'. Paper presented at the IMILCO – University of Marmara conference, 'Irregular migration, informal labour and community in Europe', Istanbul, December

Mai, N. (2001) 'Transforming Traditions: A Critical Analysis of the Trafficking and Exploitation of Young Albanian Girls in Italy' in King, R. (ed.) *The*

Mediterranean Passage: Migration and Cultural Encounters in Southern Europe, Liverpool: Liverpool University Press

Masselot, A. (2007) 'The State of Gender Equality Law in the European Union', in *European Law Journal*, March 2007, Vol. 13, No. 2, pp. 152–68

Ong, A. (1999) *Flexible Citizenship: The Cultural Logic of Transnationality*, Durham, NC: Duke University Press

Prifti, P. R. (1975) 'The Albanian Women's Struggle for Emancipation', in *Southeastern Europe*, Vol. 2, No. 2, pp. 109–29

Putnam, R. (1993) *Making Democracy Work*, Princeton NJ: Princeton University Press

Smart, J. (1997) 'Borrowed Men on Borrowed Time: Globalisation, Labour Migration and Local Economies in Alberta', in *Canadian Journal of Regional Science*, Vol. 21, No. 2, pp. 141–56

Vahanian, N. (2008) 'A Conversation with Catherine Malabou', in *Journal for Cultural and Religious Theory*, Vol. 9, pp. 1–13, accessed at: http://www.jcrt.org/archives/09.1/Malabou.pdf

Woolcock, M. (2000) 'Social Capital in Theory and Practice: Where Do We Stand?'. Paper presented at the 21st annual conference on economic issues, accessed at: http//www.world-bank.org/poverty/scapital/index.html

Young, A. (2000) *Women Who Become Men: Albanian Sworn Virgins*, Oxford: Berg

Concluding Remarks

State Security Versus the Security of Migrants: The Unbalanced Contest

Annie Phizacklea

While the chapters in this collection address very different scenarios in a range of European member states there is an underlying theme; in the face of ever increasing securitization of migration in member states (surveillance of external borders, visa restrictions, forced return of irregulars and failed asylum seekers etc.) how can the already limited access of third country nationals to protection be safeguarded?

The simple answer is that, since the 1970s and the effective construction of a 'Fortress Europe', political elites supported by large parts of the media have been more concerned with emphasizing the supposed destabilizing threats posed by migration rather than with its benefits or concern for the rights of third country nationals to protection. These issues have been addressed in the collection through analyses of political and media constructions of migrants as a 'threat' (a 'cultural threat', a 'terrorist threat', a 'criminal threat' etc.) in the work of Tsoukala, Karyotis, Campani, and Krasteva, while the tension between state securitization and the personal security of migrants is explored by Maguire, Likic-Broboric, Parusel, Schuster, Nagtegaal, Pajnik, Ribas-Mateos, Koser and Lazaridis and Konsta. There is of course no clear cut division between these approaches; for instance, Campani is also addressing the issue of the personal security of migrants through individual narratives.

Tsoukala's content analysis of two right wing and two 'leftish' upmarket daily newspapers in France and Greece has provided us with an insightful comparative analysis of contemporary media constructions of the 'migrant' and of the 'threat' posed by immigration. Her research indicates that while it is based on 'upmarket' newspapers (as opposed to the populist, sensationalist tabloids) who adopt an ambivalent stance towards the old negative stereotyping of migrants, events such as the 2005 riots in France being reported as 'a claim for fair treatment' are so marginal that they are drowned out by the right wing and political calls for a strict clampdown on 'juvenile delinquency' in the areas concerned where 'not so perfectly integrated nationals' are regarded as the enemy within. Tsoukala argues that media representation around migration issues in the late 2000s differs significantly from previous decades, that now we have a two-fold axis, desecuritizing discourses (usefulness) and securitizing discourses (the threat), the latter being highly politicized, particularly in terms of the 'cultural' threat discourses in France. But these are the 'upmarket' newspapers; have things changed that much? Is there really a two-fold axis? In the 1970s and 1980s there was a very strong alternative press in France and Greece (albeit that immigration was not an issue in Greece at the time) which did act as a counterpoint to the old stereotypes, but with economic crisis these alternative voices are muted.

Karyotis's analysis of political elite support for the security-migration nexus suggests that they are anchored in preconceived ideas about their responsibility to protect symbolic boundaries of the nation state. As each and every member state has adopted highly restrictive immigration controls since the 1970s, elites may be hesitant to challenge the apparatus of control. The clamour for greater securitization then creates not a safer society but one that lives in permanent fear from real or perceived threats. He points out that this results in the same contradictory set of unintended consequences that we have witnessed since the 1970s. First, migrants stay rather than leave for fear of losing access to labour markets, increasing the likelihood of irregularity as migrants who entered 'Fortress Europe' with papers become overstayers and therefore irregular. Second, scapegoating of migrants further marginalizes their position in society with the potential to trigger migrant anger (the enemy within), and finally it leads to the conflation of asylum seekers with a terrorist threat.

Campani has taken up these issues in the case of Italy, where a populist use of the security-migration nexus casts male irregular migrants as a criminal threat to national security, and women migrants as a moral threat. Thus, as in France, the national populist discourse combines security with identity threat discourses. She contrasts this with the narratives of migrant women in Italy whose primary concerns focus on human security; the safety of their children, their wish to be 'legal' in a state that criminalizes overstayers. We will return to these issues in our consideration of issues around the personal security and protection of third country nationals.

Togral pursues the theme of 'cultural threat' in her chapter, pursuing Martin Barker's argument proposed in the 1980s, that the old racist stereotypes had been superseded by the new 'cultural' racisms. In 1979 Margaret Thatcher (then leader of the UK Conservative opposition) warned in the lead-up to the general election that the UK was in danger of being 'swamped' by other 'cultures'. Togral suggests that so-called 'Islamification' has been reified as a destabilizing factor; a security threat to national identities and European 'values'.

Krasteva crystallizes these issues in her analysis of the burqa debate in France, her chapter (as do others) demonstrates the tension between state securitization and the personal/human security of migrants.

Maguire's chapter sums up this tension in his analysis of contemporary biometric security, drawing on historical evidence. For instance, for social reformers such as Bentham security predicates liberty; liberty is security. Maguire posits that states such as Qatar offer an example of contemporary discourses of security in which liberty is understood as security. But in the Qatar case technology is used to govern society through differentiated access to spaces and privileges, 'biometric citizenship'. A society composed of e-gates maximizes the security and liberties of some, while minimizing the risks associated with the mobility and identity of 'suspect' populations. Qatar, in common with the Gulf States as a whole is heavily dependent on migrant labour.

Likic-Brboric analyses the tension within the broader context of EU policy. She argues that the most recent phases of EU enlargement have intensified the conflict between market and socially minded politics. So, when we look at the *issue* of migration, it is framed in terms of 'partnership and solidarity', but the actual policy approach is about effective return of 'illegals' and security oriented border control. She argues that this orientation constrains the development of a right based mobility/migration regime. We may reflect that, even before the Schengen Agreement, certain member states were being identified as the 'soft underbelly' of Europe as far as irregular migration was concerned, a concern which intensified when Russia and other post-Soviet states hastily adopted the 1951 Geneva Convention on refugees in 1992. When hundreds of thousands of 'transit' migrants flowed into those (totally unprepared) countries from Afghanistan, Iraq and Somalia, the result, as one commentator put it, was that 'once they have bounced off the walls of 'fortress Europe', [they] find themselves in shaky camps in very poor living conditions' (Condagnone, 1997: 44). Little did we know 25 years ago that the elaborate system of fortification of restrictive immigration policies would be extended to asylum seekers too. Even though these policies fly in the face of international understandings which propose that asylum seekers should be assessed differently to other types of migrants, current policies are designed to prevent them arriving in the first place and, if they do, to dump them back to the last 'safe' place they can be traced to, or, worse, into the hands of the very regime they were escaping from. These issues are accentuated in the case of minors, taken up in Parusel's chapter, in which he addresses the question of how/whom can redress the current balance of affairs.

He argues that the case of unaccompanied minors strikes at the heart of the tension between on the one hand international law and the EU legislation dealing with the well-being of children and on the other the securitization approach, including tightened borders and limited access to protection for foreign nationals. His central argument is that while the migration control framework represents a distinctive obstacle to the effective protection of unaccompanied migrant children, the presence of international human rights law, advocacy work and the involvement of non-governmental actors in EU policy development may offer some opportunities (although limited) to improve the well-being of migrant children. Parusel is right to stress how limited these opportunities are in the face of the 'Europeanization' of asylum and refugee policies over the last 25 years. When the emphasis is always on the destabilizing effects of such migration to EU countries, there is rarely an attempt to point to the benefits of migration or to look to real ways of effectively protecting those migrants in need of a safe haven. It is interesting that one of the few EU reports that points to the benefits of migration constituted a justification for those existing members who did not impose restrictive regulations on the mobility of EU8 workers. The 2006 report from the European

Commission argued that despite fears to the contrary Eastern European workers had contributed to economic growth in Britain, Ireland and Sweden, filling gaps in the labour market and increasing tax revenues. The presence of so many workers on relatively low wages acting as a deflationary pressure; in contrast, the report suggests that in those countries which imposed restrictions there had been higher levels of undeclared work (CEC, 2006). So where are the official reports that point to the contribution that refugees have and continue to make to European societies or to how the rights of third country nationals should be better protected? Parusel's answer is simple; EU policy makers are intent on preventing new 'pull factors', concentrating on how to make Europe as unattractive as possible for any potential third country national.

Schuster takes up this theme in her chapter based on the experiences of a group of Afghan migrants in Paris who have fled through Greece and Italy. In order to evade any legal or moral responsibility for providing protection to those in need, member states maintain asylum seekers in a state of legal, physical and psychological insecurity by using the Dublin II agreement (identifying the first 'safe' country entered by someone claiming protection) and Eurodac (finger printing). Local politicians feed into the overriding philosophy that if protection and support is offered this simply encourages more people to come.

Nagtegaal supplies an interesting twist to this story in the Dutch case. The Netherlands regards itself as a 'tolerant and hospitable' society to those seeking protection, but in reality has gone along with the European consensus in taking 'self protective' measures against refugees who are perceived as a threat to stability. Echoing Maguire he points to the security/liberty nexus. The logic is that new technologies allow borders to be better controlled, enhancing the self protective measures of the state (securitization) while they provide protection for 'real' refugees (liberty). But, he argues that there is considerable evidence to show that the vast majority of government data collection sites are not in themselves properly secured.

Pajnik draws our attention to the vicious circle of migrant attempts to gain at least some social rights/citizenship in Slovenia, echoing the narratives of the women in Campani's research. These are the individual accounts of the migrant quest for personal security within a system that institutionalizes legal and psychological insecurity. Lazaridis and Konsta extend this theme theoretically, using the concept of abjectification. They argue that migrant women are transformed by law and through law from legal subjects to legal 'abjects' excluded from the host society. The women use social capital as a means of resisting this process of abjectification, creating securities or in practice pseudosecurities for themselves. Ribas-Meteo chooses border towns as key sites for a better understanding of the realities of border reinforcement and the vulnerability of human rights. Koser assesses the impact of the global financial crisis on migrants and migration in Europe. The chapter highlights the continuity of historical trends; recession means a fall in migration, but in the face of restrictionist immigration legislation, those already resident stay on in the hope of an economic upturn, even if this means becoming an 'overstayer', joining the ranks of the undocumented and sending little money home in the meantime. Not surprisingly, migrants become scapegoats in times of economic uncertainty, experiencing rising levels of anti-immigrant sentiment and discrimination.

All the chapters in this volume point to the same conclusion. In the face of an increased securitization-migration nexus (a securitization which becomes ever more elaborate technologically) migrants are cast as a cultural/criminal/terrorist threat, and, in these recessionary times, a threat to economic stability. Their need for protection (and this applies to migrant workers as well as those asylum seekers who desperately need such protection) becomes an unwelcome issue. The recognition and enforcement of existing rights is then interpreted as a 'pull factor', which will only encourage others to seek entry to 'Fortress Europe'. This logic informs the opposition to further mass regularization programmes for irregular migrants such as those undertaken in Greece, Italy, Portugal and Spain.

So how can what appears to be an irrevocable tide be turned? Parusel has argued that the presence of international human rights law, advocacy and the involvement of non-governmental actors in EU policy development offers a limited opportunity for improving the current situation. But 'limited' is the key word here. There has to be a real commitment in practice to a rights based mobility/migration regime in Europe rather then the lip service that is paid to such rights in current day to day realities. Given the record of practice over the last forty years, this amounts to a very uphill struggle.

References

CEC, Commission of the European Communities (2006) 'Report on the Functioning of the Transitional Arrangements set out in the 2003 Accession Treaty (period 1 May 2004–April 2006)', Brussels

Codagnone, C. (1997) 'The New Migration in Russia in the 1990s' in Koser, K. and Lutz, H. (eds) *The New Migration in Europe: Social Constructions and Social Realities*, Basingstoke, Macmillan

Index

'n' (e.g. '207n') indicates reference exclusively to a note. Where an entry has eight or more indicators, and where appropriate, major references are printed in bold (e.g. '**239**').

RESEARCH IN MIGRATION AND ETHNIC RELATIONS SERIES

Other titles in this series

Long-Distance Nationalism
Diasporas, Homelands and Identities
Zlatko Skrbiš

Education and Racism
A Cross National Inventory of Positive Effects of Education on Ethnic Tolerance
Edited by Louk Hagendoorn and Shervin Nekuee

New Chinese Migrants in Europe
The Case of the Chinese Community in Hungary
Pál Nyíri

European Nations and Nationalism
Theoretical and Historical Perspectives
Edited by Louk Hagendoorn, György Csepeli, Henk Dekker and Russell Farnen

Transitions
Russians, Ethiopians, and Bedouins in Israel's Negev Desert
Edited by Richard Isralowitz and Jonathan Friedlander

Comparative Perspectives on Racism
Edited by Jessika ter Wal and Maykel Verkuyten

Arguing and Justifying
Assessing the Convention Refugees' Choice of Moment, Motive and Host Country
Robert F. Barsky

Patriarchal Structures and Ethnicity in the Italian Community in Britain
Azadeh Medaglia

South Africa's Racial Past
The History and Historiography of Racism, Segregation, and Apartheid
Paul Maylam

Poland's Post-War Dynamic of Migration
Krystyna Iglicka

Ethnicity and Nationalism in Italian Politics
Inventing the *Padania*: Lega Nord and the Northern Question
Margarita Gómez-Reino Cachafeiro

Integration and Resistance
The Relation of Social Organisations, Global Capital, Governments and International Immigration in Spain and Portugal
Ricard Morén-Alegret

New European Identity and Citizenship
Edited by Rémy Leveau, Khadija Mohsen-Finan and Catherine Wihtol de Wenden

Globalizing Chinese Migration
Trends in Europe and Asia
Edited by Pál Nyíri and Igor Saveliev

Illegal Immigrants and Developments in Employment in the Labour Markets of the EU
Jan Hjarnø

European Encounters
Migrants, Migration and European Societies Since 1945
Edited by Rainer Ohliger, Karen Schönwälder and Triadafilos Triadafilopoulos

Identity and Integration
Migrants in Western Europe
Edited by Rosemarie Sackmann, Bernhard Peters and Thomas Faist

Integrating Immigrants in the Netherlands
Cultural versus Socio-Economic Integration
Edited by Louk Hagendoorn, Justus Veenman and Wilma Vollebergh

Governance in Multicultural Societies
Edited by John Rex and Gurharpal Singh

International Migration in the New Millennium
Global Movement and Settlement
Edited by Danièle Joly

Citizenship in European Cities
Immigrants, Local Politics and Integration Policies
Edited by Rinus Penninx, Karen Kraal, Marco Martiniello and Steven Vertovec

Migration Between States and Markets
Edited by Han Entzinger, Marco Martiniello and Catherine Wihtol de Wenden

Nationalism and Exclusion of Migrants
Cross-National Comparisons
Edited by Mérove Gijsberts, Louk Hagendoorn and Peer Scheepers

Transnational Social Spaces
Agents, Networks and Institutions
Edited by Thomas Faist and Eyüp Özveren

Roma and Gypsy-Travellers in Europe
Modernity, Race, Space and Exclusion
Angus Bancroft

International Migration Research
Constructions, Omissions and the Promises of Interdisciplinarity
Edited by Michael Bommes and Ewa Morawska

East to West Migration
Russian Migrants in Western Europe
Helen Kopnina

Migration and its Enemies
Global Capital, Migrant Labour and the Nation-State
Robin Cohen

Globalizing Migration Regimes
New Challenges to Transnational Cooperation
Edited by Kristof Tamas and Joakim Palme

Migration, Regional Integration and Human Security
The Formation and Maintenance of Transnational Spaces
Edited by Harald Kleinschmidt

Moving Lives
Narratives of Nation and Migration among Europeans in Post-War Britain
Kathy Burrell

Immigrant Women and Feminism in Italy
Wendy Pojmann

Cities and Labour Immigration
Comparing Policy Responses in Amsterdam, Paris, Rome and Tel Aviv
Michael Alexander

Diversity Management and Discrimination
Immigrants and Ethnic Minorities in the EU
John Wrench

Minority Rights Protection in International Law
The Roma of Europe
Helen O'Nions

The African Diaspora in the United States and Europe
The Ghanaian Experience
John A. Arthur

Labour Migration from Turkey to Western Europe, 1960–1974
A Multidisciplinary Analysis
Ahmet Akgündüz

Paradoxes of Cultural Recognition
Edited by Sharam Alghasi, Thomas Hylland Eriksen and Halleh Ghorashi

Irregular Migration in Europe
Myths and Realities
Edited by Anna Triandafyllidou

Inclusion and Exclusion of Young Adult Migrants in Europe
Barriers and Bridges
Edited by Katrine Fangen, Kirsten Fossan, Ferdinand Andreas Mohn

Muslim Diaspora in the West
Negotiating Gender, Home and Belonging
Edited by Haideh Moghissi and Halleh Ghorashi

Managing Ethnic Diversity
Meanings and Practices from an International Perspective
Edited by Reza Hasmath

Media in Motion
Cultural Complexity and Migration in the Nordic Region
Edited by Elisabeth Eide and Kaarina Nikunen

For Product Safety Concerns and Information please contact our EU representative GPSR@taylorandfrancis.com Taylor & Francis Verlag GmbH, Kaufingerstraße 24, 80331 München, Germany

Batch number: 10725973

Printed by Printforce, the Netherlands